Advances in Industrial Engineering

Series Editor: Gavriel Salvendy, Purdue University, West Lafayette, IN 47907, U.S.A.

Vol. 1. Performance Criteria and Incentive Systems (S. Globerson)
Vol. 2. Techniques for Multiobjective Decision Making in Systems Management (F. Szidarovszky, M. E. Gershon and L. Duckstein)
Vol. 3. Analysis and Design of Hierarchical Control Systems: With Special Reference to Steel Plant Operations (T. J. Williams, ed.)
Vol. 4. Productivity Improvement: Employee Involvement and Gainsharing Plans (E. M . Dar-El)
Vol. 5. Manufacturing Research: Organizational Issues (A. Gerstenfeld, H. J. Bullinger and H. J. Warnecke, eds.)
Vol. 6. Innovation: A Challenge to the Engineer (K. Holt)
Vol. 7. Manufacturing Research Perspectives: U.S.A.–Japan (A. Gerstenfeld, C. Ganz and T. Sata, eds.)
Vol. 8. Applications of Fuzzy Set Methodologies in Industrial Engineering (G. W. Evans, W. Karwowski and M. R. Wilhelm, eds.)
Vol. 9. People and Product Management in Manufacturing (J. A. Edosomwan, ed.)
Vol. 10. Integrative Management, Innovation and New Venturing: A Guide to Sustained Profitability (M. Terkel)
Vol. 11. Surface Modeling for CAD/CAM (B.K. Choi)

Advances in Industrial Engineering, 11

Surface Modeling for CAD/CAM

Byoung K. Choi

Korea Advanced Institute of
Science and Technology
Chongryang
Seoul, Korea

ELSEVIER
Amsterdam–Oxford–New York–Tokyo 1991

ELSEVIER SCIENCE PUBLISHERS B.V.
Sara Burgerhartstraat 25
P.O. Box 211, 1000 AE Amsterdam, The Netherlands

Distributors for the United States and Canada:

ELSEVIER SCIENCE PUBLISHING COMPANY, INC.
655 Avenue of the Americas
New York, N.Y. 10010, U.S.A.

ISBN: 0 444 88482 3

Printed in The Netherlands

CONTENTS

PREFACE

The purpose of this book is to provide a comprehensive treatment on the subject of **sculptured surface modeling.** *"Curved object manufacturing"* is an important subject in modern industry as more variety of industrial products are being designed with *"sculptured surfaces"*. The subject area is called **CAMM** (computer-aided modeling & machining) which is becoming one of the most critical issues in **CAD/CAM**. The initial draft of the book was based on the lecture notes of a Ph.D. level course (IE 675 CAM-II) offered by the author at Purdue University during his visit in 1988. Since then quite a few new (published and unpublished) results have been added to the initial draft.

There has been a gap between the theory in **CAGD** (computer-aided geometric design) and the practices in CAMM. In a way, this is a gap between *"academia"* and *"industry"*. This book is written to help reduce the gap by delivering *practical "solutions"* to engineers and at the same time by providing *"problems"* to researchers. The main focus of the book, however, is on the *"approach"* to and *"strategy"* for solving engineering problems. Solving an engineering problem requires a sound understanding of the structure or domain of the problem as well as available techniques or tools, which this book intends to deliver. The *"domain"* of surface modeling in this book is **descriptive shape models**, that is, geometric shapes that can be *"described"* (on engineering drawings or otherwise).

This is not a book on CAGD *per se*, but broad mathematical principles of the methods are covered, in addition to some details of practical implementation. Overall structure of CAMM is presented in the first chapter, and brief mathematical backgrounds are provided in the next chapter, followed by the subject of *"curve modeling"* in Chapters 3 and 4. Backgrounds for *"surface modeling"* are given in the two chapters that follow. Chapters 7 through 13 are the main body for surface modeling. Chapter 14 presents practical approaches to developing commercial CAMM systems, and various issues in developing a *unified shape modeler* for CIM are presented in the last chapter.

This book is written for engineers, but researchers and developers in CAGD may also find it useful. The book is reasonably self-contained so that an undergraduate level of algebraic calculus is enough to follow the discussions. Recent developments in CAGD research are also introduced, and references are provided to guide the reader to further researches.

ACKNOWLEDGEMENTS

It is a great pleasure and honor to dedicate this book to Professor Moshe M. Barash, my former advisor at Purdue, to whom I owe so much in many respects. He has been pioneering in the area of CAM and CIM, and is about to retire from his more than thirty years of dedication to his profession.

This book would not have been completed in time without the help from Mr. Hayong Shin, one of the best Ph.D students that I know of, who patiently has derived many of the algebraic formulas in the book and spent so much time in proof reading the manuscripts and pasting the illustrations. The invaluable help from Mr. Jung H. Park, another Ph.D student who spent so much time in editing the final draft and in producing "Mac Draw" outputs, is greatly appreciated. Many thanks are due also to other graduate students of mine in providing *"graphics outputs"* and other assistances, and to Miss Won who did such an excellent job in converting the Chi-writer version initially prepared by the author to this "camera copy ready" Texture version of the book.

Special thanks are due to Professor Salvendy of Purdue for his arrangement and encouragement for the writing of this book. I would also like to thank my good friends T. C. Chang and C. Chu of Purdue for their reviews and suggestions on earlier chapters of the book and for their support and encouragement.

Very special thanks to my wife, Yong, and my little boy, Samuel, who endured my *"absence"* so long during the writing of the book. Their loving care and encouragement kept the writing going.

Byoung K. Choi
Seoul, Korea
August, 1990

CHAPTER 1

FRAMEWORKS FOR SURFACE MODELING AND MACHINING

1.1 INTRODUCTION

Efficient manufacturing of curved objects in an important issue is modern industry as more variety of industrial products are being designed with "free formed" or **sculptured** surfaces. Products are designed with sculptured surfaces to make them look better and/or function better. Sculptured surfaces in household products such as telephones serve to enhance their appearance, while these in aero-dynamic parts such as turbine blade are to meet functional requirements of the product. The former is called an **aesthetic surface**, and the latter a **functional surface**.

The purpose of this book is to outline in a self-contained manner the major developments in *sculptured surface modeling for computer-aided design and manufacturing.* In this book, the term **modeling** is used to mean the activity of *constructing a mathematical or computer model from the description of a shape*, usually, given in the form of engineering drawing or stored in wireframe forms. This is in contrast to the term **design** which refers to the activity of creating, in most cases interactively, a geometric shape. On the other hand, the term "geometric modeling" refers to a collection of techniques or tools that may be used in both "design" and "modeling". We are specifically interested in modeling sculptured surfaces to be machined on NC (numerically controlled) machines.

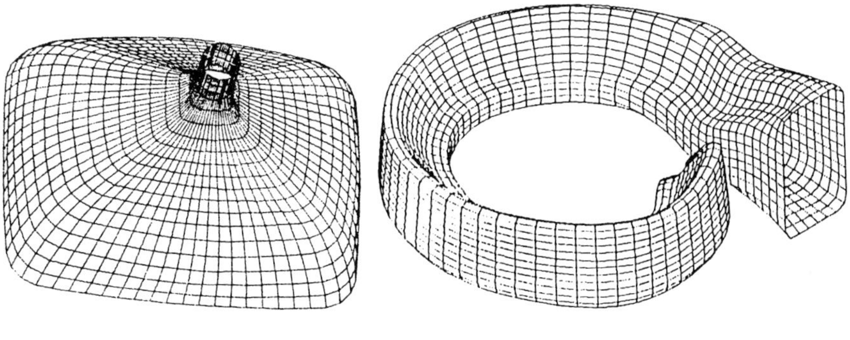

Figure 1.1 Sculptured Surface Examples (CRT funnel and pump volute).

Most of engineering disciplines are concerned with modeling a system: It may be a production system, a highway system, or a robot arm. In modeling a system, the engineer should understand the structure of the target system, that is the modeling domain, as well as the tools used in building and analyzing the model. In our case, the modeling domain is the *description* of sculptured surfaces which are detailed in blue prints or stored in computer. The surface description itself is a form of shape model, called **descriptive shape model** (DSM), which is subject to human interpretation.

Mathematical tools used in surface modeling are referred to as "computational geometry", "surface engineering geometry", or "CAGD (computer aided geometric design)". Excellent treatments of the subjects are available in the literature, e.g., Faux and Pratt (1980), Mortenson (1985), Ding and Davies (1987), and Farin (1988). But, it is not easy for a manufacturing engineer to digest the materials and apply them in constructing a computer model of a DSM. In this respect, this book attempts to present the material in a digested form.

This book is concerned with constructing a computer model of a DSM so that the curved object can be machined on NC machines (instead of copy milling). The computer model for NC machining is called a **computational geometric model** (CGM) as introduced in Choi *et al* (1988a). Basic backgrounds on algebraic and differential geometry are presented in Chapter 2, followed by comprehensive discussions on curve modeling in Chapters 3 and 4. An overview of surface patch models is provided in Chapter 5. Chapter 6 is devoted to Bezier curves and surfaces as they play a key role in surface modeling.

Chapters 7 through 13 constitute the main body for sculptured surface modeling. The topics to be discussed are:

- *Surface construction from 3D data array (Chap. 7),*
- *Surface construction from scattered 3D data (Chap. 8),*
- *Surface construction from 2D cross sections (Chap. 9),*
- *Surface construction from 3D curve-nets (Chap. 10),*
- *Construction of blending surfaces (Chap. 11),*
- *Surface intersecting and trimming (Chap. 12),*
- *Non-parametric surface modeling (Chap. 13).*

In Chapter 14, an approach to developing commercial CAMM (computer-aided modeling and machining) systems is proposed, and a futuristic discussion on the issues in developing a unified shape modeler (ie, a solid modeler with sculptured surface capability) is provided in the last chapter.

The present chapter starts with a discussion on frameworks for curved object manufacturing to provide a basis for later discussions.

1.2 FRAMEWORKS FOR CURVED OBJECT MANUFACTURING

As it is impossible to machine a sculptured surface to its exact dimension on a conventional machine tool, a model or template is always employed in machining curved objects. Thus, the manufacturing cycle for curved objects usually consists of three distinctive phases: Design phase, modeling phase, and machining phase. This section reviews the evolution of frameworks for curved object manufacturing. A view of conventional manufacturing cycle is first presented, and than the contemporary framework for curved object manufacturing is described, followed by a framework of an idealistic manufacturing cycle.

1.2.1 Conventional Cycle of Curved Object Manufacturing

Traditionally, curved objects have been machined on **copy milling** machines. As shown in Fig. 1.2, designing starts with a *conceptual shape model* of the product. The designer may make a "physical clay model" in order to better evaluate his conception about the shape or to perform necessary functional analyses. A series of modifications may be necessary to obtain a satisfactory **clay model** (not necessarily made of clay). When the final approved clay model is obtained, 3D coordinates on the surface of the clay model are measured with a coordinate measuring machine. This is a common practice in designing automobiles, footwear, and aero-dynamic parts.

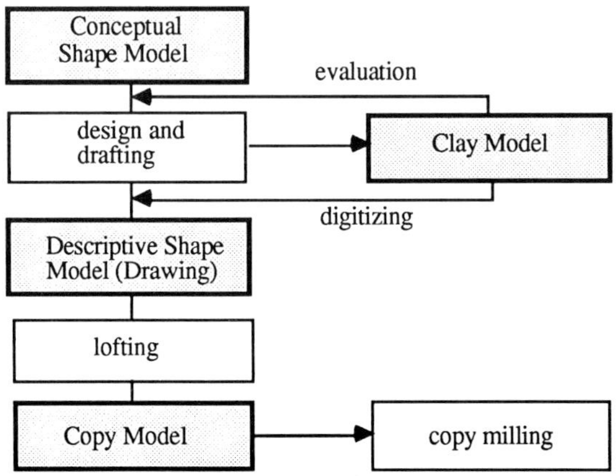

Figure 1.2 Conventional Cycle of Curved Object Manufacturing

Once the shape of the object has been finalized, draftsmen produce a set of blue prints which serve as a DSM (descriptive shape model) of the object. 3D coordinate values are provided in the blue prints if necessary (3D coordinate values of a surface are sometimes generated from engineering calculations). The activities described so far are carried out in the Design Department. If the product is to be produced by other than machining, the blue prints of the part are directed to the Die-making Department (or an outside vendor) where blue prints of the dies are produced.

The next step in the conventional manufacturing cycle of Fig. 1.2 is to make a *copy model* (of dies in most cases) according to the specifications of the blue prints. Here we need a highly skilled loftsman who manually produces the copy model, usually made of wood or plaster, from blue prints. Sometimes, a copy model may be copied directly from the clay model. Finally, sculptured surface machining (usually ball-end milling) is carried out on a copy milling machine by tracing out the copy model. To obtain a smooth surface, grinding and scraping are followed by copy milling.

In the conventional manufacturing cycle of Fig. 1.2, one may easily identify the problems that manufacturing engineers have to cope with:

- It is difficult to achieve a desired level of accuracy in the final machined shape, mainly because of the "copying process";
- It is easy to make errors, by mistake or misinterpretation, because the process involves manual handling of enormous amount of numeric data;
- It is very unproductive because the shape models are reconstructed manually and the process is serial (ie, clay model → parts drawing → copy model → copy milling).

1.2.2 Contemporary Framework for Curved Object Manufacturing

The advent of computer graphics and the progress in geometric modeling have made a significant impact on the manufacturing of curved objects. Among the activities in Fig. 1.2, following four areas have received direct benefits from the new technologies:

- *Drafting*: Engineering drawings are produced by a CADD (computer aided design and drafting) system.
- *Measured 3D data*: Mathematical models of sculptured surfaces are directly constructed from the 3D coordinate values.
- *Lofting*: Mathematical models of sculptured surfaces are constructed from the (curve-net type) description of the surface.
- *Copy milling*: Copy milling is replaced by NC machining.

If the above mentioned activities are carried out "computer-aided", the resulting manufacturing cycle would become the one shown in Fig. 1.3. For a manufacturing engineer, the main difference between the conventional manufacturing cycle of Fig. 1.2 and the contemporary framework in Fig. 1.3 is the substitution of "lofting" with "surface modeling". As a result, the "copy model" and "copy milling" are replaced by a "CGM (computational geometric model)" and "NC machining", respectively. Additional advantage is in the possibility of on-line inspection and corrective machining (grinding) of the finished part surface. It is to be noted in Fig. 1.3 that portions of CGM may as well be obtained during the CADD stage.

With the contemporary framework for design/modeling/machining shown in Fig. 1.3, most difficulties inherent in the conventional manufacturing cycle can be alleviated:

- *Machined part surfaces become more accurate and precise;*
- *Chances of making (dimensioning) errors are greatly reduced;*
- *Overall manufacturing cycle time would be reduced considerably.*

Another very important advantage of the contemporary framework is that it becomes much easier to *accommodate design changes* (which happen all the time) because all the shape models (ie, blue prints and CGM) are now stored in computer. The contemporary framework, however, demand much more careful planning for the NC machining of the curved object.

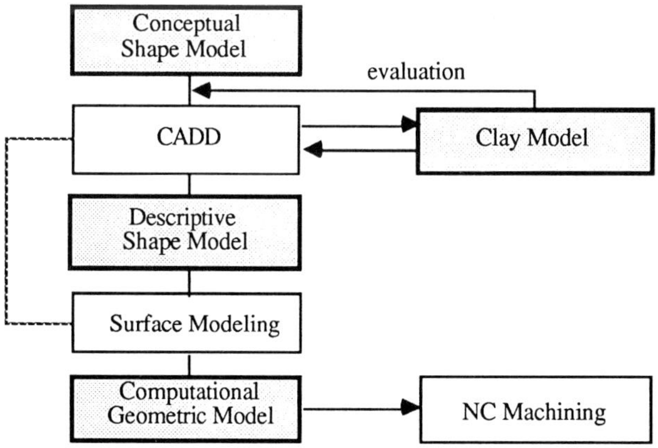

Figure 1.3 Contemporary Framework for Curved Object Manufacturing

1.2.3 A CIM Framework for Curved Object Manufacturing

As far as the serialism is concerned, the conventional cycle of Fig. 1.2 and the contemporary framework of Fig. 1.3 are almost identical. We may call the contemporary framework a modular approach or "islands of automation". Thus, it is not difficult to envision an idealistic framework for curved objects manufacturing where all the components are integrated. Such a framework is proposed in Fig. 1.4. In the proposed CIM framework, design and modelling are merged into a true CAD, and as a result, all the shape information is completely stored as a CGM once the shape of the object is finalized.

The state-of-the-art in CAGD may fully support the implementation of the contemporary framework presented in Fig. 1.3. But current industrial practices are somewhere in-between the conventional cycle and the contemporary framework mainly because of lack of effective modeling methods. The purpose of this book, in this respect, is to provide surface modeling methodologies enabling a full implementation of the contemporary framework. Currently, rough estimates of typical lead times for a car and a home electronic product are about 3 years and 0.5 years, respectively. With the CIM framework in Fig. 1.4, they could be reduced to 3 months and 0.5 months. Then why not implement it right away?

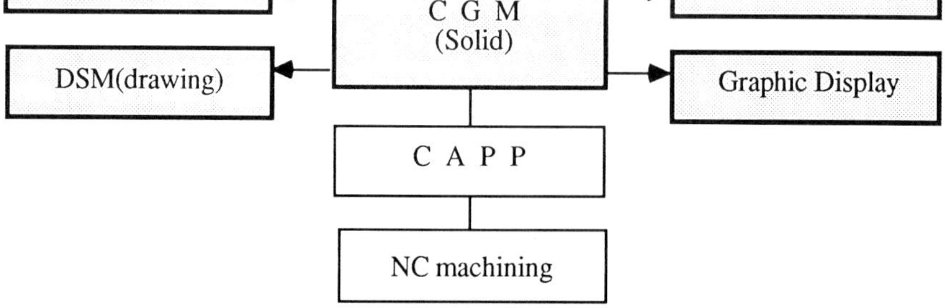

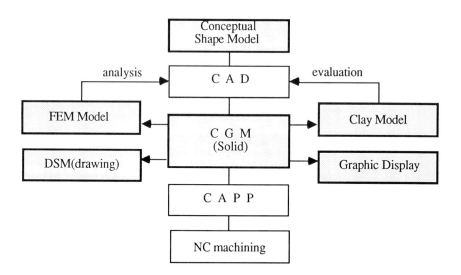

Figure 1.4 CIM Framework for Curved Object Manufacturing

The current state-of-the-art in CAGD has yet to "invent" a general purpose *unified shape modeler* that can support the CIM framework of Fig. 1.4: The unified shape modeler should be able to construct a CGM (computational geometric model) directly from a "conceptual shape model"; it should be supported by powerful graphics hardware and new rendering technologies; it should also serve other CIM functions such as design analyses, machining process planning, assembly planning, and preform design. In the mean time, however, it is essential to be able to implement the contemporary framework in its full capacity. Issues in developing such a unified shape modeler are discussed in the last chapter of this book.

1.3 ANATOMY OF SURFACE MODELING

In this section, existing methods of describing curves and surfaces in engineering drawings are closely examined, and then an overall schema for sculptured surface modeling is proposed. Our discussion, however, will be based on common practices of drafting rather than rigorous concepts of descriptive (projective) geometry. Further, we restrict ourselves to the problem of constructing a *computational geometric model* (CGM) from a describable shape, not from conceivable shapes. Namely, our modeling domain is *descriptive shape model* (DSM).

The term **descriptive shape model** is used to mean a geometric entity that can be described, in terms of simple geometric entities, on a drawing or a CRT screen. There are four types of geometric entities in a DSM:

- *Point*
- *Curve*
- *Surface*
- *Solid*

A DSM is intended to be interpreted by human, but its description should be consistent and unambiguous so that it can be converted to a unique CGM. This requires that a DSM be described in terms of exact numeric values as much as possible: A *point* is exactly specified by its coordinate values; a *curve* may be defined as a sequence of points or by its equation; a *surface* can be defined by a set of points (or a net of curves) or by its equation; a *solid* is defined by a set of bounding surfaces.

1.3.1 Curve Description Methods

If a curve is drawn on a plane it is called a *2D curve*, and a curve defined in space is called a *3D curve*. There are basically three ways of describing a **2D curve**:

- *by the use of 2D curve primitives (Fig. 1.5a);*
- *rounding (blending) between two curves (Fig. 1.5b);*
- *as a sequence of 2D points (Fig. 1.5c).*

A **3D curve**, on the other hand, may be described by using one of the four methods below:

- *as a sequence of 3D points (Fig. 1.5c);*
- *as an intersection between 3D surfaces (Fib. 1.5d);*
- *as a projection of a 2D curve onto a 3D surface (Fig. 1.5e);*
- *as a set of 2D curves on orthogonal projection planes (not shown).*

Typical methods of describing 2D and 3D curves are shown in Fig. 1.5. A simple method of describing a "curve segment" on a plane is the use of conic sections (ie, straight line, ellipse, parabola, etc.) which are unambiguously described by their "parameters" (eg, center point coordinates, radius values, focal points). We may call these conic sections "curve primitives". A composite curve is constructed by connecting a series of curve primitives as shown in Fig. 1.5a. When a smooth curve is required, "roundings" may be specified at some joins as shown in Fig. 1.5b.

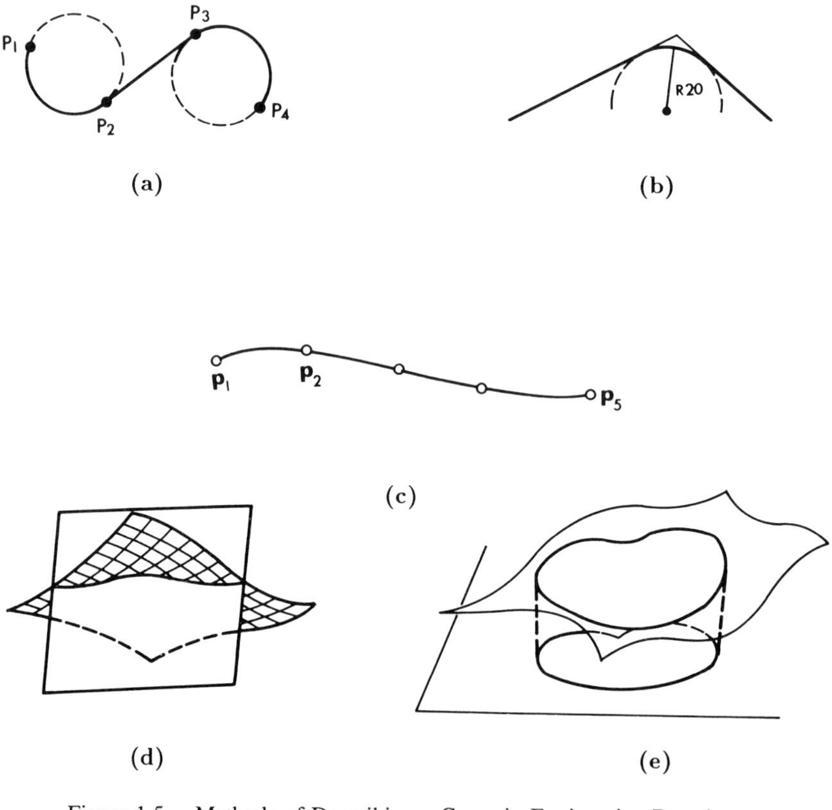

Figure 1.5 Methods of Describing a Curve in Engineering Drawing

The most popular method of describing a "free-formed" curve is to specify a sequence of points through which the curve is to pass (Fig. 1.5c). This method allows the representation of 3D space curves as well as 2D curves. An indirect method of describing a space curve is to define the curve as an intersection of two surfaces as depicted in Fig. 1.5d. In this case, the curve can not be drawn exactly. Another popular method of defining a space curve in a drawing is to draw its 2D image and then specify a "projection surface" as shown in Fig. 1.5e (it actually is a reversed projection).

1.3.2 Surface Description Methods

The geometric entity **surface** can not be *drawn* on a drawing, but it can be *described* in a drawing. There are basically five ways of describing surfaces on engineering drawings:

- *by the use of surface primitives;*
- *as a mesh of curves;*
- *as a sweeping of cross section curves;*
- *as a set of 3D points;*
- *as a blending of two or more surfaces.*

1) Surface Primitives:

A simple method of defining a surface is the use of **surface primitives**, for example quadric surfaces, which can be exactly specified by a few parameters as shown in Fig. 1.6. In order to define a useful surface, however, individual surface primitives may need to be "trimmed" and/or "compounded" (meaning that the entire surface is a Boolean sum of individual primitives).

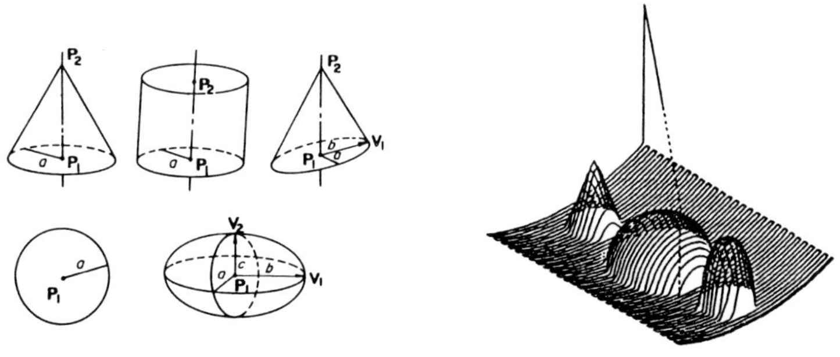

Figure 1.6 Surface Primitives and Compound Surface

2) Curve-net:

The second method of describing a surface in engineering drawing is to specify a network of curves as shown in Fig. 1.7a. The designer specifies important **feature curves** as a curve-net which is to be "filled" by smooth interpolating surfaces.

3) Sectional curves and profile curves:

A variant of the curve-net method is to define a series of cross-section **curves** together with profile curves as shown in Fig. 1.7b. Here a smooth surface is constructed by "sweeping" cross section curves along the profile curves.

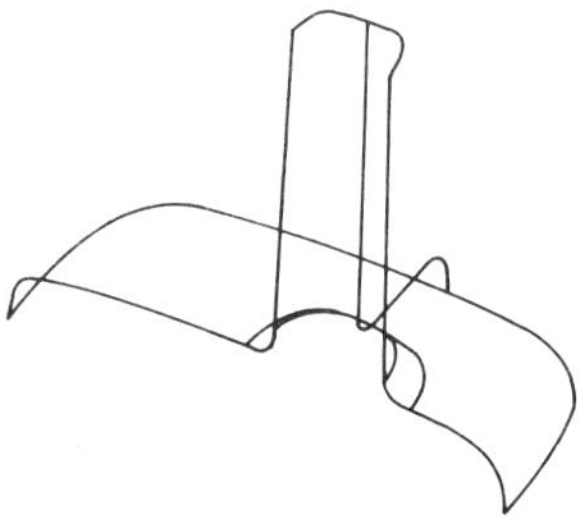

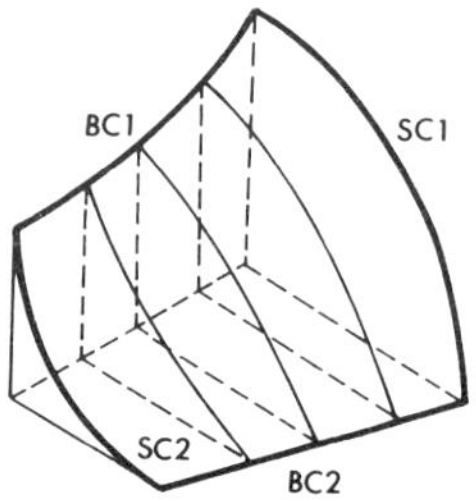

(a) (b)

Figure 1.7 Surface Defined by (a) Curve-net, and (b) Cross-sections

4) Point Data:

The fourth method of defining free-formed surfaces is to specify a **set of 3D Points**. A smooth surface is obtained by interpolating the data points. When a "clay model" is used in obtaining a final surface shape, the surface is described by a set of measured coordinate values. The point data may or may not have a regular arrangement.

5) Blending Surface:

The term *blending* is used in CAGD (computer aided geometric design) community to mean the construction of a smooth transition surface between neighboring surfaces. The term **rounding** or **filleting** is better known among engineers. A blending surface is defined by specifying a blending radius together with participating base surfaces. Examples of surface blending are given in Fig. 1.8.

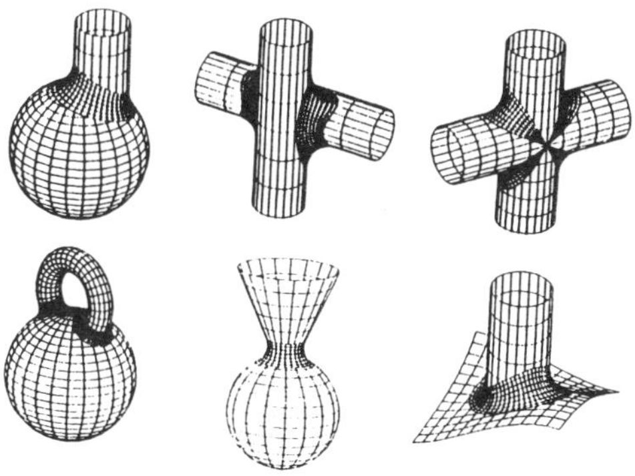

Figure 1.8 Blending Surface

1.3.3 <u>Overall Surface Modeling Schema</u>

The curve and surface description methods discussed in the previous section may be depicted as *inter-relationships among the geometric entities* as shown in Fig. 1.9. In general, a geometric entity is defined either by its **primitives** or with respect to other types of geometric entities. An arrow into a geometric entity indicates a method of defining the geometric entity.

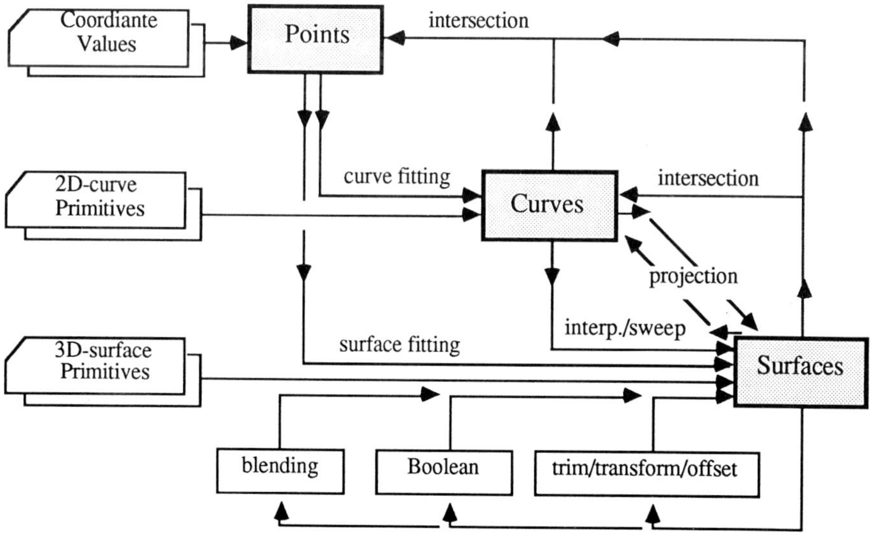

Figure 1.9 Anatomy of Surface Modeling

For example, a *point* is defined *1) by its coordinate values, 2) as an intersection of two curves, or 3) as an intersection of three surfaces.* **Methods of defining a surface** are:

- *surface primitives*
- *point data interpolation (ie, surface fitting)*
- *curve-net interpolation*
- *section curve sweeping*
- *as a blend of base surfaces.*

In addition, a predefined surface may need to be trimmed, transformed, and/or offsetted in order to obtain a new surface. To form a **compound surface** (ie, a collection of surfaces), individual surfaces may be subjected to some Boolean operations with respect to a common domain.

All of the *modeling operations* appearing in Fig. 9 will be covered in this book. That is,

- *Curve fitting method in Chap. 4,*
- *Surface fitting methods in Chaps. 7 and 8,*
- *Section curve sweeping in Chap. 9,*
- *Curve-net interpolation in Chap. 10,*
- *Blending surfaces in Chap. 11,*
- *Surface/surface intersection and trimming in Chap. 12,*
- *Surface primitives and Boolean operation in Chap. 13.*

In general, a geometric entity is constructed from lower level ones (eg, curve from points, surface from points or curves) or by specifying parameters of its primitives. The lower level geometric entities and parameters used in constructing a geometric entity are sometimes called **geometric handles** which can be "turned" to modity the shape. A polyhedron may also serve as a handle for a curved object as will be seen in Chapter 11. Handle points that are not on the curve or surface are called **control points** which are widely used in interactive design.

This book deals mainly with fixed handles (ie, existing descriptive models), but all the methods introduced in the book may be used in interactive shape design as well, if we allow the handles to be turned. In general, a CAM system is concerned with *fixed handles*, while a CAD system is used in designing an object by *turning the handles.* Practical approaches to developing commercial CAM software systems are proposed in Chapter 14, and a schema of unified shape modeler is proposed in Chapter 15. The "integrated CAM architecture" proposed in Chapter 14 is well suited for the concept of "concurrent engineering", while the "unified shape modeler" is a necessity for CIM.

1.4 ISSUES IN SCULPTURED SURFACE MACHINING

Once a computational geometric model (CGM) has been constructed (refer to Fig. 1.3), a series of NC-codes are to be generated in order to machine the surface on NC machine tools. A CGM of a surface is usually expressed as a vector valued function $\mathbf{r}(u, v)$. In die cavity machining, for example, a collection of individual CGMs have to be machined in a single setting. The following sequence of metal removal operations may be needed in obtaining a finished die surface starting from a block of raw stock:

a) *Drilling* of initial holes for subsequent endmilling.

b) *Endmilling* of initial cavity.

c) Rough cutting with larger ball-nosed endmills.

d) Finish cutting with smaller ball-nosed endmills.

e) EDM (electrical discharge machining) for sharp internal corners.

f) Grinding and scraping.

The first three steps are called **roughing operations**, and the last three are **finishing operations**.

The main issues in sculptured surface machining are of course productivity and quality. For an economical machining of sculptured surface, we need to have a good machining process plan so that machining sequences, cutter passes, and cutting conditions are all optimized. Process planning for the machining of sculptured surfaces falls into two categories: Rough cut planning and finish cut planning. Maximizing metal removal rates is the main criteria in rough cut planning, while minimizing cutter travel distances is an important measure in finish cut planning. Another very important productivity issue is the trade-off between machining and scraping. Scraping is a time consuming process because it is in most part done manually even though some progresses have been made in applying industrial robots (Lilly *et al*, 1988).

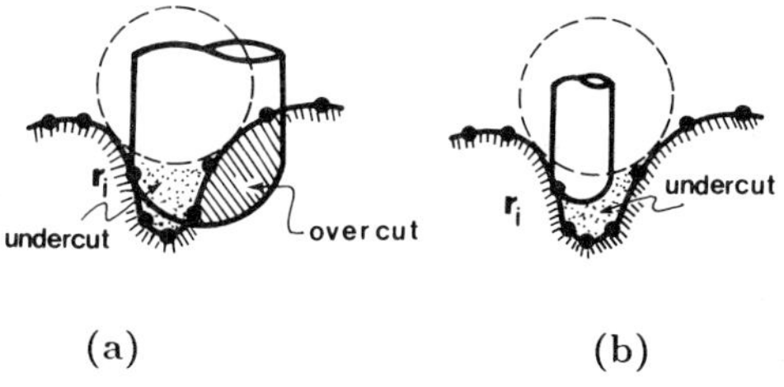

Figure 1.10 Cutter Interferences in NC Machining

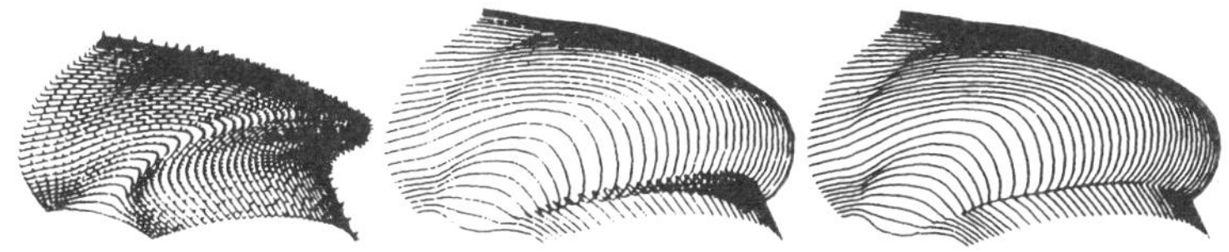

Figure 1.11 Cutter Interference Handling

The quality issue is concerned with inaccuracy and **gouging**: Machining inaccuracy comes from various sources such as tool wear, cutter deflection, improper fixturing, and presetting errors. Gouging or cutter interference is one of the most critical problems in sculptured surface machining. When machining cavity regions, the cutter easily invades a portion of the surface as depicted in Fig. 1.10. Sometimes a considerable amount of gouging comes from cutter deflection. Thus, it is essential to have mechanisms for verifying NC-codes before actual machining is carried out. An NC verification system should be able to identify possible cutter deflections as well. When gouging occurs, the cutter paths form a loop which has to be removed from the planned cutter paths. An example is given in Fig. 1.11. For more details, the reader is referred to Choi and Jun (1989).

1.5 DISCUSSIONS

This book is based on the "contemporary framework for curved object manufacturing" depicted in Fig. 1.3, in which manufacturing begins with a *descriptive shape model* (DSM). In other words, this book is concerned with surface modeling which would make the manufacturing activities *computer-aided*. However, the surface modeling methods to be introduced in this book may well be implemented in interactive CAD systems as well. In practice, a growing number of commercial CAD/CAM systems supports some of the 'surface modeling' function of Fig. 1.3. This means that a CGM of the surface may be obtained directly during the CAD stage. In this respect, this book should be the concern of both manufacturing engineer and design engineer.

CHAPTER 2

BASICS OF DIFFERENTIAL GEOMETRY AND COORDINATE TRANSFORMATION

2.1 INTRODUCTION

In this chapter we review elementary results in differential geometry and coordinate transformation that are relevant to surface modeling and machining. Most of the results presented here will be used in later chapters. Another purpose of this chapter is to introduce new terminologies in the fields. The topics to be discussed include:

- *methods of representing curves and surfaces,*
- *properties of curve such as unit tangent and curvature,*
- *properties of surface such as unit normal and normal curvature, and*
- *concept of coordinate transformations and frames.*

The reader who is familiar with the subjects may skip the entire chapter.

2.2 BASICS OF VECTOR ALGEBRA

In this section, we give a minimum set of results in vector algebra, understanding of which is essential for our discussions on the properties of curves and surfaces.

2.2.1 Vectors in Cartesian Space

Consider a point $P(x_1, y_1, z_1)$ in a Cartesian coordinate system. As shown in Fig. 2.1a, the position of P may be regarded as a point obtained as a result of a series of **orthogonal displacements** from the origin 0. That is, move the origin point in x-axis direction by x_1, move the resulting point in y-axis direction by y_1, and then move it in z-axis direction by z_1. Then, the total displacement **a** may be expressed as

$$\mathbf{a} = (x_1, y_1, z_1)$$

which is a vector representation of the point $P(x_1, y_1, z_1)$. Let **i**, **j**, **k** denote unit displacements in x-, y-, z-axis directions, respectively:

$$\mathbf{i} = (1, 0, 0)$$

$$\mathbf{j} = (0, 1, 0)$$

$$\mathbf{k} = (0, 0, 1).$$

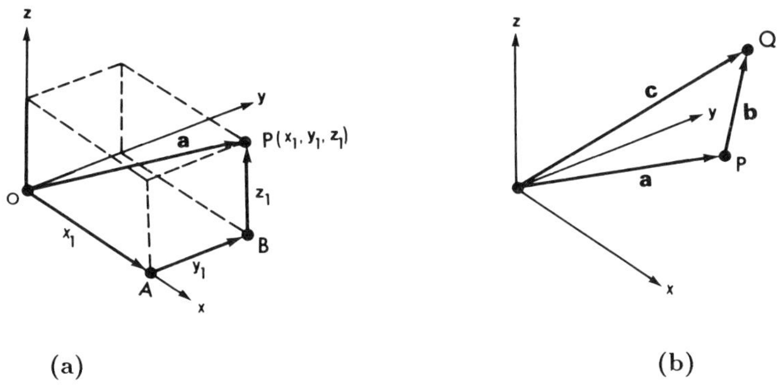

Figure 2.1 Definition of a Vector

Then, the total displacement a can be expressed as a sum of orthogonal displacements. That is,

$$\mathbf{a} = x_1\mathbf{i} + y_1\mathbf{j} + z_1\mathbf{k}.$$

The quantity **a** in the above expression is called a **vector**, and it is specified by its *direction* and *magnitude*.

Now let us move the point P to a new position Q as shown in Fig. 2.1b. Let

$$\mathbf{b} = (x_2,\ y_2,\ z_2)$$

represent the relative displacement from P to Q, then the total displacement c of Q from the origin O is the sum of the two displacements. Namely, we have

$$\begin{aligned}\mathbf{c} &= \mathbf{a} + \mathbf{b}\\ &= (x_1 + x_2,\ y_1 + y_2,\ z_1 + z_2)\\ &= (x_1 + x_2)\mathbf{i} + (y_1 + y_2)\mathbf{j} + (z_1 + z_2)\mathbf{k}.\end{aligned}$$

We have demonstrated that two vectors are added by simply adding scalar values of corresponding *elements* and that a vector can be multiplied by a scalar. The **magnitude** of a vector **a** is given as an Euclidean norm. Namely,

$$|\mathbf{a}| = (x_1^2 + y_1^2 + z_1^2)^{1/2} \tag{2.1}$$

We can *normalize* a vector by dividing it by its magnitude. That is, we have

$$\mathbf{u} = \mathbf{a}\ /\ |\mathbf{a}|$$

which is called a *unit vector* in the *direction* of **a**.

2.2.2 Vector Multiplications

Addition, subtraction, scalar multiplication, and scalar differentiation of vectors are all carried out element-wise. But multiplying two vectors is not that intuitive. Let us consider the two vectors $\mathbf{a}$, $\mathbf{b}$ arranged as in Fig. 2.2 where A is the common starting point of both vectors and θ is the angle between the two vectors.

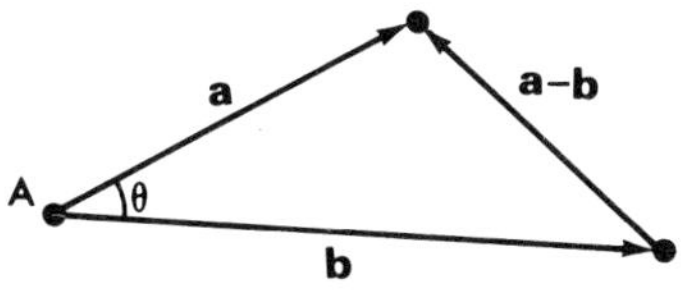

Figure 2.2 Definition of Scalar and Vector Products

The **scalar product** of $\mathbf{a}$ and $\mathbf{b}$ is defined as

$$\mathbf{a} \cdot \mathbf{b} = x_1 x_2 + y_1 y_2 + z_1 z_2. \qquad (2.2-a)$$

An application of the second cosine rule to the sides of the triangle in Fig. 2.2 will result in

$$|\mathbf{a} - \mathbf{b}|^2 = |\mathbf{a}|^2 + |\mathbf{b}|^2 - 2|\mathbf{a}||\mathbf{b}|cos\theta,$$

where θ is the angle between $\mathbf{a}$ and $\mathbf{b}$. Then, from the definition of scalr product, we have

$$\begin{aligned} 2|\mathbf{a}||\mathbf{b}|cos\theta &= |\mathbf{a}|^2 + |\mathbf{b}|^2 - |\mathbf{a} - \mathbf{b}|^2 \\ &= \mathbf{a} \cdot \mathbf{a} + \mathbf{b} \cdot \mathbf{b} - (\mathbf{a} - \mathbf{b}) \cdot (\mathbf{a} - \mathbf{b}) \\ &= (x_1^2 + y_1^2 + z_1^2) + (x_2^2 + y_2^2 + z_2^2) - \{(x_1 - x_2)^2 + (y_1 - y_2)^2 + (z_1 - z_2)^2\} \\ &= 2(x_1 x_2 + y_1 y_2 + z_1 z_2) = 2\ \mathbf{a} \cdot \mathbf{b}. \end{aligned}$$

Thus, another way of expressing the above scalar product is

$$\mathbf{a} \cdot \mathbf{b} = |\mathbf{a}|\ |\mathbf{b}|cos\theta. \qquad (2.2-b)$$

A geometric interpretation of the new scalar product form is that it represents a *projected length* of the first vector in the direction of the second vector (if the second vector is a unit vector). Further, the modulus of the vector $\mathbf{a}$ is given by

$$|\mathbf{a}|^2 = \mathbf{a} \cdot \mathbf{a}.$$

A **vector product** of two vectors a,b is a vector perpendicular to both vectors (pointing into the paper in Fig. 2.2), and it is defined as

$$\mathbf{a} \times \mathbf{b} = (y_1 z_2 - z_1 y_2)\mathbf{i} + (z_1 x_2 - x_1 z_2)\mathbf{j} + (x_1 y_2 - y_1 x_2)\mathbf{k} \qquad (2.3-a)$$

which is easier to memorize if written in the form of a determinant as:

$$\mathbf{a} \times \mathbf{b} = \begin{vmatrix} \mathbf{i} & \mathbf{j} & \mathbf{k} \\ x_1 & y_1 & z_1 \\ x_2 & y_2 & z_2 \end{vmatrix}$$

From (2.3-a) and (2.2), the modulus of a × b is evaluated as follows:

$$\begin{aligned} |\mathbf{a} \times \mathbf{b}|^2 &= (y_1 z_2 - z_1 y_2)^2 + (z_1 x_2 - x_1 z_2)^2 + (x_1 y_2 - y_1 x_2)^2 \\ &= (x_1^2 + y_1^2 + z_1^2)(x_2^2 + y_2^2 + z_2^2) - (x_1 x_2 + y_1 y_2 + z_1 z_2)^2 \\ &= |\mathbf{a}|^2 |\mathbf{b}|^2 - |\mathbf{a}|^2 |\mathbf{b}|^2 cos^2\theta \\ &= |\mathbf{a}|^2 |\mathbf{b}|^2 sin^2\theta. \end{aligned}$$

Thus, an alternative expression for vector product is obtained as

$$\mathbf{a} \times \mathbf{b} = |\mathbf{a}||\mathbf{b}| sin\theta \, \mathbf{u} \qquad (2.3-b)$$

where $\mathbf{u}$ is a unit vector in the direction of a × b. A geometric interpretation of the above vector product form is that *its magnitude represents the area of the parallelogram formed by the two vectors* a, b. A few algebraic properties of scalar and vector products are listed below:

a) $\mathbf{a} \cdot \mathbf{b} = \mathbf{b} \cdot \mathbf{a}$; $\mathbf{a} \times \mathbf{b} = -\mathbf{b} \times \mathbf{a}$
b) $\mathbf{a} \cdot (\mathbf{b} + \mathbf{c}) = \mathbf{a} \cdot \mathbf{b} + \mathbf{a} \cdot \mathbf{c}$; $\mathbf{a} \times (\mathbf{b} + \mathbf{c}) = \mathbf{a} \times \mathbf{b} + \mathbf{a} \times \mathbf{c}$
c) $\mathbf{a} \cdot \mathbf{a} = |\mathbf{a}|^2$; $\mathbf{a} \times \mathbf{a} = 0$
d) $\mathbf{i} \cdot \mathbf{i} = 1$, $\mathbf{i} \cdot \mathbf{j} = 0$, etc.(holds for any two perpendicular unit vectors)
e) $\mathbf{i} \times \mathbf{j} = \mathbf{k}$, $\mathbf{j} \times \mathbf{k} = \mathbf{i}$, $\mathbf{k} \times \mathbf{i} = \mathbf{j}$.

A **triple scalar product** is a combination of the scalar and vector products and is expressed as

$$\mathbf{c} \cdot (\mathbf{a} \times \mathbf{b}) = x_3(y_1 z_2 - z_1 y_2) + y_3(z_1 x_2 - x_1 z_2) + z_3(x_1 y_2 - y_1 x_2) \qquad (2.4-a)$$

where c = (x_3, y_3, z_3), which is easier to memorize if written in the determinant form

$$\mathbf{c} \cdot (\mathbf{a} \times \mathbf{b}) = \begin{vmatrix} x_1 & y_1 & z_1 \\ x_2 & y_2 & z_2 \\ x_3 & y_3 & z_3 \end{vmatrix} \qquad (2.4-b)$$

A triple vector product is a combination of two vector products. The following identity relation is very convenient in evaluating a triple vector product:

$$\mathbf{a} \times (\mathbf{b} \times \mathbf{c}) = (\mathbf{a} \cdot \mathbf{c})\mathbf{b} - (\mathbf{a} \cdot \mathbf{b})\mathbf{c}. \tag{2.5}$$

The above identity is easily verified by faithfully evaluating both sides of the expression.

2.3 CURVE GEOMETRY

Intuitively, a curve is defined as a trajectory of a point satisfying certain constraints. This section presents different methods of representing curves and basic properties of curves.

2.3.1 Representation of Curves

There are three ways of representing a curve. In terms of the functional form of a curve, a curve may be classified as

- *an implicit curve,*
- *an explicit (or non-parametric) curve, or*
- *a parametric curve.*

Let us consider a unit circle drawn on a sheet of paper by using compasses. In order to obtain a mathematical representation of the circle, we first need to define a coordinate system on the paper where the circle is to be drawn. Now we draw a circle with one leg fixed at the origin of the xy-coordinate system as shown in Fig. 2.3a. Since the distance between the center (0,0) of the circle and a point (x, y) on the circle is 1, we have the following relation between the two coordinate variables:

$$x^2 + y^2 = 1 \quad or \quad x^2 + y^2 - 1 = 0 \tag{2.6}$$

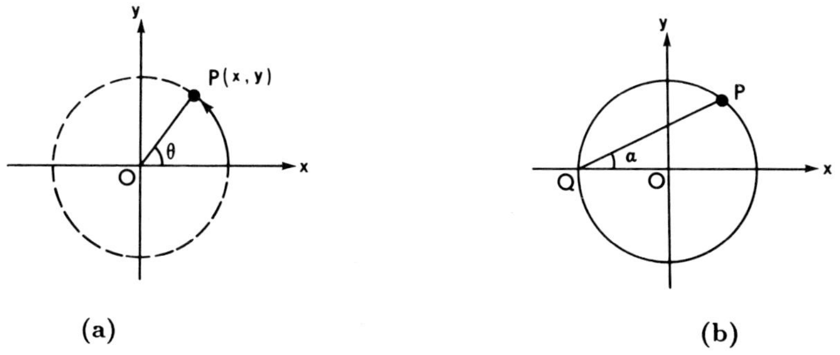

Figure 2.3 Parameterization of a Unit Circle at the Origin

If we need only upper half of the unit circle, the above expression is rewritten as

$$y = (1 - x^2)^{1/2}. \tag{2.7}$$

Equation (2.6) is in a form $g(x, y) = 0$, called **implicit equation** of a curve, and Eqn. (2.7) is said to be in an **explicit** or **nonparametric** form. Let θ denote the angle between the line PO and x-axis as shown in Fig. 2.3a, then individual coordinate values become functions of θ. That is,

$$x = x(\theta) = cos\, \theta \ ; \quad y = y(\theta) = sin\, \theta, \tag{2.8}$$

where θ is the *parameter* of the curve. The above equation is called a **parametric equation** of a curve.

Let us consider a different parameter for the unit circle. As shown in Fig. 2.3b, this time the angle α between the line PQ and x-axis is chosen, where Q is the left intersecting point of the circle with the x-axis line. Then, by definition, we have

$$tan\, \alpha = y \ / \ (x + 1).$$

From this equation and Eqn. (2.6), we immediately obtain the following:

$$x = x(t) = (1 - t^2)/(1 + t^2) \ ; \quad y = y(t) = 2t/(1 + t^2) \tag{2.9}$$

where $t = tan\, \alpha$. The above equation is also a parametric equation of the unit circle, and it is said to be in a **rational polynomial** (parametric) **form** because individual equations are defined as ratios of polynomials in the parameter t. The process of obtaining a (rational) parametric representation from an implicit polynomial equation (of curve or surface) is called *parameterization*. An excellent discussion on the subject is given in Waggenspack *et al* (1987).

A **space curve** is conveniently represented by using a parametric equation of the form

$$x = x(t), \quad y = y(t), \quad z = z(t).$$

For notational convenience, we use the following vector notation when describing a space curve in a 3D Cartesian coordinate system:

$$\mathbf{r}(t) = (x(t),\ y(t),\ z(t)).$$

With a parametric representation, a segment of a curve is easily defined by specifying the range of the parameter. A space curve can not be represented by using a single implicit (or explicit) equation. Since an implicit equation of the form $g(x, y, z) = 0$ stands for a surface, we need two such equations to define a space curve. In this case, the space curve to be defined is in fact an intersection of the two surfaces.

2.3.2 Properties of a Curve

In the discussions that follow, the term *"curve"* is used to mean a *"regular parametric representation of the curve"*. Curve properties to be discussed are

- *flow rate of the curve,*
- *unit tangent vector,*
- *curvature,*
- *principal normal vector, and*
- *radius of curvature.*

Let us consider a regular parametric representation of a curve given as

$$\mathbf{r} = \mathbf{r}(t) = (x(t),\ y(t),\ z(t)).$$

Then, the derivative of the vector valued function $\mathbf{r}(t)$ is defined as

$$\dot{\mathbf{r}}(t) = d\mathbf{r}(t)/dt = (dx/dt,\ dy/dt,\ dz/dt).$$

Higher order derivatives are defined similarly.

1) Flow rate of a curve:

The magnitude of the derivative vector $\dot{\mathbf{r}}$(t) is called **flow rate** of the curve. That is,

$$\dot{s}(t) = |\dot{\mathbf{r}}(t)|. \tag{2.10}$$

Imagine that the curve is a road and the parameter t represents time. Then flow-rate corresponds to how fast you drive a car on the road. If we take the accumulated length (ie, the mileage reading of the car) of the curve (ie, road) as its parameter s, the resulting curve equation $\mathbf{r}(s)$ becomes a *naturally parameterized* curve which has a uniform flow rate of 1. Flow rate is not a property of the curve itself but a result of parameterization.

2) Unit tangent vector:

Properties of a curve that are independent of parameterization are called **intrinsic properties** of the curve. *Unit tangent* vector (gradient) and *curvature* are the two most important intrinsic properties. Let s be the natural parameter (ie, arc length) of a curve $\mathbf{r}(t)$. That is,

$$s = \int_0^s |\dot{\mathbf{r}}(t)|dt$$

then, **unit tangent vector** of the curve $\mathbf{r}(t)$ is defined as

$$\mathbf{T} = d\mathbf{r}/ds. \tag{2.11 - a}$$

By applying a chain rule of differentiation, an alternate expression for the unit tangent vector is obtained as

$$\mathbf{T} = \dot{\mathbf{r}}(t)/|\dot{\mathbf{r}}(t)|. \tag{2.11 - b}$$

3) Curvature of a curve:

Let s and $\mathbf{T}$ respectively denote a natural parameter and unit tangent vector of a curve $\mathbf{r}(t)$, then **curvature** of the curve is defined as

$$\kappa = |d\mathbf{T}/ds|. \tag{2.12 - a}$$

By applying the chain rule of differentiation and after some algebraic manipulation, the *curvature* of the curve is given by

$$\kappa = \frac{|\dot{\mathbf{r}} \times \ddot{\mathbf{r}}|}{|\dot{\mathbf{r}}|^3} \tag{2.12 - b}$$

where, $\dot{\mathbf{r}} \equiv d\mathbf{r}(t)/dt$ and $\ddot{\mathbf{r}} \equiv d\dot{\mathbf{r}}(t)/dt$. For an explicit 2D curve of the form $y = y(x)$, the above equation simplifies to

$$\kappa = y''/(1 + y'^2)^{3/2},$$

where, $y' = dy/dx$ and $y'' = dy'/dx$. This may be easily verified by expressing $(2.12 - b)$ in its component form and using the relations $x = t$, $y = y(t)$ and $z = 0$.

4) Principal normal vector:

By differentiating the unit tangent $\mathbf{T}$ with respect to t and then normalizing the result, we obtain another unit vector $\mathbf{N}$ called **principal normal vector** of the curve. That is,

$$\mathbf{N} = (d\mathbf{T}/dt)/|d\mathbf{T}/dt| \equiv (d\mathbf{T}/ds)/|d\mathbf{T}/ds|. \tag{2.13}$$

Since $\mathbf{T}$ is a unit vector (ie, $\mathbf{T} \cdot \mathbf{T} = 1$), $\mathbf{N}$ should be perpendicular to $\mathbf{T}$ as depicted in Fig. 2.4. The plane defined by $\mathbf{T}$ and $\mathbf{N}$ is called the **osculating plane**. A third vector perpendicular to both $\mathbf{T}$ and $\mathbf{N}$ is called a **binormal vector** which is given by $\mathbf{B} = \mathbf{T} \times \mathbf{N}$.

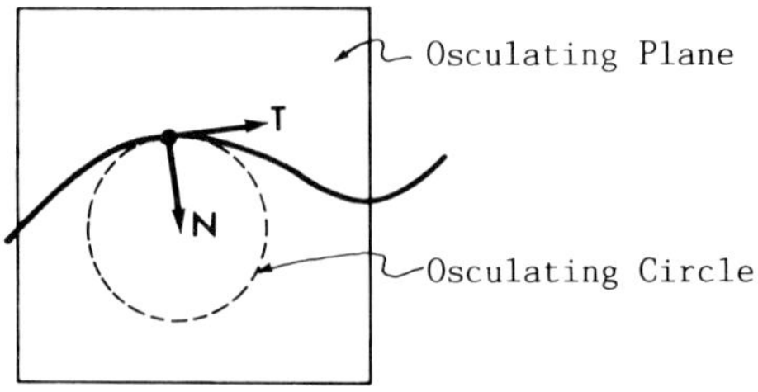

Figure 2.4 Principal Normal Vector N and Osculating Plane

5) Radius of curvature:

Let us define a circle on the osculating plane (see Fig. 2.4) such that it passes through the current point **r**(t) and its curvature is the same as that of the curve at that point. This circle is called the **osculating circle**. The radius of this circle is called the **radius of curvature** ρ of the curve, and it is given by

$$\rho = 1/\kappa \,. \tag{2.14}$$

Example 2.1 : *Find the unit tangent vector* **T** *and the radius of curvature* ρ *of the parametric curve given by* $(x = \cos\theta,\ y = \sin\theta,\ z = 0)$.

By differentiating the curve equation $\mathbf{r}(\theta) = (\cos\theta,\ \sin\theta,\ 0)$, *we obtain*

$$\dot{\mathbf{r}} = (-\sin\theta,\ \cos\theta,\ 0) \quad and \quad \ddot{\mathbf{r}} = (-\cos\theta,\ -\sin\theta,\ 0).$$

Thus, the unit tangent vector is obtained as:

$$\mathbf{T} = \dot{\mathbf{r}}/|\dot{\mathbf{r}}| = (-\sin\theta,\ \cos\theta,\ 0).$$

By applying the vector product formula $(2.3 - a)$, *we have*

$$\dot{\mathbf{r}} \times |\ddot{\mathbf{r}}| = (0,\ 0,\ 1).$$

From (2.12), *the curvature* κ *is found to be 1, and so is the radius of curvature* (2.14). *This is expected because the curve is a unit circle.* ◇

Still another property of a space curve is the **torsion** of the curve. The torsion of a curve is given by (Lipschutz, 1969)

$$\tau = -(d\mathbf{B}/ds) \cdot \mathbf{N},$$

where **B** and **N** are binormal and principal normal vectors, respectively. The principal equations concerning the properties of a space curve are called the **Serret-Frenet equations**. They are

$$d\mathbf{r}/ds = \mathbf{T},$$

$$d\mathbf{T}/ds = \kappa\mathbf{N},$$

$$d\mathbf{N}/ds = \tau\mathbf{B} - \kappa\mathbf{T},$$

$$d\mathbf{B}/ds = -\tau\mathbf{N}.$$

In the above equations, s is the natural parameter of the curve $\mathbf{r}(s)$ and κ and τ are *curvature* and *torsion* of the curve, respectively.

2.4 SURFACE GEOMETRY

The section presents basic concepts of surface representation and some useful results on surface properties. More detailed discussions on the subject may be found in Faux and Pratt (1980) or a textbook on differential geometry such as the one by Lipschutz (1969).

2.4.1 Representation of Surfaces

Any physical object is bounded by its surfaces. The mathematical equation describing a surface may be in *implicit form, parametric form, or nonparametric form.*

1) Implicit representation:

Let us take a sphere of radius 1 centered at the origin of a 3D Cartesian coordinate system. Then the inside of the sphere is a set of points satisfying the following inequality:

$$x^2 + y^2 + z^2 < 1.$$

And the equation

$$x^2 + y^2 + z^2 = 1 \quad or \quad x^2 + y^2 + z^2 - 1 = 0 \tag{2.15}$$

defines the surface of the sphere. In general, an **implicit equation** of the form $g(x, y, z) = 0$ represents the boundary (surface) of the two disjoint half spaces $g(x, y, z) > 0$ and $g(x, y, z) < 0$.

2) Parametric representation:

In differential geometry, a *surface* is defined as *"the image of a sufficiently regular mapping of a set of points in a plane (domain) into 3D space"*, and it is expressed as

$$\mathbf{r}(u, v) = (x(u, v),\ y(u, v),\ z(u, v)) \tag{2.16}$$

where u and v are **parameters** of the surface. Returning to our unit sphere, one may easily parameterize the implicit equation (2.15) by taking u and v as longitude and latitude, respectively. That is,

$$\mathbf{r}(u, v) = (cos\, v\ cos\, u,\ cos\, v\ sin\, u,\ sin\, v) \tag{2.17}$$

with $0 \leq u \leq 2\pi$ and $-\pi/2 \leq v \leq \pi/2$. As with the unit circle (2.9), a different parameterization using a rational polynomial form is possible. Such a parametric form is given in Mudur (1986) as

$$x(u, v) = \frac{(1 - u^2)(1 - v^2)}{(1 + u^2)(1 + v^2)}; \quad y(u, v) = \frac{2u(1 - v^2)}{(1 + u^2)(1 + v^2)}; \quad z(u, v) = \frac{(1 + u^2)2v}{(1 + u^2)(1 + v^2)}.$$

3) Non-parametric representation:

When the domain of a surface is taken to be the xy-plane of a Cartesian coordinate system, the parametric form (2.16) becomes a **nonparametric** equation (Note $u \equiv x,\ v \equiv y$):

$$\mathbf{r}(u,v) = (u,\ v,\ z(u,v)) \quad or \quad z = z(x,y). \tag{2.18}$$

If we are interested only in the upper hemisphere of the unit sphere, Eqn. (2.15) is expressed in an explicit form as

$$z = (1 - x^2 - y^2)^{1/2} \quad for \quad (x^2 + y^2) \leq 1. \tag{2.19}$$

If the surface is defined on a finite (bounded) domain, it is called a surface **patch** or simply a patch. An assembly of patches with prescribed interpatch continuity conditions is called a **composite** surface. The term **compound** surface is sometimes used to denote a collection of topologically unrelated surface elements specified in a domain of interest.

2.4.2 Tangent and Normal of a Surface

Let us define a 2D parametric curve $\mathbf{u}(t)$ on the domain (u, v-plane) of a parametric surface $\mathbf{r}(u,v)$ as shown in Fig. 2.5. That is,

$$\mathbf{u}(t) = [u(t)\ v(t)]^T. \tag{2.20}$$

Then, the map of $\mathbf{u}(t)$ forms a curve $\mathbf{r}(t)$ lying on the surface $\mathbf{r}(u,v)$ so that

$$\begin{aligned} \mathbf{r}(t) &= \mathbf{r}(u(t),\ v(t)) \\ &= (x(u(t),\ v(t)),\ y(u(t),\ v(t)),\ z(u(t),\ v(t))). \end{aligned} \tag{2.21}$$

A special case of (2.20) is an *isoparametric curve* in which $u(t) = t$ and $v = v^*$ or $v(t) = t$ and $u = u^*$.

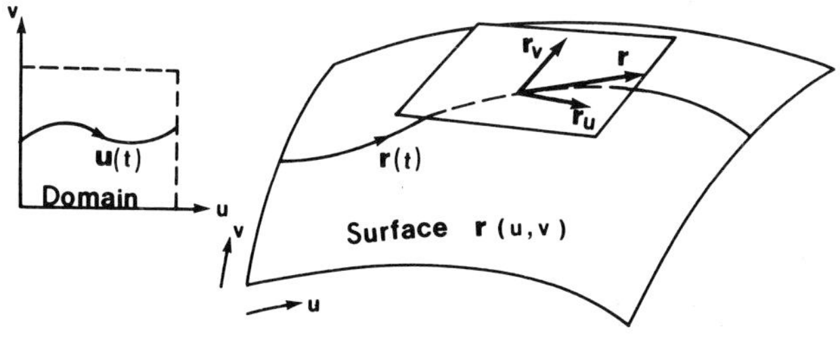

Figure 2.5 Curve on the Surface and Tangent Plane

1) Tangent vector of a surface:

To begin with, let us define partial derivatives of $\mathbf{r}(u, v)$ as

$$\mathbf{r}_u = \partial\mathbf{r}/\partial u\ ; \quad \mathbf{r}_{uv} = \partial^2\mathbf{r}/\partial u\partial v \quad ; \quad etc. \tag{2.22}$$

On differentiating (2.21) with respect to t, we have

$$\dot{\mathbf{r}} = \frac{d\mathbf{r}}{dt} = \frac{\partial\mathbf{r}}{\partial u}\cdot\frac{du}{dt} + \frac{\partial\mathbf{r}}{\partial v}\cdot\frac{dv}{dt} = \mathbf{r}_u\dot{u} + \mathbf{r}_v\dot{v}\ . \tag{2.23}$$

Note in (2.23) that $\dot{\mathbf{r}}$ is the tangent vector of $\mathbf{r}(t)$ and $\mathbf{r}_u$ and $\mathbf{r}_v$ are tangent vectors of *isoparametric curves*. The three tangent vectors $\dot{\mathbf{r}}$, $\mathbf{r}_u$, $\mathbf{r}_v$ define a plane called *tangent plane*.

2) Normal vector of a surface:

A unit vector $\mathbf{n}$ normal to the tangent plane is called the **unit normal vector** of the surface at the current point. The unit normal vector is obtained by *normalizing* the vector product of $\mathbf{r}_u$ and $\mathbf{r}_v$.

$$\mathbf{n} = (\mathbf{r}_u \times \mathbf{r}_v)/|\mathbf{r}_u \times \mathbf{r}_v|. \tag{2.24}$$

The unit normal vector is essential for surface interrogation. For example, an **offset surface** (with offset distance d) is given by

$$\mathbf{r}^{\circ}(u, v) = \mathbf{r}(u, v) + d\ \mathbf{n}(u, v).$$

The normal vector N of an implicit surface $g(x, y, z) = 0$ is given by

$$\mathrm{N} = (\partial g/\partial x,\ \partial g/\partial y,\ \partial g/\partial z) \tag{2.25}$$

and the unit normal vector $\mathbf{n}$ is obtained by normalizing N. Namely,

$$\mathbf{n} = \mathrm{N}\ /\ |\mathrm{N}|.$$

Example 2.2 : *$ax + by + cz - d = 0$ is an implicit equation of a plane. Find its unit normal vector and the distance of the plane from the origin.*

From (2.25), a normal vector of the plane is given by $\mathrm{N} = (a, b, c)$. Thus, unit normal vector is expressed as

$$\mathbf{n} = (a/e,\ b/e,\ c/e) \quad with \quad e = |\mathrm{N}| = (a^2 + b^2 + c^2)^{1/2}.$$

By dividing the original equation by e, it is expressed as

$$\mathbf{n}\cdot\mathbf{r} = d/e, \quad where \quad \mathbf{r} = (x,\ y,\ z).$$

From the definition of the scalar product (2.2), $\mathbf{n}\cdot\mathbf{r}$ becomes the length of $\mathbf{r}$ projected in the direction of the unit vector $\mathbf{n}$. Thus, the (perpendicular) distance of the plane from the origin is equal to $d/(a^2 + b^2 + c^2)^{1/2}$. ◇

3) First fundamental matrix:

The tangent vector given by Eqn. (2.23) may be expressed in a matrix form as (vectors are treated as column vectors):

$$\dot{\mathbf{r}} = \mathbf{r}_u \dot{u} + \mathbf{r}_v \dot{v} = \mathbf{A}\ \dot{\mathbf{u}}, \tag{2.26}$$

where $\mathbf{A} = [\mathbf{r}_u\ \mathbf{r}_v]$, and $\dot{\mathbf{u}} = d\mathbf{u}(t)/dt = (du/dt,\ dv/dt) = [\dot{u}\ \dot{v}]^T$. And the modulus of the tangent vector is evaluated as follows.

$$|\dot{\mathbf{r}}|^2 = \dot{\mathbf{r}} \cdot \dot{\mathbf{r}} = (\dot{\mathbf{r}})^T(\dot{\mathbf{r}}) = \dot{\mathbf{u}}^T\mathbf{A}^T\mathbf{A}\ \dot{\mathbf{u}} = \dot{\mathbf{u}}^T\mathbf{G}\ \dot{\mathbf{u}}, \tag{2.27}$$

where $\mathbf{G}$ is called the **first fundamental matrix** and is expressed as

$$\mathbf{G} = \mathbf{A}^T\mathbf{A} = \begin{bmatrix} \mathbf{r}_u \cdot \mathbf{r}_u & \mathbf{r}_u \cdot \mathbf{r}_v \\ \mathbf{r}_u \cdot \mathbf{r}_v & \mathbf{r}_v \cdot \mathbf{r}_v \end{bmatrix}. \tag{2.28}$$

Thus, the unit tangent vector $\mathbf{T}$ is expressed in terms of $\mathbf{G}$ as

$$\mathbf{T} = \dot{\mathbf{r}}/|\dot{\mathbf{r}}| = (\mathbf{A}\ \dot{\mathbf{u}})/(\dot{\mathbf{u}}^T\mathbf{G}\ \dot{\mathbf{u}})^{1/2}. \tag{2.29}$$

The following is a useful relation for computing surface areas and cross-section areas:

$$|\mathbf{r}_u \times \mathbf{r}_v| = |\mathbf{G}|^{1/2} \tag{2.30}$$

For example, the area of a parametric surface $\mathbf{r}(u,v)$ is expressed as

$$S = \int\int |\mathbf{r}_u \times \mathbf{r}_v| du dv = \int\int |\mathbf{G}|^{1/2} du dv.$$

Example 2.3 : *Find the unit tangent vector of the curve* $\mathbf{r}(t) = \mathbf{r}(u(t),\ v(t)) = \mathbf{r}(t,t)$ *on the surface of a paraboloid* $\mathbf{r}(u,v) = (u\cos v, u\sin v, u^2/2)$ *with* $0 \le u \le 2,\ 0 \le v \le 2\pi$. *Note* $\mathbf{u}(t) = (t,t)$. *All the vectors are column vectors.*

Since $\dot{\mathbf{u}} = [1\ 1]^T$, $\mathbf{r}_u = [\cos v\quad \sin v\quad u]^T$, *and* $\mathbf{r}_v = [-u\sin v\quad u\cos v\quad 0]^T$, *we have*

$$\mathbf{A}\dot{\mathbf{u}} = [\cos v - u\sin v\quad \sin v + u\cos v\quad u]^T\ ;$$

$$\mathbf{r}_u \cdot \mathbf{r}_u = (1+u^2), \quad \mathbf{r}_u \cdot \mathbf{r}_v = 0, \quad \text{and} \quad \mathbf{r}_v \cdot \mathbf{r}_v = u^2.$$

Thus, the first fundamental matrix is evaluated as

$$\mathbf{G} = \begin{bmatrix} 1+u^2 & 0 \\ 0 & u^2 \end{bmatrix}, \quad \text{and} \quad \dot{\mathbf{u}}^T\mathbf{G}\ \dot{\mathbf{u}} = 1 + 2u^2.$$

Finally, from Eqn. (2.29), *the unit tangent vector is given as*

$$\mathbf{T}(t) = [\cos t - t\sin t\quad \sin t + t\cos t\quad t]^T/(1+2t^2)^{1/2} \quad \text{since } u = v = t. \qquad \diamond$$

2.4.3 Curvature of Surface

1) Second fundamental matrix:

Let us consider the curve $\mathbf{r}(t)$ on the surface $\mathbf{r}(u,v)$ depicted in Fig. 2.5. The second derivative of $\mathbf{r}(t)$ with respect to t is evaluated as

$$\ddot{\mathbf{r}} = \dot{u}(\dot{u}\mathbf{r}_{uu} + \dot{v}\mathbf{r}_{uv}) + \ddot{u}\mathbf{r}_u + \dot{v}(\dot{v}\mathbf{r}_{vv} + \dot{u}\mathbf{r}_{uv}) + \ddot{v}\mathbf{r}_v.$$

On taking a scalar product with unit surface normal vector $\mathbf{n}$, the result is expressed as (noting that $\mathbf{r}_u \cdot \mathbf{n} = \mathbf{r}_v \cdot \mathbf{n} = 0$):

$$\begin{aligned} \ddot{\mathbf{r}} \cdot \mathbf{n} &= (\dot{u})^2 \mathbf{r}_{uu} \cdot \mathbf{n} + 2\dot{u}\dot{v}\mathbf{r}_{uv} \cdot \mathbf{n} + (\dot{v})^2 \mathbf{r}_{vv} \cdot \mathbf{n} \\ &= \dot{\mathbf{u}}^T \mathbf{D}\ \dot{\mathbf{u}}, \end{aligned} \qquad (2.31-a)$$

where $\dot{\mathbf{u}} = \begin{bmatrix} \dot{u} \\ \dot{v} \end{bmatrix}$; $\mathbf{D} = \begin{bmatrix} \mathbf{r}_{uu} \cdot \mathbf{n} & \mathbf{r}_{uv} \cdot \mathbf{n} \\ \mathbf{r}_{uv} \cdot \mathbf{n} & \mathbf{r}_{vv} \cdot \mathbf{n} \end{bmatrix}$ is called the **second fundamental matrix** of the surface.

2) Normal curvature:

From the *Serret-Frenet equations* (ie, $d\mathbf{r}/ds = \mathbf{T}$; $d\mathbf{T}/ds = \kappa\mathbf{N}$), the second derivative of $\mathbf{r}(t)$ is evaluated as

$$\begin{aligned} \ddot{\mathbf{r}} &= d\dot{\mathbf{r}}/dt \\ &= d(\dot{s}\mathbf{T})/dt \\ &= \ddot{s}\mathbf{T} + \dot{s}\dot{\mathbf{T}} \\ &= \ddot{s}\mathbf{T} + \dot{s}(\dot{s}\kappa\mathbf{N}). \end{aligned}$$

Again, by taking a scalar product with $\mathbf{n}$, we obtain (noting that $\mathbf{T} \cdot \mathbf{n} = 0$)

$$\ddot{\mathbf{r}} \cdot \mathbf{n} = (\dot{s})^2 \kappa\mathbf{N} \cdot \mathbf{n}. \qquad (2.31-b)$$

The quantity $\kappa\mathbf{N} \cdot \mathbf{n}$ in the above expression is called the *normal curvature* κn. From (2.31-a) and (2.31-b), the **normal curvature** is given by

$$\kappa n \equiv \kappa\mathbf{N} \cdot \mathbf{n} = (\dot{\mathbf{u}}^T \mathbf{D}\ \dot{\mathbf{u}})/(\dot{s})^2 = (\dot{\mathbf{u}}^T \mathbf{D}\ \dot{\mathbf{u}})/(\dot{\mathbf{u}}^T \mathbf{G}\ \dot{\mathbf{u}}). \qquad (2.32)$$

Recall that $\dot{s} = |\dot{\mathbf{r}}| = (\dot{\mathbf{u}}^T \mathbf{G}\ \dot{\mathbf{u}})^{1/2}$. The physical meaning of the normal curvature is as follows: At a current point $\mathbf{r}(u(t), v(t))$ on the surface $\mathbf{r}(u,v)$, construct a plane π containing the unit tangent vector $\mathbf{T}$ and the unit surface normal vector $\mathbf{n}$. Then, the curvature of the curve obtained by intersecting the surface with the plane π becomes the normal curvature of the surface at $\mathbf{r}(t)$ along the direction of $\dot{\mathbf{u}}$.

3) Principal curvatures:

The normal curvature given by (2.32) is a function of the direction $\dot{\mathbf{u}}$. That is,

$$\kappa_n(\dot{\mathbf{u}}) = (\dot{\mathbf{u}}^T\mathbf{D}\ \dot{\mathbf{u}})/(\dot{\mathbf{u}}^T\mathbf{G}\ \dot{\mathbf{u}}).$$

Thus, extreme values of the normal curvature can be obtained from

$$\partial\kappa_n/\partial\dot{\mathbf{u}} = 2\mathbf{D}\dot{\mathbf{u}} - 2\kappa_n\mathbf{G}\dot{\mathbf{u}} = \mathbf{0}, \tag{2.33}$$

where $\dot{\mathbf{u}} = [\dot{u} \quad \dot{v}]^T$. The extreme values of the normal curvature are known as **principal curvatures**, and they are obtained from (2.33) as

$$\kappa_{n1} = (b + (b^2 - ac)^{1/2})/a\ ; \quad \kappa_{n2} = (b - (b^2 - ac)^{1/2})/a \tag{2.34}$$

$$where, \quad a = |\mathbf{G}| = \begin{vmatrix} g_1 & h \\ h & g_2 \end{vmatrix}$$

$$c = |\mathbf{D}| = \begin{vmatrix} d_1 & e \\ e & d_2 \end{vmatrix}$$

$$b = (g_1 d_2 + g_2 d_1)/2 - eh.$$

The product of the two principal curvatures is called the **Gaussian curvature** which is sometimes used as a measure of surface *smoothness.*

Example 2.4 : *Find the normal curvature of the curve* $\mathbf{u}(t) = [u(t) \quad v(t)]^T = [t \quad t]^T$ *on the surface of the paraboloid given in Example 2.3.*

From the results in Example 2.3, the second derivatives and surface normals are evaluated as

$$\mathbf{r}_{uu} = \begin{bmatrix} 0 \\ 0 \\ 1 \end{bmatrix} ;\ \mathbf{r}_{uv} = \begin{bmatrix} -\sin v \\ \cos v \\ 1 \end{bmatrix} ;\ \mathbf{r}_{vv} = \begin{bmatrix} -u\cos v \\ -u\sin v \\ 0 \end{bmatrix} ;\ \mathbf{n} = \frac{1}{\sqrt{1+u^2}} \begin{bmatrix} -u\cos v \\ -u\sin v \\ 1 \end{bmatrix}.$$

The second fundamental matrix in $(2.31-a)$ *is then given by* $(u = v = t)$

$$\mathbf{D} = \frac{1}{\sqrt{1+t^2}} \begin{bmatrix} 1 & 0 \\ 0 & t^2 \end{bmatrix} \quad and \quad \dot{\mathbf{u}}^T\mathbf{D}\dot{\mathbf{u}} = \sqrt{1+t^2}\ .$$

Finally, the normal curvature in (2.32) *is obtained as*

$$\kappa_n = (\dot{\mathbf{u}}^T\mathbf{D}\dot{\mathbf{u}})/(\dot{\mathbf{u}}^T\mathbf{G}\dot{\mathbf{u}}) = \frac{\sqrt{1+t^2}}{(1+2t^2)}.$$ ◇

2.5 COORDINATE TRANSFORMATION

This section presents basic results in coordinate transformation. The topics to be discussed are coordinate transformations and frames.

2.5.1 2D Transformations

There are three types of 2D transformations as depicted in Fig. 2.6. They are *translation, rotation* and *scaling*. These transformations are *linear* transformations.

1) 2D translation (Fig. 2.6a):

The coordinates (x', y') of the point Q obtained by *moving* the point $P(x, y)$ by t_x in x-direction and by t_y in y-direction are expressed as

$$x' = x + t_x \; ; \quad y' = y + t_y. \tag{2.35}$$

2) 2D scaling (Fig. 2.6b):

Let s_x and s_y denote *scaling factors* in x-direction and y-direction, respectively. Then, the coordinates of the scaled point Q are given by

$$x' = s_x \cdot x \; ; \quad y' = s_y \cdot y. \tag{2.36}$$

3) 2D rotation (Fig. 2.6c):

If a point $P(x, y)$ is rotated *counter-clockwise* by the angle of θ as shown in Fig. 2.6c, the resulting point $Q(x', y')$ is expressed as

$$x' = x \cos\theta - y \sin\theta, \tag{2.37 – a}$$

$$y' = x \sin\theta + y \cos\theta. \tag{2.37 – b}$$

The rotation transformation (2.37) is easily obtained by expressing the points P, Q in *polar coordinates* and then applying a trigonometric formula.

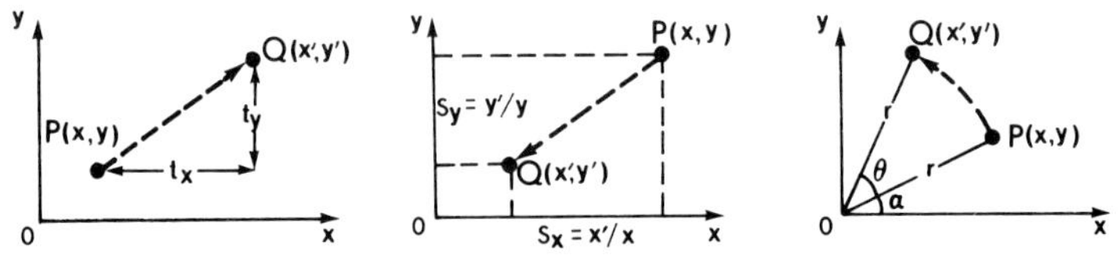

(a) (b) (c)

Figure 2.6 2D Transformations

4) Homogeneous transformation

A linear transformation in general can be expressed in a matrix form. Let us define "three components" vectors for the 2D coordinates as

$$\mathbf{h} = (x\ y\ 1)\ ; \quad \mathbf{h}' = (x'\ y'\ 1),$$

then the above 2D transformations (2.35), (2.36), and (2.37) can be expressed in matrix form as

$$\mathbf{h}' = \mathbf{h}\ \mathbf{M}, \tag{2.38}$$

where $\mathbf{M} \equiv \mathbf{T},\ \mathbf{S},\ \mathit{or}\ \mathbf{R}$;

$$\mathbf{T} = \begin{bmatrix} 1 & 0 & 0 \\ 0 & 1 & 0 \\ t_x & t_y & 1 \end{bmatrix} \ ; \quad \mathbf{S} = \begin{bmatrix} s_x & 0 & 0 \\ 0 & s_y & 0 \\ 0 & 0 & 1 \end{bmatrix} \ ; \quad \mathbf{R} = \begin{bmatrix} \cos\theta & \sin\theta & 0 \\ -\sin\theta & \cos\theta & 0 \\ 0 & 0 & 1 \end{bmatrix} .$$

The vectors $\mathbf{h}, \mathbf{h}'$ in (2.38) are (normalized) **homogeneous vectors**, and the matrix $\mathbf{M}$ is called a **homogeneous transformation matrix**. More details on homogeneous vectors may be found in §3.5.1.

2.5.2 3D Transformations

3D transformations are defined as an extension of the 2D transformations. If a 3D point $P(x, y, z)$ is translated by t_x, t_y, t_z, the coordinates (x', y', z') of the resulting point are given by

$$x' = x + t_x\ ; \quad y' = y + t_y\ ; \quad z' = z + t_z. \tag{2.39}$$

Similarly, a 3D scaling with scaling factors s_x, s_y, s_z gives the following transformations of the coordinate variables.

$$x' = s_x \cdot x\ ; \quad y' = s_y \cdot y\ ; \quad z' = s_z \cdot z. \tag{2.40}$$

As with the 2D case, the above 3D transformations can be expressed in matrix form using homogeneous vectors for the 3D points. For **4D homogeneous vectors**

$$\mathbf{h} = (x\ y\ z\ 1) \quad \mathit{and} \quad \mathbf{h}' = (x'\ y'\ z'\ 1),$$

the 3D translation (2.39) and 3D scaling (2.40) are expressed as

$$\mathbf{h}' = \mathbf{h}\ \mathbf{T}\ :\ (3D\ translation) \tag{2.41 - a}$$

$$\mathbf{h}' = \mathbf{h}\ \mathbf{S}\ :\ (3D\ scaling), \tag{2.41 - b}$$

where, $$\mathbf{T} = \begin{bmatrix} 1 & 0 & 0 & 0 \\ 0 & 1 & 0 & 0 \\ 0 & 0 & 1 & 0 \\ t_x & t_y & t_z & 1 \end{bmatrix} \ ; \quad \mathbf{S} = \begin{bmatrix} s_x & 0 & 0 & 0 \\ 0 & s_y & 0 & 0 \\ 0 & 0 & s_z & 0 \\ 0 & 0 & 0 & 1 \end{bmatrix} .$$

Since it is not easy to define a rotation around an arbitrary axis in 3D space, a general rotation is usually expressed as a concatenation of *basic rotations* around coordinate axes each of which is basically a 2D rotation similar to (2.37). The *basic rotation transformations* are given by

1) rotation around x-axis:

$$x' = x \; ; \quad y' = y \; \cos\theta - z \; \sin\theta \; ; \quad z' = y \; \sin\theta + z \; \cos\theta. \tag{2.42 - a}$$

2) rotation around y-axis:

$$x' = z \; \sin\theta + x \; \cos\theta \; ; \quad y' = y \; ; \quad z' = z \; \cos\theta + x \; \sin\theta. \tag{2.42 - b}$$

3) rotation around z-axis:

$$x' = x \; \cos\theta + y \; \sin\theta \; ; \quad y' = x \; \sin\theta + y \; \cos\theta \; ; \quad z' = z. \tag{2.42 - c}$$

It can be shown that the homogeneous transformation matrices for the transformations in (2.42) are given by

$$\mathbf{R}(x,\theta) = \begin{bmatrix} 1 & 0 & 0 & 0 \\ 0 & C & S & 0 \\ 0 & -S & C & 0 \\ 0 & 0 & 0 & 1 \end{bmatrix} ; \mathbf{R}(y,\theta) = \begin{bmatrix} C & 0 & -S & 0 \\ 0 & 1 & 0 & 0 \\ S & 0 & C & 0 \\ 0 & 0 & 0 & 1 \end{bmatrix} ; \mathbf{R}(z,\theta) = \begin{bmatrix} C & S & 0 & 0 \\ -S & C & 0 & 0 \\ 0 & 0 & C & 0 \\ 0 & 0 & 0 & 1 \end{bmatrix} \tag{2.43}$$

where $C = \cos\theta$ and $S = \sin\theta$. In general, a 3D transformation is expressed, in terms of a **homogeneous transformation matrix H**, as follows:

$$(x' \; y' \; z' \; 1) = (x \; y \; z \; 1)\mathbf{H}, \tag{2.44}$$

$$where \quad \mathbf{H} = \begin{bmatrix} r_{11} & r_{12} & r_{13} & 0 \\ r_{21} & r_{22} & r_{23} & 0 \\ r_{31} & r_{32} & r_{33} & 0 \\ t_x & t_y & t_z & 1 \end{bmatrix} = \begin{bmatrix} & \mathbf{R} & & 0 \\ & & & 0 \\ & & & 0 \\ & \mathbf{t} & & 1 \end{bmatrix}.$$

Alternatively, the transformation expression (2.44) can be rewritten as

$$(x' \; y' \; z') = (x \; y \; z) \; \mathbf{R} + \mathbf{t}. \tag{2.45}$$

If the transformation (2.44) consists of rotations and translations only (without scaling), it can be shown that the "rotation" matrix **R** in (2.45) is an **orthogonal matrix** (Paul, 1981). That is, let us define row vectors of **R** as

$$\mathbf{n} = (r_{11} \; r_{12} \; r_{13}); \quad \mathbf{o} = (r_{21} \; r_{22} \; r_{23}); \quad \mathbf{a} = (r_{31} \; r_{32} \; r_{33}), \tag{2.46}$$

then they are orthogonal unit vectors satisfying

$$\mathbf{n} \times \mathbf{o} = \mathbf{a} \; ; \quad \mathbf{o} \times \mathbf{a} = \mathbf{n} \; ; \quad \mathbf{a} \times \mathbf{n} = \mathbf{o} \; ;$$

$$|\mathbf{n}| = |\mathbf{o}| = |\mathbf{a}| = 1.$$

2.5.3 Coordinate Frames

Another interpretation of the transformation expression (2.44) is that a "copy" of the reference coordinate system itself is transformed into the 3D space to form a **moving coordinate frame.** Let $\mathbf{i}_h$, $\mathbf{j}_h$, $\mathbf{k}_h$ denote **homogeneous axis vectors** of the reference coordinate system. That is,

$$\mathbf{i}_h = (1\ 0\ 0\ 0) \qquad (2.47-a)$$

$$\mathbf{j}_h = (0\ 1\ 0\ 0) \qquad (2.47-b)$$

$$\mathbf{k}_h = (0\ 0\ 1\ 0) \qquad (2.47-c)$$

By applying the transformation (2.44) to the above homogeneous vectors, we get

$$\mathbf{i}'_h = \mathbf{i}_h\ \mathbf{H} = (1\ 0\ 0\ 0)\ \mathbf{H} = (\mathbf{n}\ 0) \qquad (2.48-a)$$

$$\mathbf{j}'_h = \mathbf{j}_h\ \mathbf{H} = (0\ 1\ 0\ 0)\ \mathbf{H} = (\mathbf{o}\ 0) \qquad (2.48-b)$$

$$\mathbf{k}'_h = \mathbf{k}_h\ \mathbf{H} = (0\ 0\ 1\ 0)\ \mathbf{H} = (\mathbf{a}\ 0) \qquad (2.48-c)$$

where $\mathbf{n}$, $\mathbf{o}$, $\mathbf{a}$ are as defined in (2.46). The above results tell that the orthogonal vectors $\mathbf{n}$, $\mathbf{o}$, $\mathbf{a}$ in $\mathbf{H}$ become the axis vectors of the moving coordinate frame transformed by (2.44). The origin of the moving coordinate frame is similarly expressed as

$$\begin{aligned} \mathbf{P}'_h &= (0\ 0\ 0\ 1)\ \mathbf{H} \\ &= (t_x\ t_y\ t_z\ 1) \qquad (2.49) \\ &= (\mathbf{t}\ 1) \end{aligned}$$

The moving coordinate frame $\mathbf{H}$ is depicted in Fig. 2.7.

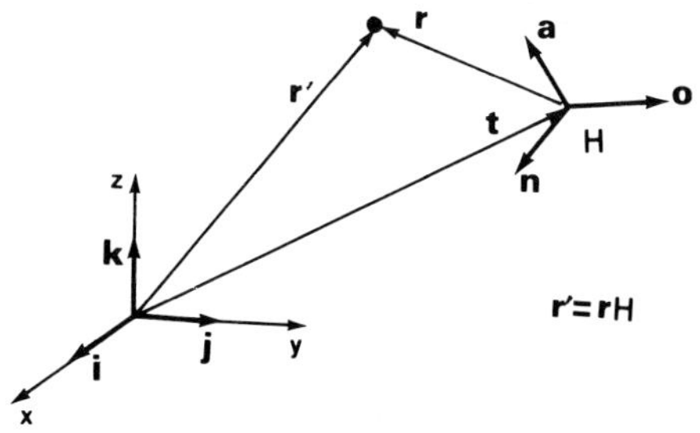

Figure 2.7 Moving Coordinate Frame and Transformation

For this reason, the homogeneous transformation matrix **H** itself is called a **coordinate frame**. Thus, the transformation (2.44) can be interpreted as follows (refer to Fig. 2.7):

a) $\mathbf{r} = (x\ y\ z)$ represents coordinate values of a *point* w.r.t. the local *(moving)* frame **H**, and

b) the "transformed point" $\mathbf{r}' = (x'\ y'\ z')$ corresponds to the coordinate values of the same point w.r.t. the reference coordinate system.

Rewriting (2.44), we have

$$\mathbf{h}' = \mathbf{h}\,\mathbf{H} \quad or \quad \mathbf{h} = \mathbf{h}'\,\mathbf{H}^{-1},$$

$$where \quad \mathbf{h} = (\mathbf{r}\ 1) = (x\ y\ z\ 1),$$

$$\mathbf{h}' = (\mathbf{r}'\ 1) = (x'\ y'\ z'\ 1),$$

$$\mathbf{H} = \begin{bmatrix} n_x & n_y & n_z & 0 \\ o_x & o_y & o_z & 0 \\ a_x & a_y & a_z & 0 \\ t_x & t_y & t_z & 1 \end{bmatrix}.$$

One may easily verify that the inverse of **H** is expressed as

$$\mathbf{H}^{-1} = \begin{bmatrix} n_x & o_x & a_x & 0 \\ n_y & o_y & a_y & 0 \\ n_z & o_z & a_z & 0 \\ -\mathbf{n}\cdot\mathbf{t} & -\mathbf{o}\cdot\mathbf{t} & -\mathbf{a}\cdot\mathbf{t} & 1 \end{bmatrix}, \tag{2.50}$$

$$where \quad \mathbf{n} = (n_x\ n_y\ n_z),$$

$$\mathbf{o} = (o_x\ o_y\ o_z),$$

$$\mathbf{a} = (a_x\ a_y\ a_z),$$

$$\mathbf{t} = (t_x\ t_y\ t_z).$$

CHAPTER 3

POLYNOMIAL CURVE MODELS

3.1 INTRODUCTION

Presented in this chapter are *mathematical models for a curve segment.* In general, a *curve* is modeled as a **composite curve** composed of a number of curve segments, possibly, of different types. For each curve segment, a set of mathematical functions is employed in order to specify the functional relationships among the coordinate variables (x, y, z). Any type of mathematical function can be used as a *curve model.* In practice, however, **polynomial curve models** are widely used, as they are easier to work with and yet flexible enough to represent most curves in engineering artifacts.

As discussed in the previous chapter, a *curve model* may be in *implicit, explicit, or parametric* forms. *Implicit* and *explicit* curve models are used in representing 2D curves only. In each form, polynomial curve models are expressed as follows:

1) Implicit polynomial curve model (2D curve only):

$$g(x,y) = \sum_{i=0}^{m} \sum_{j=0}^{n} c_{ij}\, x^i\, y^j = 0 \quad for\ some\ integers\ m, n.$$

2) Explicit polynomial curve model (2D curve only):

$$y = f(x) = a + bx + cx^2 + ... \ (Cartesian\ coordinates).$$

$$r = h(\theta) = \alpha + \beta\theta + \gamma\theta^2 + ... \ (Polar\ coordinates),$$

$$with\ x = r \cos\theta \ ; \quad y = r \sin\theta.$$

3) Parametric polynomial curve model:

$$\mathbf{r}(t) \equiv (x(t),\ y(t),\ z(t)) = \mathbf{a} + \mathbf{b}t + \mathbf{c}t^2 + ...$$

It is in theory possible to convert one form of curve model to another. The process of converting an implicit polynomial to a parametric (rational) polynomial is called **parametrization,** and the reverse process is called **implicitization.** Parametric curve models that are widely used are

- *standard polynomial curve model,*
- *Ferguson curve,*
- *Bezier curve,*
- *Uniform B-spline curve, and*
- *Non-uniform B-spline (NUB) curve.*

Topics to be discussed in this chapter are as follows:

a) Implicit polynomial curve (conic section) in §3.2.

b) Explicit polynomial curves in §3.2.

c) Ferguson, Bezier, and uniform B-spline curves in §3.3.

d) Non-uniform B-spline curves in §3.4.

e) Rational Bezier curves in §3.5.

3.2 2D CURVE MODELS

In conventional engineering drawings, 2D curve segments are the major building blocks for describing 3D shapes. A 2D curve segment may be represented by an implicit polynomial, an explicit polynomial, or a parametric polynomial. In this section, however, only implicit and explicit polynomial models will be discussed.

3.2.1 Implicit Polynomial Curve Models

An implicit function of the form $g(x, y) = 0$ represents a 2D curve on the x,y-plane. For example, a *straight line* and a *circle* are represented by

$$ax + by + c = 0 \quad and \quad (x - a)^2 + (y - b)^2 - r^2 = 0,$$

where a, b, c, r are coefficients of the implicit polynomials.

Implicit curves in general have some convenient features, for example: 1) It is easy to compute curve tangent and normal vectors, and 2) it is very simple to determine whether a point (x, y) is located inside or outside of the region formed by an implicit curve. The **tangent equation** of $g(x, y) = 0$ at a point $P(x_1, y_1)$ on the curve is given by

$$g_x(x_1, y_1)\,(x - x_1) + g_y(x_1, y_1)\,(y - y_1) = 0 \tag{3.1}$$

where, $g_x \equiv \partial g/\partial x$ and $g_y \equiv \partial g/\partial y$. And the **normal equation** is given by

$$g_y(x_1, y_1)\,(x - x_1) - g_x(x_1, y_1)\,(y - y_1) = 0. \tag{3.2}$$

A point $P(x_2, y_2)$ is said to be

- "inside" the curve $g(x, y) = 0 \quad if \; g(x_2, y_2) < 0$;
- "outside" of the curve if $g(x_2, y_2) > 0$.

However, it should be noted that the designation of the "side" is arbitrary because $g(x, y) = 0$ and $-g(x, y) = 0$ represent the same curve.

A *quadratic polynomial function g(x,y)* represents a **conic section curve**. A conic section has a clear geometric meaning, and as a result, it is relatively easy to control their shapes by setting a few *parameters*. Widely used conic section curves are *ellipse, parabola,* and *hyperbola* which have the following standard forms:

a) *Ellipse* : $x^2/a^2 + y^2/b^2 - 1 = 0,$

b) *Parabola* : $y^2 - 4ax = 0,$

c) *Hyperbola* : $x^2/a^2 - y^2/b^2 - 1 = 0,$

where a, b are *shape parameters*. These curves are called *conic sections* because they represent *intersection curves between a right cone and a plane* as depicted in Figs. 3.1, 3.2, and 3.3. Also shown in the figures are geometric meanings of the shape parameters a, b in the above conic section equations. More details on conic sections are given in §4.4.2.

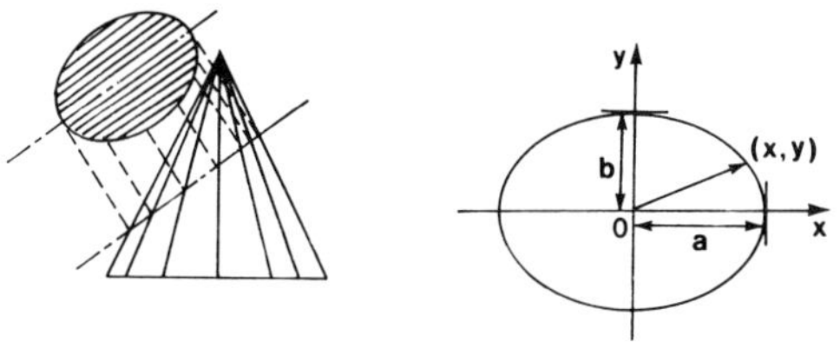

Figure 3.1 Definition of Ellipse

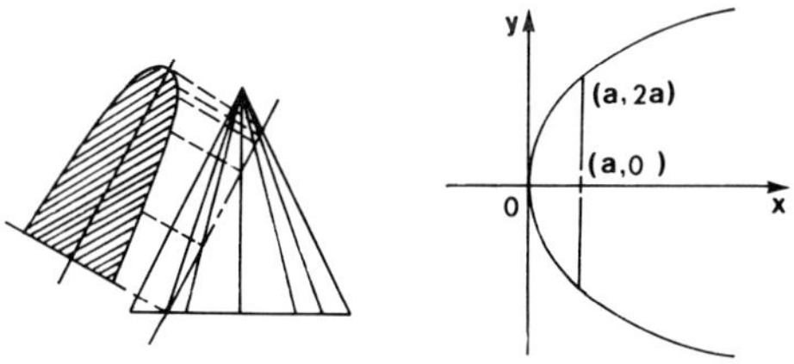

Figure 3.2 Definition of Parabola

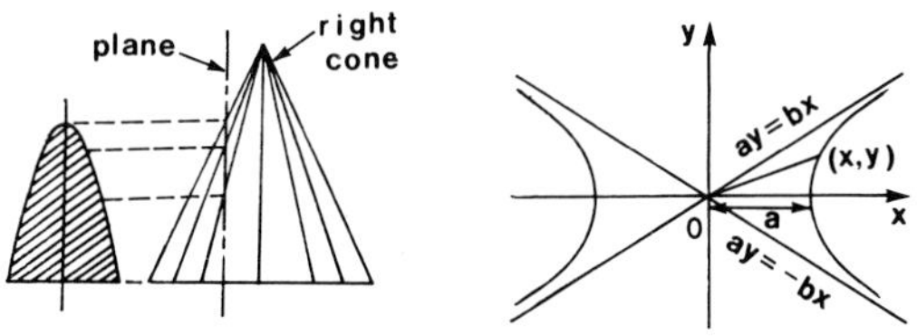

Figure 3.3 Definition of Hyperbola

The major disadvantage of implicit representation is that it is difficult to trace the curve in a sequential manner. Sequential tracing of a curve is essential in computer graphics and NC machining. Thus, in geometric modeling, parametrically represented conic sections are widely used. Methods of *parametrizing* conic section curves will be given in the next chapter (§4.4). Implicit polynomial curves of degrees higher than two are rarely used in practice.

3.2.2 Explicit Polynomial Curve Models

An explicit function of the form $y = f(x) = a + bx + cx^2 ...$ represents a curve on the x, y-plane. If $f(x)$ is a polynomial of degree two, it becomes a *parabola* which is a special case of conic section. Examples of explicit polynomial curves of degrees two and three are shown in Fig. 3.4.

A nice feature of explicit polynomial representation is that it can be converted to either an implicit polynomial or a parametric polynomial. That is, $y = f(x)$, where $f(x)$ is a polynomial in x, is equivalent to either

$$g(x, y) \equiv y - f(x) = 0$$

or

$$x(t) = t \; ; \quad y(t) = f(t), \tag{3.3}$$

where t is the parameter of the curve. Thus, the explicit representation has the advantages of both implicit and parametric representations. Namely, it is

a) easy to compute curve tangent and normal vectors,
b) easy to determine whether a point is located above or below the curve,
c) easy to trace the curve in a sequential manner.

Figure 3.4 Explicit Polynomial Curves

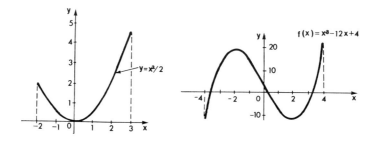

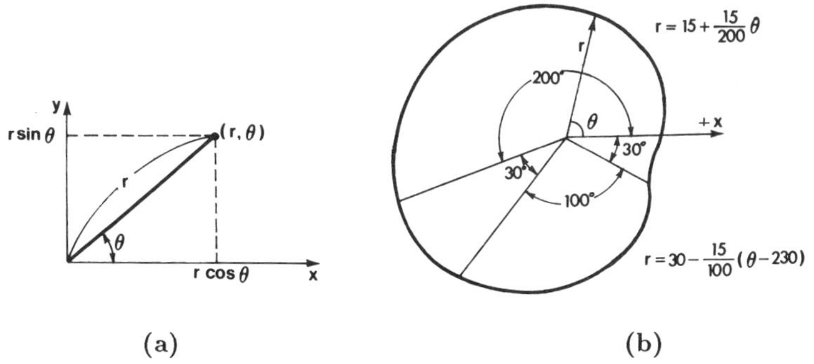

Figure 3.5 Polar Coordinate and Profile of a Cam

The major drawback of explicit representation, however, is that it can not handle a curve having loops or vertical segments. Explicit representation (or a parametric representation of the form (3.3)) is sometimes called *non-parametric* representation as discussed in the previous chapter.

In a polar coordinate system, a curve on the x, y-plane is represented as an explicit (polynomial) function of the form

$$r = h(\theta) = \alpha + \beta\theta + \gamma\theta^2 + \ldots \; : \quad x = r\cos\theta\,; \quad y = r\sin\theta$$

where, r and θ are as defined in Fig. 3.5a. For a linear function $f(\theta)$, we have (for some constants a,b)

$$r = f(\theta) = a + b\theta \tag{3.4}$$

which is sometimes called the *Archimedes curve.* The curve equation (3.4) is suitable for defining the profile of a mechanical cam whose function is to convert a constant speed rotational motion to a constant speed linear motion. An example of a cam profile is given in Fig. 3.5b.

3.3 PARAMETRIC POLYNOMIAL MODELS FOR CURVE SEGMENT

Presented in this section are parametric polynomial curve models that can be used in representing a curve segment specified by its *end conditions,* namely, two end points P_0, P_1 and two end tangents $\mathbf{t}_0$, $\mathbf{t}_1$. Since the curve is defined by four vectors (two position vectors and two tangent vectors), it can be modeled by a cubic vector-valued polynomial function. Cubic polynomials are widely used in ordinary (ie, non-rational) curve models because cubic is the minimum degree polynomial that provide just enough flexibility for constructing a space curve. The cubic polynomial models to be discussed are *Ferguson curve, Bezier curve,* and *uniform B-spline curve. Non-uniform B-spline curve* will be discussed in the next section.

3.3.1 Standard Polynomial Curve Models

A standard polynomial function is easy to define and efficient to evaluate. Let us consider a vector-valued cubic polynomial function of the form

$$\begin{aligned}\mathbf{r}(u) &= (\ x(u),\ y(u),\ z(u)\ ) \\ &= \mathbf{a} + \mathbf{b}\ u + \mathbf{c}\ u^2 + \mathbf{d}\ u^3 \\ &= [1\ u\ u^2\ u^3]\begin{bmatrix}\mathbf{a}\\ \mathbf{b}\\ \mathbf{c}\\ \mathbf{d}\end{bmatrix} \\ &= \mathbf{U}\ \mathbf{A},\quad with\quad 0 \le u \le 1\end{aligned} \tag{3.5}$$

In (3.5), $\mathbf{U}=\ [1\ u\ u^2\ u^3]$ is called the **power basis vector** and $\mathbf{A} = [\mathbf{a}\ \mathbf{b}\ \mathbf{c}\ \mathbf{d}]^{\mathrm{T}}$ is the **coefficient vector**.

The cubic function (3.5) does not give much geometric meaning in its form, but it may be used in constructing a smooth curve passing through four data points $P_i(i = 1, ...4)$. One way of fitting the four data points using the cubic polynomial model (3.5) is as follows: Let d_i denote the *chord-length* of P_i and P_{i+1} such that

$$d_i = |P_{i+1} - P_i| \quad for \quad i = 1, 2, 3.$$

Then, the values u_i's of the parameter at the data points P_i's can be assigned as

$$\begin{aligned} u_1 &= 0\ ; \\ u_2 &= d_1/\Sigma d_j\ ; \\ u_3 &= (d_1 + d_2)/\Sigma d_j\ ; \\ u_4 &= 1. \end{aligned}$$

In order for the cubic curve (3.5) to pass through the data points, we whould have

$$\mathbf{r}(u_i) = P_i \quad for \quad i = 1, 2, 3, 4$$

which is solved for the unknown coefficients. When the coefficients of (3.5) are determined from the above relations, it is sometimes called a **four points curve.** In general, a degree-n polynomial curve

$$\mathbf{r}(u) = \sum_{i=0}^{n} \mathbf{a}_i u^i$$

can be used fit $(n + 1)$ data points. This method of constructing curves of higher degrees has been implemented in earlier CAD/CAM systems.

3.3.2 Ferguson Curve Model

Ferguson (1964) introduced a different way of using equation (3.5). He was interested in constructing a curve segment (see Fig. 3.6)

a) joining two **end points** P_0 *and* P_1 ;
b) having specified **end tangents** $\mathbf{t}_0$ *and* $\mathbf{t}_1$.

In order for the cubic curve (3.5) to meet the **end conditions** P_0, P_1, $\mathbf{t}_0$, $\mathbf{t}_1$, the following relations should hold:

$$P_0 = \mathbf{r}(0) \; ; \quad P_1 = \mathbf{r}(1) \; ; \quad \mathbf{t}_0 = \dot{\mathbf{r}}(0) \; ; \quad \mathbf{t}_1 = \dot{\mathbf{r}}(1) \tag{3.6}$$

where $\dot{\mathbf{r}}(u) \equiv d\mathbf{r}(u)/du = \mathbf{b} + 2u\mathbf{c} + 3u^2\mathbf{d}$. Upon evaluating the cubic curve (3.5) at its ends, we have

$$P_0 = \mathbf{r}(0) = \mathbf{a} \; ; \quad P_1 = \mathbf{r}(1) = \mathbf{a} + \mathbf{b} + \mathbf{c} + \mathbf{d} \; ;$$
$$\mathbf{t}_0 = \dot{\mathbf{r}}(0) = \mathbf{b} \; ; \quad \mathbf{t}_1 = \dot{\mathbf{r}}(1) = \mathbf{b} + 2\mathbf{c} + 3\mathbf{d} \; ;$$

which are easily solved for the unknown coefficients **a,b,c,d** to give

$$\mathbf{a} = P_0 \; ;$$
$$\mathbf{b} = \mathbf{t}_0 \; ;$$
$$\mathbf{c} = -3P_0 + 3P_1 - 2\mathbf{t}_0 - \mathbf{t}_1 \; ;$$
$$\mathbf{d} = 2P_0 - 2P_1 + \mathbf{t}_0 + \mathbf{t}_1 .$$

The above solution is more conveniently expressed in a matrix form as

$$\mathbf{A} \equiv \begin{bmatrix} \mathbf{a} \\ \mathbf{b} \\ \mathbf{c} \\ \mathbf{d} \end{bmatrix} = \begin{bmatrix} 1 & 0 & 0 & 0 \\ 0 & 0 & 1 & 0 \\ -3 & 3 & -2 & -1 \\ 2 & -2 & 1 & 1 \end{bmatrix} \begin{bmatrix} P_0 \\ P_1 \\ \mathbf{t}_0 \\ \mathbf{t}_1 \end{bmatrix} \equiv \mathbf{C}\ \mathbf{S}. \tag{3.7}$$

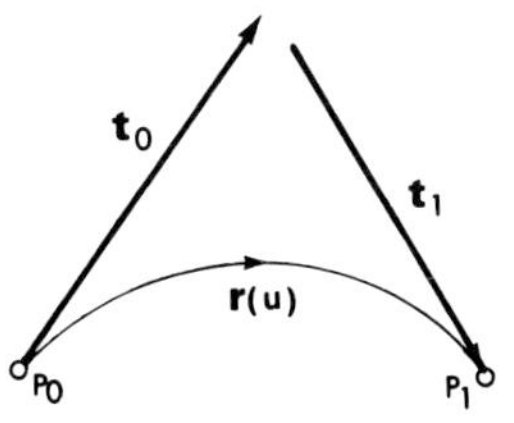

Figure 3.6 Construction of a Ferguson Curve Segment

By combining the results in (3.5) and (3.7), the equation of the curve segment $\mathbf{r}(u)$ is expressed in terms of its end conditions $P_0, P_1, \mathbf{t}_0, \mathbf{t}_1$ as

$$\mathbf{r}(u) = \mathbf{U\ A} = \mathbf{U\ C\ S} \quad with \quad 0 \leq u \leq 1 \tag{3.8}$$

$$where \quad \mathbf{C} = \begin{bmatrix} 1 & 0 & 0 & 0 \\ 0 & 0 & 1 & 0 \\ -3 & 3 & -2 & -1 \\ 2 & -2 & 1 & 1 \end{bmatrix} \ ; \quad \mathbf{S} = \begin{bmatrix} P_0 \\ P_1 \\ \mathbf{t}_0 \\ \mathbf{t}_1 \end{bmatrix}.$$

The above curve is called **Ferguson curve**. In practical applications, it is not easy to specify the *magnitudes of* the end tangents. It is a common practice to set the magnitudes of *end tangent vectors* equal to the chord-length, that is, $|\mathbf{t}_0| = |\mathbf{t}_1| = |P_1 - P_0|$. This choice gives a reasonable curve shape.

The expressions (3.5) and (3.8) are both in a **power basis form** whose evaluation is quite efficient because it allows to use *nested multiplications* (Conte and deBoor, 1980). Expressing (3.8) in a different way, we have

$$\begin{aligned} \mathbf{r}(u) &= (\mathbf{U\ C})\ \mathbf{S} \\ &= (1 - 3u^2 + 2u^3)\ P_0 + (3u^2 - 2u^3)\ P_1 + (u - 2u^2 + u^3)\ \mathbf{t}_0 + (-u^2 + u^3)\ \mathbf{t}_1 \\ &= H_0^3(u)\ P_0 + H_1^3(u)\ \mathbf{t}_0 + H_2^3(u)\ \mathbf{t}_1 + H_3^3(u)\ P_1, \end{aligned} \tag{3.9}$$

$$\begin{aligned} where \quad H_0^3(u) &= (1 - 3u^2 + 2u^3), \\ H_1^3(u) &= (u - 2u^2 + u^3), \\ H_2^3(u) &= (-u^2 + u^3), \\ H_3^3(u) &= (3u^2 - 2u^3). \end{aligned}$$

The functions $H_i^3(u)$ in (3.9) are called (cubic) **Hermite blending functions** which, evaluated at $u = 0, 1$, meet the following requirements:

$$\begin{aligned} & H_0^3(0) = H_3^3(1) = \dot{H}_1^3(0) = \dot{H}_2^3(1) = 1, \\ & H_0^3(1) = H_3^3(0) = \dot{H}_1^3(1) = \dot{H}_2^3(0) = 0, \quad and \\ & \dot{H}_0^3(j) = \dot{H}_3^3(j) = H_1^3(j) = H_2^3(j) = 0 \quad for\ all \quad j = 0, 1. \end{aligned}$$

Thus, one may easily verify that (3.9) satisfies the end conditions (3.6). In fact, (3.9) is the original definition of a *Hermite blended (or lofted) curve*, and the conversion from (3.9) to (3.8) is called a *basis translation* (from Hermite basis to power basis).

3.3.3 Bezier Curve Model

There are many ways of defining a Bezier curve. In this section, we want to construct a **cubic Bezier curve** from the Ferguson curve equation (3.8). Recall that a Ferguson curve is defined in terms of its *end conditions* P_0, P_1, $\mathbf{t}_0$, $\mathbf{t}_1$. As shown in Fig. 3.7a, let us consider the four points V_0, V_1, V_2, V_3 satisfying the following conditions:

a) V_0*: start point of the curve segment,* P_0.
b) V_1*: one-third point on starting tangent vector, ie* $(V_0 + \mathbf{t}_0/3)$.
c) V_2*: two-thirds point on ending tangent vector, ie* $(V_3 - \mathbf{t}_1/3)$.
d) V_3*: end point of the curve segment,* P_1.

That is, the Bezier points V_i's are expressed in terms of the end conditions as

$$V_0 = P_0 \ ; \quad V_1 = (V_0 + \mathbf{t}_0/3) \ ; \quad V_2 = (V_3 - \mathbf{t}_1/3) \ ; \quad V_3 = P_1.$$

Alternatively, the Ferguson input vectors (ie, P_0, P_1, $\mathbf{t}_0$, $\mathbf{t}_1$) are expressed in terms of the **control vertices** V_i's as:

$$\begin{aligned} P_0 &= V_0, \\ P_1 &= V_3, \\ \mathbf{t}_0 &= 3(V_1 - V_0), \\ \mathbf{t}_1 &= 3(V_3 - V_2). \end{aligned}$$

These relations are expressed in a matrix form as

$$\mathbf{S} \equiv \begin{bmatrix} P_0 \\ P_1 \\ \mathbf{t}_0 \\ \mathbf{t}_1 \end{bmatrix} = \begin{bmatrix} 1 & 0 & 0 & 0 \\ 0 & 0 & 0 & 1 \\ -3 & 3 & 0 & 0 \\ 0 & 0 & -3 & 3 \end{bmatrix} \begin{bmatrix} V_0 \\ V_1 \\ V_2 \\ V_3 \end{bmatrix} \equiv \mathbf{L}\ \mathbf{R} \tag{3.10}$$

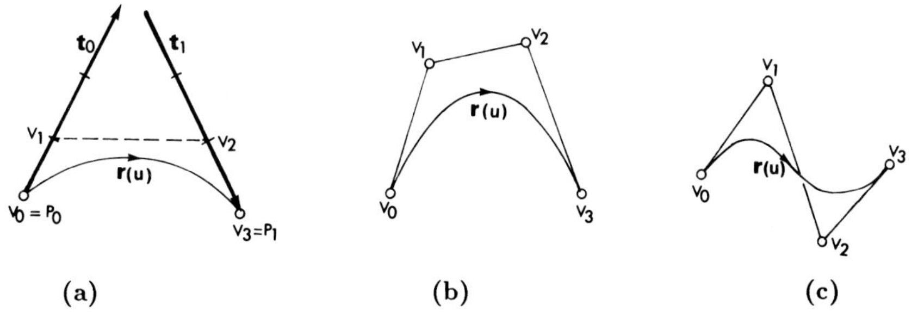

Figure 3.7 Cubic Bezier Curve Examples

Finally, the result in (3.10) is substituted into the *Ferguson curve* (3.8) to obtain the following **cubic Bezier curve** equation.

$$\begin{aligned}\mathbf{r}(u) &= \mathbf{U\ C\ S} \\ &= \mathbf{U\ C\ (L\ R)} \\ &= \mathbf{U\ (C\ L)\ R} = \mathbf{U\ M\ R} \quad with \quad 0 \le u \le 1,\end{aligned} \tag{3.11}$$

$$where \quad \mathbf{M} = \begin{bmatrix} 1 & 0 & 0 & 0 \\ -3 & 3 & 0 & 0 \\ 3 & -6 & 3 & 0 \\ -1 & 3 & -3 & 1 \end{bmatrix} \; ; \quad \mathbf{R} = \begin{bmatrix} V_0 \\ V_1 \\ V_2 \\ V_3 \end{bmatrix}.$$

It is called *Bezier curve* in honor of the earlier work by Bezier (1972). A nice feature of Bezier curve is that the shape of the curve resembles that of the *polygon* formed by the control vertices as shown in Fig. 3.7. As with the Ferguson curve, the Bezier curve (3.11) can also be expressed in a polynomial form as

$$\begin{aligned}\mathbf{r}(u) &= (\mathbf{U\ M})\ \mathbf{R} \\ &= B_0^3(u)\ V_0 + B_1^3(u)\ V_1 + B_2^3(u)\ V_2 + B_3^3(u)\ V_3 \\ &= \sum_{i=0}^{3} B_i^3(u)\ V_i,\end{aligned} \tag{3.12}$$

$$\begin{aligned} where \quad B_0^3(u) &= (1-u)^3, \\ B_1^3(u) &= 3u(1-u)^2, \\ B_2^3(u) &= 3u^2(1-u), \\ B_3^3(u) &= u^3. \end{aligned}$$

$B_i^3(u)$ are called *cubic* **Bernstein polynomials** (or Bernstein blending functions).

The Bernstein polynomials correspond to the terms in *binomial expansion* $(u+v)^n$. With $v = 1-u$, the general form of Bernstein polynomials of *degree* n is expressed as

$$B_i^n(u) = \frac{n!}{(n-i)!\ i!}\ u^i(1-u)^{n-i}. \tag{3.13 - a}$$

The Bernstein polynomials (3.13-a) are called **Bezier basis functions** which are used in defining a *degree-n Bezier curve* (3.13-b) by blending $n+1$ control vertices.

$$\mathbf{r}(u) = \sum_{i=0}^{n} B_i^n(u)\ V_i \quad with \quad 0 \le u \le 1 \tag{3.13 - b}$$

On evaluating the general Bezier equation and its derivative at $u = 0, 1$, we have

$$\begin{aligned} \mathbf{r}(0) &= V_0, \\ \mathbf{r}(1) &= V_n, \\ \dot{\mathbf{r}}(0) &= n(V_1 - V_0), \\ \dot{\mathbf{r}}(1) &= n(V_n - V_{n-1}). \end{aligned} \tag{3.14}$$

Example 3.1 : *Construct a quadratic Bezier curve using the results in* (3.14).

The end points of the quadratic Bezier curve are: $\mathbf{r}(0) = V_0$; $\mathbf{r}(1) = V_2$, *and end tangents are given by* $\dot{\mathbf{r}}(0) = 2(V_1 - V_0)$ *and* $\dot{\mathbf{r}}(1) = 2(V_2 - V_1)$. *By evaluating the quadratic polynomial* $\mathbf{r}(u) = \mathbf{a}+\mathbf{b}u+\mathbf{c}u^2$ *at* $u = 0, 1$, *and equating the results with the end conditions, we have*

$$\mathbf{r}(0) = \mathbf{a} = V_0; \quad \mathbf{r}(1) = \mathbf{a} + \mathbf{b} + \mathbf{c} = V_2;$$

$$\dot{\mathbf{r}}(0) = \mathbf{b} = 2(V_1 - V_0); \quad \dot{\mathbf{r}}(1) = \mathbf{b} + 2\mathbf{c} = 2(V_2 - V_1);$$

On substituting the above results into the quadratic equation, we get

$$\begin{aligned} \mathbf{r}(u) &= V_0 + 2(V_1 - V_0)u + (V_0 - 2V_1 + V_2)u^2 \\ &= (1-u)^2 V_0 + 2u(1-u)V_1 + u^2 V_2 \\ &= \sum_{i=0}^{2} B_i^2(u)\; V_i \quad \textit{with} \quad 0 \le u \le 1. \end{aligned}$$

The above quadratic Bezier curve is expressed in a matrix form as

$$\mathbf{r}(u) = [1\ u\ u^2] \begin{bmatrix} 1 & 0 & 0 \\ -2 & 2 & 0 \\ 1 & -2 & 1 \end{bmatrix} \begin{bmatrix} V_0 \\ V_1 \\ V_2 \end{bmatrix}. \qquad \diamond$$

In fact, the standard definition of a (cubic) Bezier curve is the *Bezier basis function form* (3.12) which gives a better geometric interpretation than the *power basis* (ie matrix) *form* (3.11). For example, a Bezier curve may be sub-divided into two or its degree may be elevated. The main advantage of the power basis form is in its computational efficiency (and convenience). It is always possible to translate the Bezier basis form into the matrix form.

The main merit of Bezier representation is in its ability to modify the shape of the curve by moving the control vertices. The polygon obtained by joining successive control vertices is called the **characteristic polygon**. It can be shown that a Bezier curve is confined inside the convex polygon formed by its control vertices. This property is called the **convex hull property**. See Chapter 6 for more details.

3.3.4 Uniform B-spline Curve Model

In general, a B-spline curve is a algebraically defined function. In order to help understand geometrical properties of a (cubic) B-spline curve, this subsection introduces a geometric construction of a **cubic B-spline curve**. Consider the four control vertices $V_0, ..., V_3$ shown in Fig. 3.8. As depicted in the figure, let us define

$M_0 = (V_0 + V_2)/2$: *middle point of V_0 and V_2.*

$M_1 = (V_1 + V_3)/2$: *middle point of V_1 and V_3.*

$P_0 = (2V_1 + M_0)/3$: *one-third position on the line joining V_1 and M_0.*

$P_1 = (2V_2 + M_1)/3$: *one-third position on the line joining V_2 and M_1.*

We want to construct a cubic curve segment $\mathbf{r}(u)$ satisfying the following:

a) It starts from P_0 and ends at P_1.
b) The end tangent vector $\mathbf{t}_0$ at P_0 is equal to $(M_0 - V_0)$.
c) The end tangent vector $\mathbf{t}_1$ at P_1 is equal to $(M_1 - V_1)$.

Thus, the *end points* P_0, P_1 of the curve segments are expressed in terms of the *control vertices* as follows:

$$P_0 \equiv \mathbf{r}(0) = [4V_1 + (V_0 + V_2)]/6 \qquad (3.15-a)$$

$$P_1 \equiv \mathbf{r}(1) = [4V_2 + (V_1 + V_3)]/6. \qquad (3.15-b)$$

The *end tangents* $\mathbf{t}_0$, $\mathbf{t}_1$ are also expressed similarly as

$$\mathbf{t}_0 \equiv \dot{\mathbf{r}}(0) = (V_2 - V_0)/2\,; \quad \mathbf{t}_1 \equiv \dot{\mathbf{r}}(1) = (V_3 - V_1)/2. \qquad (3.16)$$

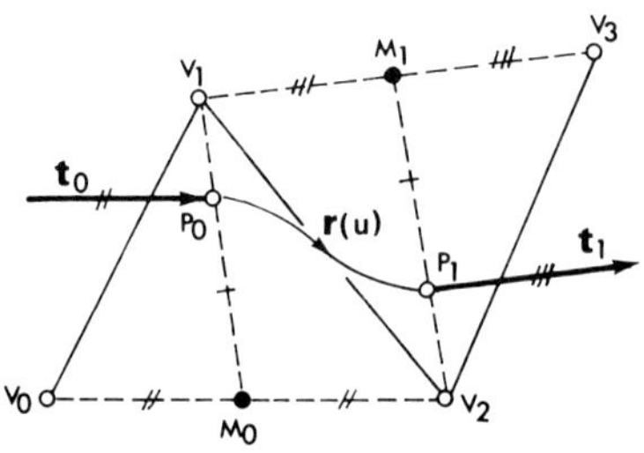

Figure 3.8 Construction of (Uniform) Cubic B-spline Curve Segment

The relations (3.15) and (3.16) are concisely expressed in a matrix form as

$$\mathbf{S} \equiv \begin{bmatrix} P_0 \\ P_1 \\ \mathbf{t}_0 \\ \mathbf{t}_1 \end{bmatrix} = \frac{1}{6} \begin{bmatrix} 1 & 4 & 1 & 0 \\ 0 & 1 & 4 & 1 \\ -3 & 0 & 3 & 0 \\ 0 & -3 & 0 & 3 \end{bmatrix} \begin{bmatrix} V_0 \\ V_1 \\ V_2 \\ V_3 \end{bmatrix} \equiv \mathbf{K}\ \mathbf{R}$$

On substituting the above result into the Ferguson curve equation (3.8), we obtain the following expression for **cubic uniform B-spline curve**:

$$\begin{aligned} \mathbf{r}(u) &= \mathbf{U}\ \mathbf{C}\ \mathbf{S} \\ &= \mathbf{U}\ \mathbf{C}\ (\mathbf{K}\ \mathbf{R}) \\ &= \mathbf{U}\ (\mathbf{CK})\ \mathbf{R} \\ &= \mathbf{U}\ \mathbf{N}\ \mathbf{R} \quad with \quad 0 \le u \le 1, \end{aligned} \tag{3.17}$$

where $\mathbf{U} = [1\ u\ u^2\ u^3]$,

$\mathbf{C} =$ *Ferguson's coefficient matrix*,

$$\mathbf{N} = \frac{1}{6} \begin{bmatrix} 1 & 4 & 1 & 0 \\ -3 & 0 & 3 & 0 \\ 3 & -6 & 3 & 0 \\ -1 & 3 & -3 & 1 \end{bmatrix} \quad : \quad cubic\ B\text{-}spline\ Coeff.\ matrix,$$

$\mathbf{R} = [V_0\ V_1\ V_2\ V_3]^T$.

This equation has the same form as the Bezier equation (3.11).

As with the Bezier curve, a general form for a "degree-n" (uniform) B-spline curve can easily be obtained as discussed in Versprille (1975). For example, a **quadratic (uniform) B-spline** curve is expressed in a matrix form as

$$\mathbf{r}(u) = \mathbf{U}\ \mathbf{N_q}\ \mathbf{R} \quad with \quad 0 \le u \le 1, \tag{3.18 – a}$$

where $\mathbf{U} = [1\ u\ u^2]$,

$$\mathbf{N_q} = \frac{1}{2} \begin{bmatrix} 1 & 1 & 0 \\ -2 & 2 & 0 \\ 1 & -2 & 1 \end{bmatrix},$$

$\mathbf{R} = [V_0\ V_1\ V_2]^T$.

On evaluating the above quadratic B-spline curve equation at its ends, one may easily verify that the followings hold:

$$\mathbf{r}(0) = (V_0 + V_1)/2,$$
$$\mathbf{r}(1) = (V_1 + V_2)/2,$$
$$\dot{\mathbf{r}}(0) = (V_1 - V_0),$$
$$\dot{\mathbf{r}}(1) = (V_2 - V_1).$$

And a **quartic B-spline curve** is expressed as

$$\mathbf{r}(u) = \mathbf{U}\ \mathbf{N}\ \mathbf{R} \quad with \quad 0 \le u \le 1, \tag{3.18 - b}$$

$$where \quad \mathbf{U} = [1\ u\ u^2\ u^3\ u^4],$$

$$\mathbf{N} = \frac{1}{24}\begin{bmatrix} 1 & 11 & 11 & 1 & 0 \\ -4 & -12 & 12 & 4 & 0 \\ 6 & -6 & -6 & 6 & 0 \\ -4 & 12 & -12 & 4 & 0 \\ 1 & -4 & 6 & -4 & 1 \end{bmatrix}$$

$$\mathbf{R} = [V_0\ V_1\ V_2\ V_3\ V_4]^T.$$

When the cubic uniform B-spline curve equation (3.17) is expressed in a polynomial form, we have an expression similar to the Bezier equation (3.12):

$$\mathbf{r}(u) = (\mathbf{U}\ \mathbf{N})\ \mathbf{R}$$
$$= \sum_{i=0}^{3} N_i^3(u)\ V_i, \tag{3.19}$$

$$where \quad N_0^3(u) = (1 - 3u + 3u^2 - u^3)/6,$$
$$N_1^3(u) = (4 - 6u^2 + 3u^3)/6,$$
$$N_2^3(u) = (1 + 3u + 3u^2 - 3u^3)/6,$$
$$N_3^3(u) = u^3/6.$$

The polynomial functions $N_i^3(u)$ in (3.19) are called *uniform cubic B-splines* (or B-spline blending functions). It is called cubic because its *degree* is 3 (*order* 4). In this book, a degree-n B-spline function is denoted by $N_i^n(u)$, but the notation $N_{n+1,i}(u)$ is commonly used in the literature.

Other aspects of B-spline curves will be discussed in more detail in the next section where non-uniform B-spline curve models are introduced.

3.4 NON-UNIFORM B-SPLINE CURVE MODELS

Presented in this section is a mathematical definition of NUB (*non-uniform B-spline*) curves. It is also shown in this section that the Bezier and uniform B-spline curves introduced in the previous section are in fact special cases of NUB-curves.

3.4.1 B-spline Basis Function

Let us consider the following *recursive scalar function* $L_i^n(t)$ defined with respect to a non-decreasing sequence of **knot points** $\{t_i\}$:

$$L_i^n(t) = \frac{(t - t_i)}{(t_{i+n-1} - t_i)} L_i^{n-1}(t) + \frac{(t_{i+n} - t)}{(t_{i+n} - t_{i+1})} L_{i+1}^{n-1}(t), \qquad (3.20)$$

$$where \quad L_i^1(t) = \begin{cases} 1, & t \in [t_i,\ t_{i+1}] \quad and \quad t_i < t_{i+1} \\ 0, & otherwise. \end{cases}$$

The above recurrence function is known as the **Cox-deBoor recursive function** which is the standard method of defining a **B-spline basis function** (of degree $n-1$). To understand geometric properties of the function, we need to evaluate it from the beginning. Let $n = 2$ in (3.20), then the first iteration is carried out as follows:

$$L_i^2(t) = \frac{(t - t_i)}{(t_{i+1} - t_i)} L_i^1(t) + \frac{(t_{i+2} - t)}{(t_{i+2} - t_{i+1})} L_{i+1}^1(t)$$

$$= \begin{cases} (t - t_i)/(t_{i+1} - t_i), & t \in [t_i,\ t_{i+1}] \\ (t_{i+2} - t)/(t_{i+2} - t_{i+1}), & t \in [t_{i+1},\ t_{i+2}] \\ 0\ , & otherwise. \end{cases}$$

In order to simplify algebraic manipulations, a *difference operator* ∇ is introduced:

$$\nabla_i = (t_{i+1} - t_i) \qquad (3.21-a)$$

$$\nabla_i^k = \nabla_i + ... + \nabla_{i+k-1} = (t_{i+k} - t_i) \qquad (3.21-b)$$

The above difference operator in fact defines a **knot span**. Using (3.21), the Cox-deBoor function with $n = 2$ is expressed as

$$L_i^2(t) = \begin{cases} (t - t_i)/(\nabla_i), & t \in [t_i,\ t_{i+1}] \\ (t_{i+2} - t)/(\nabla_{i+1}), & t \in [t_{i+1},\ t_{i+2}] \\ 0\ , & otherwise. \end{cases}$$

Using the difference operators, the next iteration with $n = 3$ is expressed as:

$$L_i^3(t) = \frac{(t - t_i)}{\nabla_i^2} L_i^2(t) + \frac{(t_{i+3} - t)}{\nabla_{i+1}^2} L_{i+1}^2(t)$$

$$= \begin{cases} (t - t_i)^2/(\nabla_i^2 \nabla_i), & t \in [t_i,\ t_{i+1}] \\ (t - t_i)^2/(\nabla_i^2 \nabla_i) - \nabla_i^3 (t - t_{i+1})^2/(\nabla_{i+1}^2 \nabla_{i+1} \nabla_i), & t \in [t_{i+1},\ t_{i+2}] \\ (t_{i+3} - t)^2/(\nabla_{i+1}^2 \nabla_{i+2}), & t \in [t_{i+2},\ t_{i+3}] \\ 0\ , & otherwise. \end{cases} \tag{3.22}$$

The function (3.22) is the B-spline of degree 2 (order 3). Functional shapes of B-splines of degrees up to 3 are plotted in Fig. 3.9.

Since a B-spline basis function has different forms on different intervals in the parameter space, the basis function in the kth interval is identified by the second subscript $[k]$. That is,

$$L_{i[k]}^n(t) \equiv L_i^n(t) \quad for \quad t \in [t_{i+k-1},\ t_{i+k}] \ : \quad k = 1, 2, ..., n. \tag{3.23}$$

According to the above definition, for example, *the first interval of* (3.22) can be rewritten as follows:

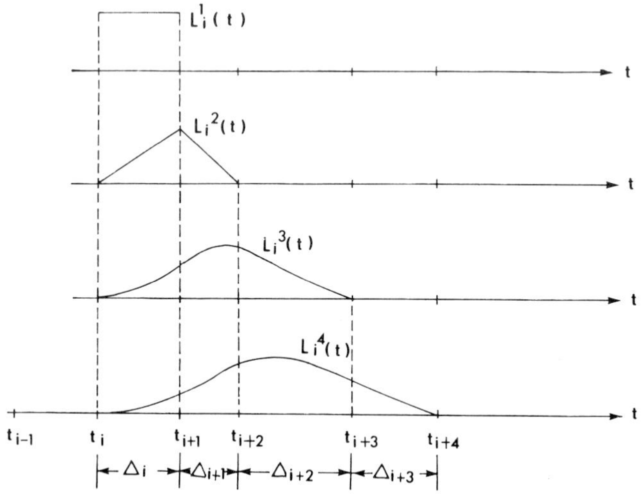

Figure 3.9 Non-uniform B-spline Functions

$$L^3_{i[1]}(t) \equiv L^3_i(t) \; for \; t \in [t_i, \; t_{i+1}] = (t - t_i)^2/(\nabla^2_i \nabla_i).$$

Let us define the following linear transformation between u and t:

$$u = (t - t_i)/(t_{i+1} - t_i) = (t - t_i)/\nabla_i. \tag{3.24}$$

As depicted in Figure 3.10, only three B-spline functions of degree 3 are non-zero in the interval $t \in [t_i, \; t_{i+1}]$. They are $L^3_{i-2[3]}(t)$, $L^3_{i-1[2]}(t)$, and $L^3_{i[1]}(t)$. These B-spline basis functions are easily obtained from (3.22) with index "i" shifted accordingly so that the range of the parameter t is always $[t_i, \; t_{i+1}]$:

$$\begin{aligned} L^3_{i-2[3]}(t) &= (t_{i+1} - t)^2/(\nabla^2_{i-1}\nabla_i) \\ &= (\nabla_i/\nabla^2_{i-1}) + u(-2\nabla_i/\nabla^2_{i-1}) + u^2(\nabla_i/\nabla^2_{i-1}) \qquad (3.25-a) \\ &\equiv N^2_0(u) \quad with \quad 0 \le u \le 1 \end{aligned}$$

$$\begin{aligned} L^3_{i-1[2]}(t) &= (t - t_{i-1})^2/(\nabla^2_{i-1}\nabla_{i-1}) - \nabla^3_{i-1}(t - t_i)^2/(\nabla^2_i\nabla_i\nabla_{i-1}) \\ &= (\nabla_{i-1}/\nabla^2_{i-1}) + u(2\nabla_i/\nabla^2_{i-1}) + u^2\frac{\nabla_i}{\nabla_{i-1}}\left\{\frac{\nabla_i}{\nabla^2_{i-1}} - \frac{\nabla^3_{i-1}}{\nabla^2_i}\right\} \qquad (3.25-b) \\ &\equiv N^2_1(u) \quad with \quad 0 \le u \le 1 \end{aligned}$$

$$L^3_{i[1]}(t) = (t - t_i)^2/(\nabla^2_i\nabla_i) = (u\nabla_i)^2/(\nabla^2_i\nabla_i) \equiv N^2_2(u) \quad with \quad 0 \le u \le 1. \qquad (3.25-c)$$

Shown in Figure 3.10 are the three basis functions (3.25) that are non-zero in $t \in [t_i, \; t_{i+1}]$.

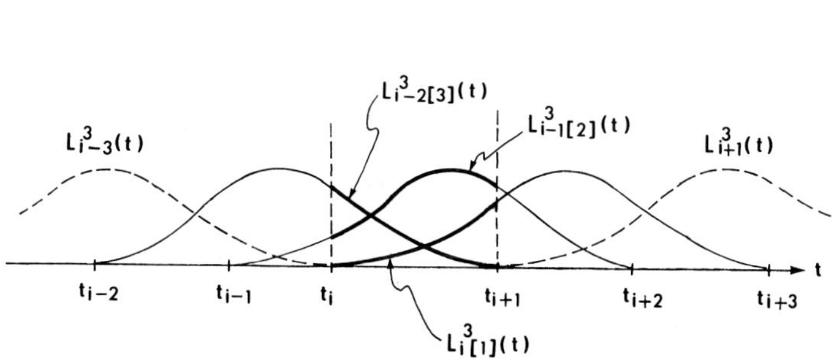

Figure 3.10 Non-zero Quadratic B-spline Basis Functions in $[t_i, \; t_{i+1}]$

3.4.2 Non-uniform B-spline Curve Models

For a given sequence of 3D points $\{P_j\}$, let us construct the following vector valued function of $t \in [t_i,\ t_{i+1}]$ by using the quadratic B-spline basis functions $L_j^3(t)$ defined in (3.22).

$$\mathbf{r}(t) = \Sigma\ P_j\ L_j^3(t) \quad : \quad t \in [t_i,\ t_{i+1}]. \tag{3.26}$$

As shown in Fig. 3.10 the blending functions $L_j^3(t)$ in (3.26) are non-zero only when $j = i-2,\ i-1,\ i.$ Let

$$V_0 = P_{i-2},$$
$$V_1 = P_{i-1},$$
$$V_2 = P_i,$$

then, from the results in (3.22) and (3.25), the vector valued function (3.26) is expressed as:

$$\begin{aligned}
\mathbf{r}(t) &= \sum_j P_j\ L_j^3(t) \quad : \quad t \in [t_i,\ t_{i+1}] \\
&= P_{i-2}\ L_{i-2}^3(t) + P_{i-1}\ L_{i-1}^3(t) + P_i\ L_i^3(t) \quad : \quad t \in [t_i,\ t_{i+1}] \\
&= V_0\ L_{i-2[3]}^3(t) + V_1\ L_{i-1[2]}^3(t) + V_2\ L_{i[1]}^3(t) \\
&= V_0\ N_0^2(u) + V_1\ N_1^2(u) + V_2\ N_2^2(u) \\
&= \mathbf{U}\ \mathbf{N}_q\ \mathbf{R} \equiv \mathbf{r}(u) \quad : \quad u \in [0, 1]
\end{aligned} \tag{3.27}$$

$$where, \quad \mathbf{U} = [1\ u\ u^2]\ ; \quad \mathbf{R} = [V_0\ V_1\ V_2]^T,$$

$$\mathbf{N}_q = \begin{bmatrix} (\nabla_i/\nabla_{i-1}^2) & (\nabla_{i-1}/\nabla_{i-1}^2) & 0 \\ (-2\nabla_i/\nabla_{i-1}^2) & (2\nabla_i/\nabla_{i-1}^2) & 0 \\ (\nabla_i/\nabla_{i-1}^2) & \frac{\nabla_i}{\nabla_{i-1}}\{\frac{\nabla_i}{\nabla_{i-1}^2} - \frac{\nabla_{i-1}^3}{\nabla_i^2}\} & \frac{\nabla_i}{\nabla_i^2} \end{bmatrix}$$

$$\nabla_i = (t_{i+1} - t_i), \quad \nabla_i^2 = \nabla_i + \nabla_{i+1}, \quad etc.$$

The curve segment given by (3.27) is called a **quadratic NUB-curve** segment. It is to be noted that the quadratic NUB-curve is *supported* by six knots, t_{i-2} through t_{i+3}, even though its parameter range is $t \in [t_i,\ t_{i+1}]$ as can seen in Fig. 3.10. However, the end knots, t_{i-2} and t_{i+3}, are redundant because this information is not used in computing the curve. Thus, the quadratic NUB-curve segment is completely specified by the three *knot spans* $\nabla_{i-1},\ \nabla_i,\ \nabla_{i+1}$ and the three *control vertices* $V_0,\ V_1,\ V_2$.

From the results of (3.22), the recursive relation (3.20) can be evaluated one step further to obtain

$$L_i^4(t) = L_{i[k]}^4(t) \quad for \quad t \in [t_{i+k-1},\ t_{i+k}] \ : \quad k = 1,2,3,4$$

$$with \begin{cases} L_{i[1]}^4(t) = (t - t_i)^3/(\nabla_i^3 \nabla_i^2 \nabla_i); \\ L_{i[2]}^4(t) = L_{i[1]}^4(t) - \nabla_i^4 (t - t_{i+1})^3/(\nabla_{i+1}^3 \nabla_{i+1}^2 \nabla_{i+1} \nabla_i); \\ L_{i[3]}^4(t) = L_{i[4]}^4(t) - \nabla_i^4 (t_{i+3} - t)^3/(\nabla_{i+2} \nabla_{i+1}^3 \nabla_{i+1}^2 \nabla_i^3); \\ L_{i[4]}^4(t) = (t_{i+4} - t)^3/(\nabla_{i+3} \nabla_{i+2}^2 \nabla_{i+1}^3). \end{cases}$$

By following the same steps of the quadratic case above, a cubic NUB-curve segment $\mathbf{r}(u)$ is evaluated as follows.

$$\begin{aligned} \mathbf{r}(t) &= \Sigma_j \ L_j^4(t)\ P_j \quad : \quad t \in [t_i,\ t_{i+1}] \\ &= L_{i-3}^4(t)\ P_{i-3} + L_{i-2}^4(t)\ P_{i-2} + L_{i-1}^4(t)\ P_{i-1} + L_i^4(t)\ P_i \\ &= L_{i-3[4]}^4(t)\ V_0 + L_{i-2[3]}^4(t)\ V_1 + L_{i-1[2]}^4(t)\ V_2 + L_{i[1]}^4(t)\ V_3 \\ &= \mathbf{U}\ \mathbf{N_c}\ \mathbf{R} \equiv \mathbf{r}(u) \quad : \quad u \in [0,1], \end{aligned} \tag{3.28}$$

$$where \quad \mathbf{U} = [1\ u\ u^2\ u^3], \quad \mathbf{R} = [V_0\ V_1\ V_2\ V_3]^T,$$

$$\mathbf{N}_c = \begin{bmatrix} \frac{(\nabla_i)^2}{\nabla_{i-1}^2 \nabla_{i-2}^3} & (1 - n_{11} - n_{13}) & \frac{(\nabla_{i-1})^2}{\nabla_{i-1}^3 \nabla_{i-1}^2} & 0 \\ -3n_{11} & (3n_{11} - n_{23}) & \frac{3\nabla_i \nabla_{i-1}}{\nabla_{i-1}^3 \nabla_{i-1}^2} & 0 \\ 3n_{11} & -(3n_{11} + n_{33}) & \frac{3(\nabla_i)^2}{\nabla_{i-1}^3 \nabla_{i-1}^2} & 0 \\ -n_{11} & (n_{11} - n_{43} - n_{44}) & n_{43} & \frac{(\nabla_i)^2}{\nabla_i^3 \nabla_i^2} \end{bmatrix}$$

$$n_{43} = -\{\frac{1}{3}\ n_{33} + n_{44} + (\nabla_i)^2/(\nabla_i^2 \nabla_{i-1}^3)\},$$

$$n_{i,j} = element\ in\ row\text{-}i,\ column\text{-}j,$$

$$\nabla_i^k \equiv \nabla_i + \nabla_{i+1} + \ldots + \nabla_{i+k-1}.$$

The derivation of the matrix form (3.28) requires a considerable amount of algebraic manipulation. When a matrix representation for a higher degree NUB curve is needed, the *basis translation* method proposed in Choi *et al* (1990a) may be used. An evaluation of a NUB curve (and surface) may be carried out by

a) directly evaluating the Cox-deBoor form (3.26),

b) using the deBoor algorithm (deBoor, 1972*),*

c) evaluating the matrix form (3.28), *or*

d) converting it into a Bezier curve (Boehm, 1981*).*

As with the evaluation of Bezier curves, the main advantage of the matrix form (ie, power basis form) is in its computational efficiency.

3.4.3 Special Cases of NUB-Curves

The cubic NUB curve (3.28) has the same form as the uniform cubic B-spline curve (3.19), but its coefficient matrix $\mathbf{N}_c$ depends on the knot spacing. Thus, for the same set of control vertices, quite different curve shapes may be obtained by changing the spacings between knot points.

When all the knot points $\{t_i\}$ are taken on successive integers, we have a uniform knot spacing. That is, the *knot spans* may be set to

$$\nabla_i = 1 \quad for \quad all \quad i.$$

Then, the NUB coefficient matrix $\mathbf{N}_c$ in (3.28) becomes the coefficient matrix $\mathbf{N}$ of the cubic uniform B-spline curve (3.17). That is, since $\nabla_i = 1$, $\nabla_i^2 = 2$, etc., $\mathbf{N}_c$ in (3.28) is trivially evaluated to give

$$\mathbf{N}_c = \begin{bmatrix} 1/6 & 2/3 & 1/6 & 0 \\ -1/2 & 0 & 1/2 & 0 \\ 1/2 & -1 & 1/2 & 0 \\ -1/6 & 1/2 & -1/2 & 1/6 \end{bmatrix} = \mathbf{N}$$

Thus, the uniform cubic B-spline curve (3.17) is obtained from the general B-spline curve (3.28) when the knot spacing becomes *uniform*.

We consider another special knot spacing with which the NUB curves would become Bezier curves. Let us define a knot sequence as

$$t_{i-2} = t_{i-1} = t_i = 0 \quad ; \quad t_{i+1} = t_{i+2} = t_{i+3} = 1$$

then, the corresponding *knot spans* are given by

$$\nabla_i = 1 \quad ; \quad \nabla_j = 0 \quad for \quad all \quad j \neq i$$

which makes the coefficient matrix $\mathbf{N}_c$ in (3.28) degenerate to the coefficient matrix $\mathbf{M}$ of the cubic Bezier curve (3.11). In summary, we have shown that both Bezier curve and uniform B-spline curve are special cases of a NUB-curve.

3.5 RATIONAL CURVE MODELS

A *rational function* is defined as a ratio of two polynomial functions. A rational curve model gives more flexibility in curve shapes than an ordinary (ie, non-rational) polynomial curve model does. A rational curve model has an ordinary polynomial form if it is expressed in **homogeneous coordinates**. This section presents rational extensions of the Bezier curve models of §3.3.3.

3.5.1 Homogeneous Coordinates

Earlier in Chapter 2, it was shown that a parametric equation of a unit circle (2.9) is given by

$$\begin{aligned}\mathbf{r}(u) &= (x(u),\ y(u),\ z(u)) \\ &= ((1-u^2)/(1+u^2),\ 2u/(1+u^2),\ 0/(1+u^2)).\end{aligned}$$

Since each element of the 3D vector in the above equation has a common denominator, we may employ a "four elements" vector which has its first three elements equal to the numerators of the 3D coordinates and its fourth element equal to the common denominator. The four-elements vector representation of the unit circle is then expressed as

$$\mathbf{R}(u) = ((1-u^2),\ 2u,\ 0,\ (1+u^2)) = (X(u),\ Y(u),\ Z(u),\ h(u)).$$

The vector $\mathbf{R}(u)$ is called a *homogeneous vector* and its elements become *homogeneous coordinates* of the 3D point $\mathbf{r}(u)$.

Every time a **homogeneous vector** $\mathbf{R}=(X,\ Y,\ Z,\ h)$ is encountered, its Cartesian coordinates $x,\ y,\ z$ are recovered by dividing the first three elements $X,\ Y,\ Z,$ by the last element h. That is, $(X,\ Y,\ Z,\ h)$ is converted to $(X/h,\ Y/h,\ Z/h,\ 1)$ when a geometric interpretation of the homogeneous vector is needed. This conversion is called **normalization**. A geometric meaning of normalization is that the *4D vector* $\mathbf{R}$ *is projected onto the plane* $h=1$ *in the four dimensional space*. By definition, two homogeneous vectors $(x,y,z,1)$ and (hx,hy,hz,h) represent the same *3D* point (x,y,z) if $h \neq 0$. In rational curve models, each control vertex $V_i = (x_i,\ y_i,\ z_i)$ is defined as a **homogeneous control point**

$$H_i = (w_i x_i,\ w_i y_i,\ w_i z_i,\ w_i)$$

whose $3D$ coordinates are $(x_i,\ y_i,\ z_i)$. The *weight* w_i provides additional flexibility in shape control.

The homogeneous representation of Cartesian coordinates is widely used in coordinate transformations (See §3.5) in the fields of computer graphics and robotics. For more details on coordinate transformations based on homogeneous coordinates, the reader is referred to, for example, Rogers and Adams (1976) and Paul (1981).

3.5.2 Rational Quadratic Curve Models

If all the 3D vectors (ie, control points) of an ordinary Bezier curve is replaced by corresponding homogeneous vectors, a rational Bezier curve model is obtained. In Example 3.1 of §3.3.3, a quadratic Bezier curve was given by

$$\mathbf{r}(u) = (x(u),\ y(u),\ z(u)) = \sum_{i=0}^{2} B_i^2(u)\ V_i$$
$$= [1\ u\ u^2] \begin{bmatrix} 1 & 0 & 0 \\ -2 & 2 & 0 \\ 1 & -2 & 1 \end{bmatrix} \begin{bmatrix} V_0 \\ V_1 \\ V_2 \end{bmatrix} \quad : \quad 0 \le u \le 1. \tag{3.29}$$

By appending 1 to each control vertex, $V_i = (x_i,\ y_i,\ z_i)$, normalized homogeneous vectors are obtained. Namely,

$$V_i^h = (x_i,\ y_i,\ z_i,\ 1) \quad for \quad i = 0, 1, 2.$$

Let H_i denote **homogeneous control vertices** such that

$$H_i = w_i V_i^h = (w_i x_i,\ w_i y_i,\ w_i z_i,\ w_i) \quad ; \quad w_i \neq 0. \tag{3.30}$$

Then, the Cartesian coordinates of a homogeneous control vertex H_i in (3.30) are the same as those of a 3D control vertex V_i regardless of the value of the **weight** w_i (as long as it is non-zero).

By using the homogeneous control vertices in (3.30), the **rational quadratic Bezier curve** is expressed in a homogeneous form as

$$\mathbf{R}(u) = (X(u),\ Y(u),\ Z(u),\ h(u)) = \sum_{i=0}^{2} B_i^2(u)\ H_i$$
$$= [1\ u\ u^2] \begin{bmatrix} 1 & 0 & 0 \\ -2 & 2 & 0 \\ 1 & -2 & 1 \end{bmatrix} \begin{bmatrix} H_0 \\ H_1 \\ H_2 \end{bmatrix} \quad : \quad 0 \le u \le 1, \tag{3.31}$$
$$where \quad H_i = (w_i x_i,\ w_i y_i,\ w_i z_i,\ w_i).$$

We would like to investigate the shape of the homogeneous curve (3.31). The "end conditions" of the rational curve are easily computed by evaluating (3.31) and its derivative at $u = 0$ and $u = 1$.

$$\mathbf{R}(0) = H_0 = w_0 V_0^h = (w_0 x_0,\ w_0 y_0,\ w_0 z_0,\ w_0) \tag{3.32 – a}$$

$$\mathbf{R}(1) = H_2 = w_2 V_2^h = (w_2 x_2,\ w_2 y_2,\ w_2 z_2,\ w_2) \tag{3.32 – b}$$

$$\dot{\mathbf{R}}(0) = 2(H_1 - H_0) = 2(w_1 V_1^h - w_0 V_0^h) \tag{3.32 – c}$$

$$\dot{\mathbf{R}}(1) = 2(H_2 - H_1) = 2(w_2 V_2^h - w_1 V_1^h). \tag{3.32 – d}$$

It may easily be seen from (3.32-a) and (3.32-b) that the homogeneous Bezier curve (3.31) starts from V_0 and ends at V_2 as does the ordinary Bezier curve (3.29). Note also that the relations in (3.32) are the same as those for an ordinary Bezier curve (see Eqn. (3.14) for $n = 2$).

Let $\mathbf{r}^h(u)$ denote a normalized homogeneous equation obtained by dividing R(u) by h(u). That is,

$$\mathbf{r}^h(u) \equiv \mathbf{R}(u)/h(u) = (x(u),\ y(u),\ x(u),\ 1) \tag{3.33}$$

Then, by differentiating (3.33) with respect to u, we obtain

$$\dot{\mathbf{r}}^h(u) = -\{\mathbf{R}(u)\dot{h}(u)\}/\{h(u)\}^2 + \dot{\mathbf{R}}(u)/h(u) \tag{3.34}$$

In general, the nth-derivative of the normalized homogeneous curve (3.33) is given by

$$\frac{d^n\{\mathbf{r}^h(u)\}}{du^n} = \sum_{i=0}^{n} \binom{n}{i} \frac{d^i\{\mathbf{R}(u)\}}{du^i} \frac{d^{n-i}\{1/h(u)\}}{du^{n-i}}.$$

Since the fourth elements of the homogeneous vectors in (3.32) are expressed as

$$h(0) = w_0\ ; \quad h(1) = w_2\ ; \quad \dot{h}(0) = 2(w_1 - w_0)\ ; \quad \dot{h}(1) = 2(w_2 - w_1),$$

the derivative (3.34) is evaluated at $u = 0, 1$ as

$$\dot{\mathbf{r}}^h(0) = 2(V_1^h - V_0^h) \cdot \frac{w_1}{w_0} \quad ; \quad \dot{\mathbf{r}}^h(1) = 2(V_2^h - V_1^h) \cdot \frac{w_1}{w_2}.$$

The above results show that the homogeneous Bezier curve (3.31) has the same end gradients as the ordinary Bezier curve (3.29), but the magnitudes of its end tangents are scaled by the amounts of w_1/w_0 and w_1/w_2 as depicted in Fig. 3.11. Compare the rational quadratic Bezier curve of Fig. 3.11b with the ordinary Bezier curve shown in Fig. 3.11a.

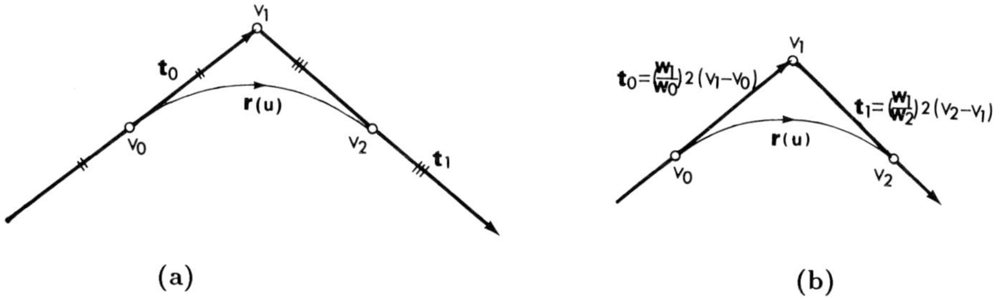

Figure 3.11 Properties of a Rational Bezier Curve

The homogeneous Bezier equation (3.31) is expressed in a **component form** as:

$$\mathbf{R}(u) = (X(u),\ Y(u),\ Z(u),\ h(u))$$
$$= (\sum_{i=0}^{2} B_i^2(u)\ w_i x_i,\ \sum_{i=0}^{2} B_i^2(u)\ w_i y_i,\ \sum_{i=0}^{2} B_i^2(u)\ w_i z_i,\ \sum_{i=0}^{2} B_i^2(u)\ w_i).$$

Thus, the Cartesian coordinates are recovered as

$$x(u) = X(u)/h(u) = \frac{(1-u)^2 w_0 x_0 + 2u(1-u)w_1 x_1 + u^2 w_2 x_2}{(1-u)^2 w_0 + 2u(1-u)w_1 + u^2 w_2}$$

$$y(u) = Y(u)/h(u) = \frac{(1-u)^2 w_0 y_0 + 2u(1-u)w_1 y_1 + u^2 w_2 y_2}{(1-u)^2 w_0 + 2u(1-u)w_1 + u^2 w_2}$$

$$z(u) = Z(u)/h(u) = \frac{(1-u)^2 w_0 z_0 + 2u(1-u)w_1 z_1 + u^2 w_2 z_2}{(1-u)^2 w_0 + 2u(1-u)w_1 + u^2 w_2}$$

which are more concisely expressed in a **rational form** as

$$\mathbf{r}(u) = (x(u),\ y(u),\ z(u))$$
$$= \frac{w_0(1-u)^2 V_0 + w_1 2u(1-u)V_1 + w_2 u^2 V_2}{w_0(1-u)^2 + w_1 2u(1-u) + w_2 u^2} \tag{3.35}$$
$$= \left(\sum_{i=0}^{2} B_i^2(u)\ w_i V_i\right) \Big/ \left(\sum_{i=0}^{2} B_i^2(u)\ w_i\right),$$

where $V_i = (x_i,\ y_i,\ z_i)$: *Bezier control vertices,*

w_i : *weights.*

Thus, we have shown that the same rational Bezier curve is expressed either in a homogeneous form (3.31) or in a rational form (3.35). In either case, it is obvious that the rational quadratic Bezier curve degenerates to the ordinary quadratic Bezier curve (3.29) when $w_i = 1$ for all i. Further, it can be shown that the *convex hull property* holds also for a rational Bezier curve for positive weights. That is, the rational quadratic Bezier curve lies inside the polygon formed by the three control vertices $V_0,\ V_1,\ V_2$ when $w_i > 0$. This rational quadratic curve model is widely used in parametrizing conic sections as will be seen in the next chapter. A rational quadratic B-spline curve is similarly constructed.

3.5.3 Rational Cubic Curve Models

We can easily construct rational curve models for higher degree Bezier curves and for B-spline curves. A rational cubic Bezier curve has exactly the same **homogeneous form** as the ordinary cubic Bezier curve of Eqn. (3.11):

$$\begin{aligned} \mathbf{R}(u) &= (X(u),\ Y(u),\ Z(u),\ h(h)) \\ &= \mathbf{U\ M\ H} \end{aligned} \tag{3.36}$$

$$where,\quad \mathbf{U} = [1\ u\ u^2\ u^3],$$

$$\mathbf{M} = \begin{bmatrix} 1 & 0 & 0 & 0 \\ -3 & 3 & 0 & 0 \\ 3 & -6 & 3 & 0 \\ -1 & 3 & -3 & 1 \end{bmatrix} \quad \mathbf{H} = \begin{bmatrix} H_0 \\ H_1 \\ H_2 \\ H_3 \end{bmatrix},$$

$$H_i = (w_i x_i,\ w_i y_i,\ w_i z_i,\ w_i),$$

$$V_i = (x_i,\ y_i,\ z_i)\ :\ Bezier\ control\ vertices,$$

$$w_i\ :\ weights.$$

The above homogeneous form is equivalent to the following **rational form**:

$$\mathbf{r}(u) = \Big(\sum_{i=0}^{3} B_i^3(u)\ w_i V_i\Big) \Big/ \Big(\sum_{i=0}^{3} B_i^3(u)\ w_i\Big)$$

$$where,\quad B_i^3(u) = \frac{3!}{(3-i)!\ i!} u^i (1-u)^{3-i},$$

$$V_i\ :\ Bezier\ control\ vertices,$$

$$w_i\ :\ weights.$$

By following the same steps for the case of quadratic Bezier curve, that is, equations (3.32) through (3.34), it can be shown that the following hold for the rational cubic Bezier curve $\mathbf{r}(u)$:

$$\mathbf{r}(0) = V_0\ ;\quad \mathbf{r}(1) = V_3\ ;$$

$$\dot{\mathbf{r}}(0) = 3(V_1 - V_0)(\frac{w_1}{w_0})\ ;$$

$$\dot{\mathbf{r}}(1) = 3(V_3 - V_2)(\frac{w_2}{w_3}).$$

Thus, the rational cubic Bezier curve starts from the first control vertex V_0 and ends at the last control vertex V_3. The end tangents of the rational Bezier curve and the ordinary Bezier curve have the same directions, but their magnitudes are scaled by the amounts of w_1/w_0 and w_2/w_3.

A rational curve model provides more degrees of freedom in defining curve shape. By providing different values of the weights w_i, the rational cubic Bezier curve can be made to pass anywhere inside the characteristic polygon. But, having too much freedom is not always a virtue. In practice, higher than quadratic rational curves are rarely used.

3.6 DISCUSSIONS

Reviewed in this chapter are *four* types of elementary polynomial curve models: They are

- *implicit polynomial curves,*
- *explicit polynomial curves,*
- *parametric polynomial curves, and*
- *rational parametric curves.*

Implicit curves are sometimes called *analytic curves*, while explicit curves are called *non-parametric* curves. Quadratic implicit curves represent *conic sections* which are easily parametrized either by using trigonometric functions or as rational (quadratic) polynomials as will be shown in the next chapter. Implicit polynomial curves have exclusively been used in traditional engineering drawings. The *(non-rational)* parametric polynomial curve models introduced are

- standard cubic polynomial model known as *four points curve,*
- *Ferguson* curve model,
- *(cubic) Bezier* curve model,
- *(cubic) uniform B-spline* curve model, and
- *(cubic) non-uniform B-spline* curve model.

These curve models are used in representing a (space) curve segment of cubic precision, but they are easily extended to higher degrees. By joining individual cubic curve segments, a *composite* curve having up to C^2-continuity is easily constructed.

The shape of non-uniform B-spline is controlled by *knot spacing* as well as by *control vertices.* The shape of a rational parametric curve is controlled also by *weights* (in addition to control vertices). In most practical application, rational parametric curve models are used in representing conic sections. A *NURB* (non-uniform rational B-spline) curve is very flexible because its shape is controlled by *control vertices, knot spacing,* and *weights.* More details on NURB curves and surfaces are provided in the last chapter of this book. The curve models discussed in this chapter are easily extended to *(tensor product)* surface models as will be seen in Chapter 5.

CHAPTER 4

COMPOSITE CURVE FITTING AND 2D CURVE MODELING

4.1 INTRODUCTION

Presented in this chapter are methods of constructing *composite curves* from a sequence of data points and special techniques for constructing 2D curve models. The curve fitting problem is to construct a *smooth* parametric curve $\mathbf{r}(t)$ passing through a sequence of data points $\{P_i : i = 0, ..., n\}$.

Any of the curve models introduced in the previous chapter may be employed in fitting a curve from a given sequence of data points. In practice, however, the Ferguson model and B-spline curve models are the popular ones. The choice of a specific curve model depends on such factors as

- *ease of implementation,*
- *computational efficiency,*
- *mathematical continuity,*
- *aesthetic smoothness, and*
- *uniformness of flow rate.*

A curve fitting problem is basically the one of solving a system of linear equations. By imposing a *second order continuity* condition at each data point P_i, a linear equation system for the unknown coefficients of a cubic curve model is obtained. Thus, for curve fitting, it is essential to have a suitable set of *continuity conditions* and to be able to solve a system of *linear equations*. The main issue in 2D curve modeling is *parametrization*. The topics to be discussed in this chapter are as follows:

1) Curve fitting of evenly spaced data points (§4.2)*:*

- *Cubic spline fitting.*
- *Uniform B-spline fitting.*

2) Curve fitting of unevenly spaced data points (§4.3)*:*

- *Chord-length spline fitting.*
- *Non-uniform B-spline fitting.*

3) 2D curve modeling (§4.4)*:*

- *Biarc curve modeling.*
- *Conic section curve modeling.*
- *parametrization of conic section curves.*

4.2 CURVE FITTING FOR EVENLY SPACED POINT DATA

This section presents two different methods of constructing smooth curves from a sequence of evenly spaced point data: One based on the Ferguson curve model (3.8), and the other based on the uniform cubic B-spline curve model (3.17). When the physical spacing of the point data becomes uneven, curves obtained by these methods tend to show local flatness and kinks.

4.2.1 Parametric Continuity Condition

Let us consider the two curve segments $\mathbf{r}^a(u)$ and $\mathbf{r}^b(u)$ in Fig. 4.1, each defined on the unit interval $u \in [0,1]$. In order for the two curve segments to be connected, the following **position continuity condition** should hold:

$$\mathbf{r}^a(1) = P_1 = \mathbf{r}^b(0). \tag{4.1 - a}$$

The composite curve in Fig. 4.1 is said to be C^1-continuous if the first derivatives of both curves at the common join are equal. That is,

$$\dot{\mathbf{r}}^a(1) = \mathbf{t}_1 = \dot{\mathbf{r}}^b(0). \tag{4.1 - b}$$

And the composite curve is said to be *second order continuous* if

$$\ddot{\mathbf{r}}^a(1) = \ddot{\mathbf{r}}^b(0). \tag{4.1 - c}$$

The conditions in (4.1) are collectively called a **parametric $\mathbf{C^2}$-condition.**

The composite curve in Fig. 4.1 is required to pass through the three points P_0, P_1, P_2, and the *end tangents* $\mathbf{t}_0$, $\mathbf{t}_2$ are assumed to be given. If the (unknown) tangent vector $\mathbf{t}_1$ at the common join is also given, each of the two curve segments can be represented as a Ferguson curve because a Ferguson curve Eqn. (3.8) is completely specified by its *end conditions*. Thus, the problem here is to determine the unknown tangent $\mathbf{t}_1$ so that the two curve segments are C^2-continuous at the common join P_1. That is, we need to compute $\mathbf{t}_1$ from the input data P_0, P_1, P_2, $\mathbf{t}_0$, $\mathbf{t}_2$ by using the C^2-condition (4.1).

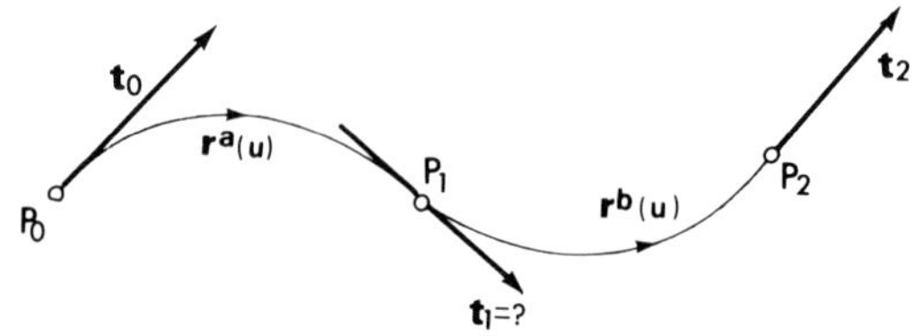

Figure 4.1 Parametric Continuity at the Common Join

4.2.2 Cubic Spline Fitting

Let us first consider the problem of determining the common tangent $\mathbf{t}_1$ of the two curve segments $\mathbf{r}^a(u)$ and $\mathbf{r}^b(u)$ in Fig. 4.1 so that the C^2-condition (4.1) is satisfied. Ferguson curve equations for the two curve segments are expressed as (See Eqn. (3.8)):

$$\mathbf{r}^a(u) = \mathbf{U}\ \mathbf{C}\ \mathbf{S}^a \quad and \quad \mathbf{r}^b(u) = \mathbf{U}\ \mathbf{C}\ \mathbf{S}^b \tag{4.2}$$

$$where, \quad \mathbf{U} = [\ 1\ u\ u^2\ u^3\ ],$$

$$\mathbf{C} = \begin{bmatrix} 1 & 0 & 0 & 0 \\ 0 & 0 & 1 & 0 \\ -3 & 3 & -2 & -1 \\ 2 & -2 & 1 & 1 \end{bmatrix},$$

$$\mathbf{S}^a = [P_0\ P_1\ \mathbf{t}_0\ \mathbf{t}_1]^T,$$

$$\mathbf{S}^b = [P_1\ P_2\ \mathbf{t}_1\ \mathbf{t}_2]^T.$$

The second derivative of the first curve segment $\mathbf{r}^a(u)$ at $u = 1$ is evaluated as

$$\begin{aligned} \ddot{\mathbf{r}}^a(1) &= \ddot{\mathbf{U}}\ \mathbf{C}\ \mathbf{S}^a|_{u=1} \\ &= [\ 0\ 0\ 2\ 6\ ]\mathbf{C}\ \mathbf{S}^a \\ &= 6P_0 - 6P_1 + 2\mathbf{t}_0 + 4\mathbf{t}_1. \end{aligned} \tag{4.3 - a}$$

And the second derivative of $\mathbf{r}^b(u)$ at $u = 0$ is similarly given by

$$\begin{aligned} \ddot{\mathbf{r}}^b(0) &= \ddot{\mathbf{U}}\ \mathbf{C}\ \mathbf{S}^b|_{u=0} \\ &= -6P_1 + 6P_2 - 4\mathbf{t}_1 - 2\mathbf{t}_2. \end{aligned} \tag{4.3 - b}$$

Since the C^2-condition (4.1) requires that the two second derivatives be equal, the following relation is obtained by equating (4.3-a) with (4.3-b).

$$\mathbf{t}_0 + 4\mathbf{t}_1 + \mathbf{t}_2 = 3(P_2 - P_0). \tag{4.4}$$

From (4.4), the unknown tangent $\mathbf{t}_1$ is determined as

$$\mathbf{t}_1 = (3P_2 - 3P_0 - \mathbf{t}_0 - \mathbf{t}_2)/4.$$

Now we consider the problem of constructing a C^2-continuous curve passing through a sequence $n + 1$ data points as shown in Fig. 4.2. It is assumed that end tangents $\hat{\mathbf{t}}_0$ and $\hat{\mathbf{t}}_n$ are given, in addition to the $n + 1$ data points $\{P_i\}$. Let $\mathbf{t}_i$ be the tangent vector at the data point P_i, then a linear equation of the form (4.4) holds for each pair of neighboring curve segments

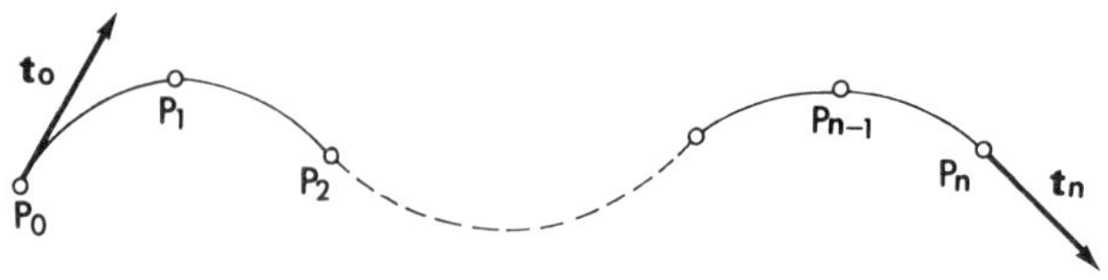

Figure 4.2 Data Point Sequence to be Interpolated

$\mathbf{r}^{i-1}(u)$ and $\mathbf{r}^{i}(u)$. That is, the following set of linear equations are obtained:

$$\mathbf{t}_{i-1} + 4\mathbf{t}_i + \mathbf{t}_{i+1} = 3(P_{i+1} - P_{i-1}) \quad for \quad i = 1, 2, ..., n-1. \tag{4.5}$$

Remember that the end tangents are assumed to be known, namely

$$\mathbf{t}_0 = \hat{\mathbf{t}}_0 \quad and \quad \mathbf{t}_n = \hat{\mathbf{t}}_n.$$

The above linear equation system has the following matrix form

$$\mathbf{A} \ \mathbf{X} = \mathbf{B},$$

where $\mathbf{A}$ is a $(n+1) \times (n+1)$ *coefficient matrix* and $\mathbf{X}$ is the unknown vector. That is, the **cubic spline fitting problem** is expressed in a matrix form as

$$\begin{bmatrix} 1 & 0 & 0 & 0 & & & \\ 1 & 4 & 1 & 0 & & & \\ 0 & 1 & 4 & 1 & & & \\ & & & \cdot & & & \\ & & & & \cdot & & \\ & & & & 1 & 4 & 1 \\ & & & & 0 & 0 & 1 \end{bmatrix} \begin{bmatrix} \mathbf{t}_0 \\ \mathbf{t}_1 \\ \mathbf{t}_2 \\ \cdot \\ \cdot \\ \mathbf{t}_{n-1} \\ \mathbf{t}_n \end{bmatrix} = \begin{bmatrix} \hat{\mathbf{t}}_0 \\ 3(P_2 - P_0) \\ 3(P_3 - P_1) \\ \cdot \\ \cdot \\ 3(P_n - P_{n-2}) \\ \hat{\mathbf{t}}_n \end{bmatrix} \tag{4.6}$$

A general solution to the linear system is given by $\mathbf{X} = \mathbf{A}^{-1}\mathbf{B}$, but (4.6) can be solved without explicitly inverting the coefficient matrix $\mathbf{A}$. Since $\mathbf{A}$ is a **tridiagonal matrix**, the solution is obtained by one forward substitution followed by a backward substitution (see Appendix A for details).

Having determined all the tangent vectors $\mathbf{t}_i$, each Ferguson curve segment $\mathbf{r}^i(u)$ at a span $[P_i,\ P_{i+1}]$ is expressed as

$$\mathbf{r}^i(u) = \mathbf{U} \ \mathbf{C} \ \mathbf{S}^i \quad for \quad i = 0, 1, 2, ..., n-1, \tag{4.7}$$

where $\mathbf{S}^i = [P_i \ \ P_{i+1} \ \ \mathbf{t}_i \ \ \mathbf{t}_{i+1}]^T$, and $\mathbf{U}$, $\mathbf{C}$ are as defined in (4.2).

The composite curve equation (4.7) is called **cubic spline curve** because it is defined by "cubic" polynomial functions and the resulting curve is similar to the one obtained from a physical "spline" where the data points P_i correspond to "ducks".

In most practical applications, the end tangents $\hat{\mathbf{t}}_0$, $\hat{\mathbf{t}}_n$ are not given. In this case, they need to be estimated by some means. Popular methods of estimating end tangents are to specify

- *circular end condition,*
- *polynomial end condition, or*
- *free-end condition.*

1) Circular End Condition:

In **circular end condition**, an end tangent is estimated by fitting a circle through the three points at the end. Shown in Fig. 4.3 are the first three data points P_0, P_1, P_2. As depicted in the figure, let us define

Q : *center of the circle to be fitted,*
$\mathbf{r} = Q - P_0$,
$\mathbf{a} = P_1 - P_0$; $\mathbf{b} = P_2 - P_0$; $\mathbf{c} = \mathbf{a} \times \mathbf{b}$.

The direction of the end tangent $\mathbf{t}_0$ should be perpendicular to both $\mathbf{r} = Q - P_0$ and $\mathbf{c} = \mathbf{a} \times \mathbf{b}$, and its magnitude may be set to the length of the chord P_0P_1. Thus, the *end tangent under the circular end condition* is given by

$$\hat{\mathbf{t}}_0 = |\mathbf{a}| \ (\mathbf{r} \times \mathbf{c}) \ / \ |\mathbf{r} \times \mathbf{c}| \tag{4.8}$$

It can be shown that the unknown vector $\mathbf{r} = Q - P_0$ in (4.8) is expressed as:

$$\mathbf{r} = \{|\mathbf{a}|^2(\mathbf{b} \times \mathbf{c}) + |\mathbf{b}|^2(\mathbf{c} \times \mathbf{a})\} \ / \ (2|\mathbf{c}|^2), \tag{4.9}$$

where $\mathbf{a} = P_1 - P_0$, $\mathbf{b} = P_2 - P_0$ *and* $\mathbf{c} = \mathbf{a} \times \mathbf{b}$.

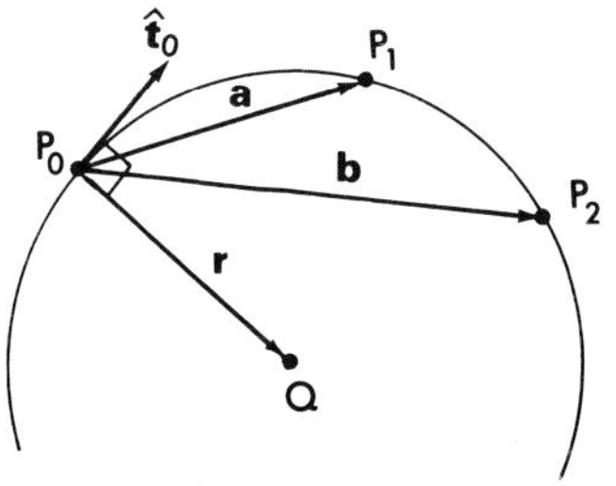

Figure 4.3 End-Tangent Estimation under Circular End Condition

Example 4.1: *Derive the expression* (4.9) *for the "radius vector"* $\mathbf{r}$.

The three planes $\pi_i\{\mathbf{r}\cdot\mathbf{n}_i = d_i\ ;\quad i = 1,2,3\}$ *intersect at a point*

$$\mathbf{r}^* = \frac{d_1(\mathbf{n}_2\times\mathbf{n}_3) + d_2(\mathbf{n}_3\times\mathbf{n}_1) + d_3(\mathbf{n}_1\times\mathbf{n}_2)}{\mathbf{n}\cdot(\mathbf{n}_2\times\mathbf{n}_3)}$$

because $\mathbf{r}^*$ *satisfies the three plane equations. Since we can set* $P_0 = (0,\ 0,\ 0)$ *without loss of generality, the three planes in our case are*

$\mathbf{r}\cdot\mathbf{c} = 0 \quad : \quad \pi_1$ *(containing* $P_0,\ P_1,\ P_2$*),*

$\mathbf{r}\cdot(\mathbf{a}/|\mathbf{a}|) = \frac{1}{2}|\mathbf{a}| : \pi_2$ *(passing through the mid-point of* P_0 *and* P_1*),*

$\mathbf{r}\cdot(\mathbf{b}/|\mathbf{b}|) = \frac{1}{2}|\mathbf{b}| : \pi_3$ *(passing through the mid-point of* P_0 *and* P_2*).*

By substituting the following into the expression for $\mathbf{r}^*$, *(4.9) is obtained:*

$$\mathbf{n}_1 = \mathbf{c},\ \mathbf{n}_2 = \mathbf{a},\ \mathbf{n}_3 = \mathbf{b},\ d_1 = 0,\ d_2 = \frac{1}{2}|\mathbf{a}|^2,\ d_3 = \frac{1}{2}|\mathbf{b}|^2. \qquad \diamond$$

2) Polynomial End Condition:

Under the **polynomial end condition**, the end tangents $\hat{\mathbf{t}}_0,\ \hat{\mathbf{t}}_n$ are obtained by fitting a *standard polynomial curve* at each end. For example, the end tangent $\hat{\mathbf{t}}_0$ may be estimated by fitting

- *a four point curve (see* §3.3.1*) through* $P_0,\ P_1,\ P_2,\ P_3$, *or*
- *a quadratic curve from the three points* $P_0,\ P_1,\ P_2$.

In either case, the end tangent $\hat{\mathbf{t}}_0$ is simply set to $\dot{\mathbf{r}}(0)$.

3) Free End Condition:

Free end condition is obtained by setting the curvatures at the end points $P_0,\ P_n$ to zero. This condition corresponds to the situation where the composite curve is not subjected to any external load at its ends. Since the curvature of a curve at a point is zero when the second derivative at that point is zero, the free end condition is satisfied by

$$\ddot{\mathbf{r}}^0(0) = \mathbf{0} \quad \textit{and}$$
$$\ddot{\mathbf{r}}^{n-1}(1) = \mathbf{0}$$

which in turn, from (4.3-a) and (4.3-b), result in the following:

$$2\mathbf{t}_0 + \mathbf{t}_1 = 3(P_1 - P_0) \quad \textit{and}$$
$$2\mathbf{t}_n + \mathbf{t}_{n-1} = 3(P_n - P_{n-1}).$$

These two linear equations are added to (4.5) to form a system of $n+1$ linear equations for the $n+1$ unknowns $\mathbf{t}_0, ..., \mathbf{t}_n$. A matrix representation of the linear system for the free end condition case is given as

$$\begin{bmatrix} 2 & 1 & 0 & 0 & & & \\ 1 & 4 & 1 & 0 & & & \\ 0 & 1 & 4 & 1 & & & \\ & & & \cdot & & & \\ & & & & \cdot & & \\ & & & & 1 & 4 & 1 \\ & & & & 0 & 1 & 2 \end{bmatrix} \begin{bmatrix} \mathbf{t}_0 \\ \mathbf{t}_1 \\ \mathbf{t}_2 \\ \cdot \\ \cdot \\ \mathbf{t}_{n-1} \\ \mathbf{t}_n \end{bmatrix} = \begin{bmatrix} 3(P_1 - P_0) \\ 3(P_2 - P_0) \\ 3(P_3 - P_1) \\ \cdot \\ \cdot \\ 3(P_n - P_{n-2}) \\ 3(P_n - P_{n-1}) \end{bmatrix} \tag{4.10}$$

which is easily solved by using the method of Appendix A.

Regardless of the end conditions used, the resulting *cubic spline curve* is composed of Ferguson curves each of which is easily converted to a cubic Bezier curve.

Example 4.2: *Convert the Ferguson curve* (4.7) *to a cubic Bezier curve.*

Since both the Bezier curve (3.11) *and the Ferguson curve equation* (4.7) *represent the same cubic curve, we have* $\mathbf{r}(u) = \mathbf{U\ M\ R} = \mathbf{U\ C\ S}$ *for* $0 \le u \le 1$. *This identity relation is satisfied when* $\mathbf{M\ R} = \mathbf{C\ S}$, *which leads to*

$$\mathbf{R} = \mathbf{M}^{-1}\mathbf{C\ S},$$

$$where \quad \mathbf{R} = \begin{bmatrix} V_0 \\ V_1 \\ V_2 \\ V_3 \end{bmatrix}; \quad \mathbf{S} = \begin{bmatrix} P_{i-1} \\ P_i \\ \mathbf{t}_{i-1} \\ \mathbf{t}_i \end{bmatrix},$$

$$\mathbf{C} = \begin{bmatrix} 1 & 0 & 0 & 0 \\ 0 & 0 & 1 & 0 \\ -3 & 3 & -2 & -1 \\ 2 & -2 & 1 & 1 \end{bmatrix},$$

$$\mathbf{M} = \begin{bmatrix} 1 & 0 & 0 & 0 \\ -3 & 3 & 0 & 0 \\ 3 & -6 & 3 & 0 \\ -1 & 3 & -3 & 1 \end{bmatrix}.$$

Thus, the Bezier control vertices V_i *in* (3.11) *are evaluated to be*

$$V_0 = P_{i-1}\ ;$$

$$V_1 = P_{i-1} + \mathbf{t}_{i-1}/3\ ;$$

$$V_2 = P_i - \mathbf{t}_i/3\ ;$$

$$V_3 = P_i. \qquad \diamond$$

4.2.3 Uniform B-spline Curve Fitting

Presented in this subsection is another method of constructing a smooth curve passing through the same data points $P_0, P_1, .., P_n$ by using the B-spline curve model (3.17). The fitted curve is a C^2-continuous curve composed of cubic B-spline curve segments. However, it should be noted that both composite curves, *the cubic spline and B-spline curves*, are the same cubic curves because both curves came from the same "four points curve" (3.5).

If a new control vertex V_3^b is appended to the B-spline curve segment of Fig. 3.8, a new curve segment is created as shown in Fig. 4.4. From the definition of a B-spline curve given by Eqn. (3.17), the two curve segments in Fig. 4.4 are expressed as:

$$\mathbf{r}^a(u) = \mathbf{U}\ \mathbf{N}\ \mathbf{R}^a \quad and \quad \mathbf{r}^b(u) = \mathbf{U}\ \mathbf{N}\ \mathbf{R}^b \tag{4.11}$$

$$where, \quad \mathbf{U} = [\ 1\ u\ u^2\ u^3\ ],$$

$$\mathbf{N} = \frac{1}{6}\begin{bmatrix} 1 & 4 & 1 & 0 \\ -3 & 0 & 3 & 0 \\ 3 & -6 & 3 & 0 \\ -1 & 3 & -3 & 1 \end{bmatrix},$$

$$\mathbf{R}^a = \begin{bmatrix} V_0^a \\ V_1^a \\ V_2^a \\ V_3^a \end{bmatrix} \quad ; \quad \mathbf{R}^b = \begin{bmatrix} V_0^b \\ V_1^b \\ V_2^b \\ V_3^b \end{bmatrix}.$$

As shown in the figure, the control vertices of the two curve segments are overlapped as:

$$V_0^b \equiv V_1^a \ ; \quad V_1^b \equiv V_2^a \ ; \quad V_2^b \equiv V_3^a .$$

By evaluating the B-spline curves in (4.11), one may easily verify that

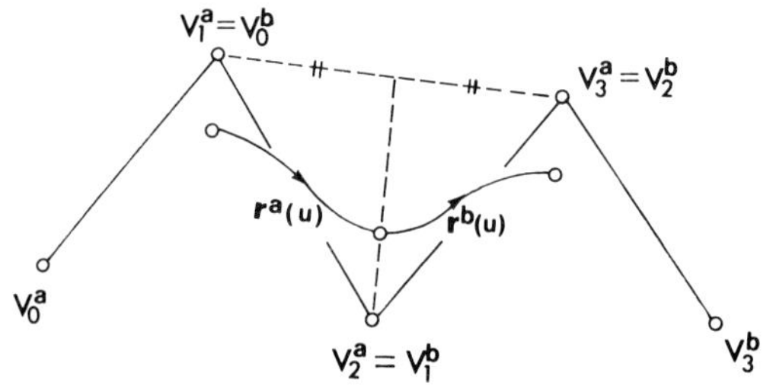

Figure 4.4 Construction of Composite B-spline Curve

$$\mathbf{r}^a(1) = (V_1^a + 4V_2^a + V_3^a)/6$$
$$\equiv (V_0^b + 4V_1^b + V_2^b)/6 \tag{4.12}$$
$$= \mathbf{r}^b(0).$$

Similarly, the following is obtained from the first derivatives of (4.11):

$$\dot{\mathbf{r}}^a(1) = (V_3^a - V_1^a)/2$$
$$\equiv (V_2^b - V_0^b)/2 \tag{4.13 - a}$$
$$= \dot{\mathbf{r}}^b(0).$$

Further, on evaluating the second derivatives of (4.11) at the common join, the following is obtained

$$\ddot{\mathbf{r}}^b(0) = V_0^b - 2V_1^b + V_2^b$$
$$\equiv V_1^a - 2V_2^a + V_3^a \tag{4.13 - b}$$
$$= \ddot{\mathbf{r}}^a(1).$$

Thus, from the results in (4.12) and (4.13), we immediately have the following continuity condition which is identical to the parametric C^2-condition (4.1).

$$\mathbf{r}^b(0) = \mathbf{r}^a(1)\ ; \quad \dot{\mathbf{r}}^b(0) = \dot{\mathbf{r}}^a(1)\ ; \quad \ddot{\mathbf{r}}^b(0) = \ddot{\mathbf{r}}^a(1).$$

In other words, a composite (uniform) B-spline curve automatically guarantees (parametric) C^2-continuity.

Now we consider the problem of constructing a composite B-spline curve from a sequence of data points shown in Fig. 4.2. Constructing a composite B-spline curve from $n+1$ data points $\{P_i : i = 0, 1, ..., n\}$ is equivalent to determining $(n+3)$ control vertices $\{V_i : i = 0, 1, ..., n+2\}$. For each data point P_i, the following relation is directly obtained from the *position continuity condition* (4.12):

$$V_i + 4V_{i+1} + V_{i+2} = 6P_i\ ; \quad i = 0, 1, ..., n. \tag{4.14}$$

Two additional linear equations have to be provided in order to determine the $n+3$ control vertices in (4.14). When the end tangents $\hat{\mathbf{t}}_0$, $\hat{\mathbf{t}}_n$ are "fixed", the two additional equations are obtained from (4.13-a) as

$$V_2 - V_0 = 2\hat{\mathbf{t}}_0 \quad (\equiv\ 2\dot{\mathbf{r}}^0(0)) \tag{4.15 - a}$$
$$V_{n+2} - V_n = 2\hat{\mathbf{t}}_n \quad (\equiv\ 2\dot{\mathbf{r}}^{n-1}(1)). \tag{4.15 - b}$$

The system of linear equations in (4.14) and (4.15) is then solved for the control vertices V_i for $i = 0, 1, ..., n+2$. A matrix representation for the linear system under the **fixed end condition** is given as

$$\begin{bmatrix} -1 & 0 & 1 & 0 & & & \\ 1 & 4 & 1 & 0 & & & \\ 0 & 1 & 4 & 1 & & & \\ & & & \cdot & & & \\ & & & & \cdot & & \\ & & & & 1 & 4 & 1 \\ & & & & -1 & 0 & 1 \end{bmatrix} \begin{bmatrix} V_0 \\ V_1 \\ V_2 \\ \cdot \\ \cdot \\ V_{n+1} \\ V_{n+2} \end{bmatrix} = \begin{bmatrix} 2\hat{\mathbf{t}}_0 \\ 6P_0 \\ 6P_1 \\ \cdot \\ \cdot \\ 6P_n \\ 2\hat{\mathbf{t}}_n \end{bmatrix}. \tag{4.16}$$

If the end tangents are not given, they may be estimated from the *circular end condition* or the *polynomial end condition* introduced in §4.2.2.

If **free-end condition** is desired, the curvatures (See Eqn. (2.13)) at both ends of the composite curve are set to zero. That is,

$\dot{\mathbf{r}}^0(0) \times \ddot{\mathbf{r}}^0(0) = \mathbf{0}$: *zero curvature at the start of the curve,*

$\dot{\mathbf{r}}^{n-1}(1) \times \ddot{\mathbf{r}}^{n-1}(1) = \mathbf{0}$: *zero curvature at the end of the curve.*

From the results of (4.12) and (4.13), the first of the above vector product forms becomes

$$\dot{\mathbf{r}}^0(0) \times \ddot{\mathbf{r}}^0(0) = \frac{1}{2}(V_2 - V_0) \times (V_2 - 2V_1 + V_0) = \mathbf{0}.$$

The above relation is satisfied by one of the followings:

$$(V_2 - V_0) = \mathbf{0},$$

$$(V_2 - 2V_1 + V_0) = \mathbf{0}, \quad or$$

$$(V_2 - V_0) = (V_2 - 2V_1 + V_0).$$

Among the three solutions, the last one is commonly adopted. Thus, we have a *multiple vertices case*. That is, at the beginning of the curve, we set

$$V_0 = V_1.$$

Similarly, the following multiple vertices case is obtained at the other end.

$$V_{n+1} = V_{n+2}.$$

The two linear equations necessary to solve (4.14) for the $n+3$ unknowns are then provided by the above *multiple vertices conditions*. The combined linear equation system is then expressed in a matrix form as

$$\begin{bmatrix} 1 & -1 & 0 & 0 & & & \\ 1 & 4 & 1 & 0 & & & \\ 0 & 1 & 4 & 1 & & & \\ & & & \cdot & & & \\ & & & & \cdot & & \\ & & & & 1 & 4 & 1 \\ & & & & 0 & -1 & 1 \end{bmatrix} \begin{bmatrix} V_0 \\ V_1 \\ V_2 \\ \cdot \\ \cdot \\ V_{n+1} \\ V_{n+2} \end{bmatrix} = \begin{bmatrix} 0 \\ 6P_0 \\ 6P_1 \\ \cdot \\ \cdot \\ 6P_n \\ 0 \end{bmatrix}. \tag{4.17}$$

Once the control vertices $\{V_i\}$ have been obtained from either (4.16) or (4.17), the B-spline curve segment spanning $[P_i,\ P_{i+1}]$ is expressed as

$$\mathbf{r}^i(u) = \mathbf{U}\ \mathbf{N}\ \mathbf{R}^i \quad for\ i = 0, 1, 2, ..., n-1 \tag{4.18}$$

$$where, \quad \mathbf{U} = [\ 1\ u\ u^2\ u^3\ ],$$

$$\mathbf{N} = \frac{1}{6}\begin{bmatrix} 1 & 4 & 1 & 0 \\ -3 & 0 & 3 & 0 \\ 3 & -6 & 3 & 0 \\ -1 & 3 & -3 & 1 \end{bmatrix},$$

$$\mathbf{R}^i = [V_i\ \ V_{i+1}\ V_{i+2}\ V_{i+3}]^T.$$

The resulting composite curve consists of n curve segments defined by $n+3$ control vertices $\{V_i :\ i = 0, 1, ..., n+2\}$.

In summary, fitting a uniform B-spline curve is equivalent to solving a system of linear equations for the control vertices $\{V_i\}$. That is,

- (4.16) *is solved when the end tangents* $\hat{\mathbf{t}}_0$, $\hat{\mathbf{t}}_n$ *are given, or*
- (4.17) *is solved if the end tangents are unknown.*

A B-spline curve allows *local modifications* (by adjusting the control vertices), but the curve fitting procedure is a global scheme meaning that a change in a data point P_i affects the entire curve.

In both the *cubic spline fitting and uniform B-spline fitting*, the same *parametric* C^2-*condition* (4.1) is employed. This means that, at a common join P_i, the same tangent vector $\mathbf{t}_i$ is used as the end tangents of both curve segments. Namely

$$\dot{\mathbf{r}}^{i-1}(1) = \mathbf{t}_i = \dot{\mathbf{r}}^i(0).$$

An implication of this is that the resulting *composite curve* would tend to show *local flatness or bulging* when the physical spacing of data points becomes uneven. Thus, if the physical spacing is uneven, the curve fitting methods of the next section should be used.

4.3 CURVE FITTING FOR UNEVENLY SPACED POINT DATA

This section presents two curve fitting methods in which physical spacings (ie, *chord-lengths*) of data points are taken into account. They are

- *chord-length spline fitting and*
- *non-uniform B-spline curve fitting.*

In the first method, the cubic spline fitting method of §4.2.2 is modified so that *chord-length* information is taken into account in determining *tangent vectors* $\mathbf{t}_i$ at data points P_i. Thus, it is called a **chord-length spline**.

The second method is based on the non-uniform B-spline curve model (3.28), and is called a **non-uniform B-spline** fitting. In this method, each chord-length is accounted for as a *knot span* for the respective curve segment. Both methods are equally suitable for constructing a smooth composite curve, which is free of local flatness and kinks, from a sequence of **unevenly spaced** data points. A standard *cubic spline* curve and a *chord-length spline* curve for the same set of unevenly spaced data points are shown in Fig. 4.5b.

(a) (b)

Figure 4.5 Effects of End-tangent Magnitude in Cubic Spline Curve

4.3.1 Geometric Continuity Condition

The two adjoining curve segments $\mathbf{r}^a(u)$ and $\mathbf{r}^b(u)$ satisfying $\mathbf{r}^a(1) = \mathbf{r}^b(0)$ are said to be *gradient continuous* at the common join if the following holds:

$$\dot{\mathbf{r}}^a(1)/|\dot{\mathbf{r}}^a(1)| = \dot{\mathbf{r}}^b(0)/|\dot{\mathbf{r}}^b(0)| = \mathbf{T} \tag{4.19}$$

where $\mathbf{T}$ is the *unit tangent* at the join. The above relation is called a *geometric C^1-condition* or simply *G^1-condition*. Recall from §2.3.2 that *curvature* κ of a curve is defined as

$$d\mathbf{T}/ds = \kappa\mathbf{N},$$

where $\mathbf{N}$ is the *principal normal vector* and s is an *arc-length parameter.*

Let $\mathbf{r}(u)$ be a *regular parametric curve.* Then, its derivatives are evaluated as (using the *chain rule of differentiation*)

$$\begin{aligned}\dot{\mathbf{r}} &= d\mathbf{r}/du \\ &= (d\mathbf{r}/ds)(ds/du) \\ &\equiv \mathbf{T}\ \dot{s},\end{aligned}$$

and

$$\begin{aligned}\ddot{\mathbf{r}} &= d\dot{\mathbf{r}}/du \\ &= \dot{s}(d\mathbf{T}/du) + (d\dot{s}/du)\mathbf{T} \\ &= \dot{s}^2(d\mathbf{T}/ds) + \ddot{s}\mathbf{T} \\ &= \dot{s}^2\kappa\mathbf{N} + \ddot{s}\mathbf{T}.\end{aligned}$$

By taking a vector product of the above two derivatives, we have

$$\begin{aligned}\dot{\mathbf{r}} \times \ddot{\mathbf{r}} &= \dot{s}\mathbf{T} \times \dot{s}^2\kappa\mathbf{N} \\ &= \dot{s}^3\kappa(\mathbf{T} \times \mathbf{N}) \\ &\equiv |\dot{\mathbf{r}}|^3\kappa\mathbf{B},\end{aligned}$$

where $\mathbf{B} = \mathbf{T} \times \mathbf{N}$ is the *binormal vector* of the curve. Thus, the curvature of a parametric curve $\mathbf{r}(u)$ can be expressed as

$$\kappa\mathbf{B} = (\dot{\mathbf{r}} \times \ddot{\mathbf{r}})/|\dot{\mathbf{r}}|^3. \qquad (4.20-a)$$

In order for the two adjoining curve segments $\mathbf{r}^a(u)$ and $\mathbf{r}^b(u)$ to be *curvature continuous* at their common join $\mathbf{r}^a(1) = \mathbf{r}^b(0)$, the following condition should hold :

$$\frac{\dot{\mathbf{r}}^a(1) \times \ddot{\mathbf{r}}^a(1)}{|\dot{\mathbf{r}}^a(1)|^3} = \frac{\dot{\mathbf{r}}^b(0) \times \ddot{\mathbf{r}}^b(0)}{|\dot{\mathbf{r}}^b(0)|^3} \qquad (4.20-b)$$

On substituting the results of (4.19) into the above relation, we have

$$\mathbf{T} \times \ddot{\mathbf{r}}^b(0) = (\beta/\alpha)^2(\mathbf{T} \times \ddot{\mathbf{r}}^a(1)),$$

where $\alpha = |\dot{\mathbf{r}}^a(1)|$ and $\beta = |\dot{\mathbf{r}}^b(0)|$. A solution to this vector equation is

$$\ddot{\mathbf{r}}^b(0) = (\beta/\alpha)^2\ddot{\mathbf{r}}^a(1). \qquad (4.21)$$

This relationship between the second derivatives is called a *geometric C^2-condition* or simply *G^2-condition.*

4.3.2 Chord-length Spline Fitting

A smooth composite curve is to be fitted through three data points $P_0,\ P_1,\ P_2$ as depicted in Fig. 4.6. Also given are the *end tangent vectors* $\mathbf{t}_0,\ \mathbf{t}_2$ of the composite curve. We want to construct a *composite Ferguson curve* by employing the G^2*-condition* (4.21).

Let $\mathbf{r}^a(u)$ and $\mathbf{r}^b(u)$ denote the two Ferguson curve segments as shown in the figure. The problem is to determine the unknown tangent vector $\mathbf{t}_1$ at the common join. Let $\mathbf{t}_1$ be equal to the end tangent at the end of the first curve segment, namely

$$\mathbf{t}_1 = \dot{\mathbf{r}}^a(1) \qquad (4.22-a)$$

then, in order to meet the G^2-condition (4-19), the end tangent at the beginning of the second curve segment should be given by

$$\dot{\mathbf{r}}^b(0) = \omega\ \mathbf{t}_1 \qquad (4.22-b)$$

$$\textit{where}\quad \omega = \alpha/\beta : \ \textit{tangent magnitude ratio},$$
$$\alpha = |\dot{\mathbf{r}}^a(1)|,$$
$$\beta = |\dot{\mathbf{r}}^b(0)|.$$

In chord-length spline fitting, the *magnitudes* of the end tangents at the common join are set to the *chord-lengths* of the curve segments. That is,

$$\alpha = |P_1 - P_0|\ ; \quad \beta = |P_2 - P_1|.$$

Thus, the *tangent magnitude ratio* ω now becomes a *chord-length ratio.* Using the *chord-length ratio,* the G^2-condition (4.21) is rewritten as

$$\ddot{\mathbf{r}}^b(0) = \omega^2\ \ddot{\mathbf{r}}^a(1), \qquad (4.23)$$

where $\quad \omega = |P_2 - P_1|\ /\ |P_1 - P_0|.$

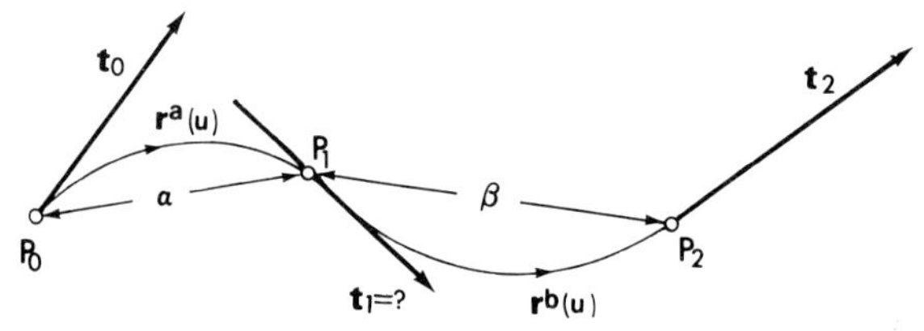

Figure 4.6 Construction of Chord-length Spline Curve

Rewriting the Ferguson curve equation (3.8), the curve segments of Fig. 4.6 are expressed as

$$\mathbf{r}^a(u) = \mathbf{U}\ \mathbf{C}\ \mathbf{S}^a \qquad (4.24-a)$$

$$\mathbf{r}^b(u) = \mathbf{U}\ \mathbf{C}\ \mathbf{S}^b \qquad (4.24-b)$$

$$where,\quad \mathbf{U} = [\ 1\ u\ u^2\ u^3\ ],$$

$$\mathbf{C} = \begin{bmatrix} 1 & 0 & 0 & 0 \\ 0 & 0 & 1 & 0 \\ -3 & 3 & -2 & -1 \\ 2 & -2 & 1 & 1 \end{bmatrix},$$

$$\mathbf{S}^a = [P_0\ P_1\ \mathbf{t}_0\ \mathbf{t}_1]^T,$$

$$\mathbf{S}^b = [P_1\ P_2\ \omega\mathbf{t}_1\ \mathbf{t}_2]^T.$$

On evaluating the second derivatives of (4.24), the G^2-condition (4.23) is expressed as

$$(-6P_1 + 6P_2 - 4\omega\mathbf{t}_1 - 2\mathbf{t}_2) \ = \ \omega^2(6P_0 - 6P_1 + 2\mathbf{t}_0 + 4\mathbf{t}_1) \qquad (4.25)$$

from which the *unknown tangent vector* $\mathbf{t}_1$ is trivially determined.

Now we consider the problem of fitting a G^2*-continuous* curve through a sequence of n+1 data points. The curve fitting problem may be stated as:

Input data (see Figure 4.2):

$\{P_i:\ i = 0, 1, ..., n\}$: *3D data points, and*

$\mathbf{t}_0,\ \mathbf{t}_n$*: end tangent vectors of the composite curve.*

Determine:

$\omega_i = |P_{i+1} - P_i|/|P_i - P_{i-1}|$ *for* $i = 1, 2, ..., n-1$ *(with* $\omega_0 = 1$*), and*

$\{\mathbf{t}_i:\ i = 1, 2, ...n-1\}$*: tangent vectors at the joins.*

Let $\mathbf{r}^i(u)$ denote the Ferguson curve for the segment spanning $[P_i,\ P_{i+1}]$, then from (4.24), it is expressed as *(for* $i = 0, 1, ..., n-1$):

$$\mathbf{r}^i(u) = \mathbf{U}\ \mathbf{C}\ \mathbf{S}^i \qquad (4.26)$$

$$where,\quad \mathbf{S}^i = [P_i\ P_{i+1}\ \omega_i\mathbf{t}_i\ \mathbf{t}_{i+1}]^T,$$

$$\omega_0 = 1,$$

$$\omega_i = |P_{i+1} - P_i|\ /\ |P_i - P_{i-1}| \quad for\ i = 1, 2, ..., n-1.$$

At each common join P_i, the G^2-condition (4.21) is expressed as

$$\ddot{\mathbf{r}}^i(0) = (\omega_i)^2\ \ddot{\mathbf{r}}^{i-1}(1).$$

By substituting the second derivatives of the Ferguson curve equation (4.26) into the above G^2 condition, we obtain the following set of $n-1$ linear equations ($for\ i = 1, 2, ..., n-1$):

$$\{\omega_{i-1}\omega_i^2\mathbf{t}_{i-1} + 2\omega_i(1+\omega_i)\mathbf{t}_i + \mathbf{t}_{i+1}\} = 3\{P_{i+1} + (\omega_i^2 - 1)P_i - \omega_i^2 P_{i-1}\} \tag{4.27}$$

$$where,\quad \omega_0 = 1,$$

$$\omega_i = |P_{i+1} - P_i| / |P_i - P_{i-1}| \quad for\ i = 1, 2, ..., n-1.$$

Assuming that the end tangents $\hat{\mathbf{t}}_0, \hat{\mathbf{t}}_n$ are given, the linear system (4.27) for **chord-length spline fitting** is expressed in a matrix form as

$$\begin{bmatrix} 1 & 0 & 0 & 0 & & & \\ \omega_1^2 & 2\omega_1 + 2\omega_1^2 & 1 & 0 & & & \\ 0 & \omega_1\omega_2^2 & 2\omega_2 + 2\omega_2^2 & 1 & & & \\ & & & \cdot & & & \\ & & & & \cdot & & \\ & & & & \omega_{n-2}\omega_{n-1}^2 & 2\omega_{n-1} + 2\omega_{n-1}^2 & 1 \\ & & & & 0 & 0 & 1 \end{bmatrix} \begin{bmatrix} \mathbf{t}_0 \\ \mathbf{t}_1 \\ \mathbf{t}_2 \\ \cdot \\ \cdot \\ \mathbf{t}_{n-1} \\ \mathbf{t}_n \end{bmatrix} = \begin{bmatrix} \hat{\mathbf{t}}_0 \\ \mathbf{b}_1 \\ \mathbf{b}_2 \\ \cdot \\ \cdot \\ \mathbf{b}_{n-1} \\ \hat{\mathbf{t}}_n \end{bmatrix} \tag{4.28}$$

$$where,\quad \mathbf{b}_i = 3\{P_{i+1} + (\omega_i^2 + 1)P_i - \omega_i^2 P_{i-1}\},$$

$$\omega_i = |P_{i+1} - P_i| / |P_i - P_{i-1}| \ : for\ i = 1, ..., n-1.$$

Notice that the linear equation (4.28) degenerates to the *cubic spline fitting case* (4.6) when $\omega_i = 1$ for all i. In other words, the chord-length spline curve becomes a cubic spline curve if all the chord-lengths are equal.

The linear system (4.28) is easily solved by the substitution method (see Appendix A). Methods of determining (or estimating) the end tangents are as given in §4.2.2:

- *circular end condition,*
- *polynomial end condition, and*
- *free-end condition.*

Under the free-end condition, however, the first and last rows in (4.28) should be changed to those of (4.10). The composite curve consisting of individual Ferguson curves in (4.26) is called a *chord-length spline* curve because chord-lengths are taken into account in defining the cubic spline curve (Kjellander and Bjorkenstan, 1983).

4.3.3 Non-uniform B-spline Fitting

This subsection presents a method of constructing a smooth NUB (non-uniform B-spline) curve passing through a sequence of 3D points. As depicted in Fig. 4.7, a composite curve composed of n cubic NUB-curve segments $\{\mathbf{r}^i(u) : i = 0, ..., n-1\}$ is defined by

- $n+3$ *control vertices* $\{V_i : i = 0, ..., n+2\}$, *and*
- $n+4$ *knot spans* $\{\nabla_i = (t_{i+1} - t_i) : i = -2, ..., n+1\}$.

Thus, the NUB-curve fitting problem may be formally stated as follows:

Input data:
$\{P_i : i = 0, 1, ..., n\}$: *3D data points, and*
$\mathbf{t}_0,\ \mathbf{t}_n$: *end tangent vectors of the composite curve.*

Determine:
$\{\nabla_i : -2, -1, ..., n+1\}$: *knot spans, and*
$\{V_i : i = 0, 1, ..., n+2\}$: *control vertices.*

The overall curve fitting procedure is similar to that of constructing a uniform B-spline curve discussed in §4.2.3. Here the cubic NUB-curve model (3.28) is employed instead of the uniform B-spline curve model (3.17). The process of fitting a B-spline curve from a sequence of point data is sometimes called the *inversion process*. The reader is advised to examine closely the relationships among the control vertices V_i, knot spans ∇_i, and data points P_i in Fig. 4.7 because the convention in this book may be different from the ones in other references.

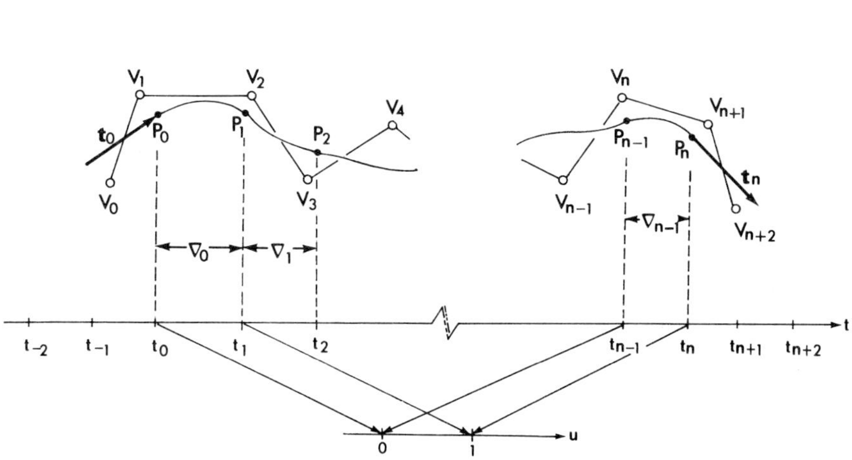

Figure 4.7 Non-uniform Cubic B-spline Curve Fitting

The first step in NUB-curve fitting is to determine the **knot spans** $\{\nabla_i\}$. As depicted in Fig. 4.7, the composite NUB-curve is supported by n knot spans, ∇_0 through ∇_{n-1}, which we call *supporting knot spans.* The remaining knot spans are called *extended knot spans.* A reasonable choice for the **supporting knot spans** is to make them equal to corresponding chord-lengths. That is,

$$\nabla_i = |P_{i+1} - P_i| \quad for\ i = 0, 1, ..., n-1. \qquad (4.29-a)$$

The choice for the **extended knot spans** does not affect the quality of the resulting NUB-curve. They could be set to zero as in (4.29-b) or be assigned a uniform value as in (4.29-c). That is,

$$\nabla_{-2} = \nabla_{-1} = \nabla_{n+1} = \nabla_n = 0, \ or \qquad (4.29-b)$$

$$\nabla_{-2} = \nabla_{-1} = \nabla_0\ ; \quad \nabla_{n+1} = \nabla_n = \nabla_{n-1} \qquad (4.29-c)$$

The zero knot spans in (4.29-b) are called **multiple knots.**

Having determined the knot spans, the next step is to build a linear equation system for the unknown vertices. For this, we need a curve equation for each cubic NUB-curve segment in Fig. 4.7. Rewriting the cubic NUB-curve model (3.28), we have

$$\mathbf{r}^i(u) = \mathbf{U}\ \mathbf{N}_c^i\ \mathbf{R}^i; \quad 0 \le u \le 1; \quad i = 0, 1, ..., n-1 \qquad (4.30)$$

$$where, \quad \mathbf{U} = [1\ u\ u^2\ u^3],$$

$$\mathbf{R}^i = [V_i\ V_{i+1}\ V_{i+2}\ V_{i+3}]^T,$$

$$\mathbf{N}_c^i = \begin{bmatrix} \frac{(\nabla_i)^2}{\nabla_{i-1}^2 \nabla_{i-2}^3} & (1 - n_{11} - n_{13}) & \frac{(\nabla_{i-1})^2}{\nabla_{i-1}^3 \nabla_{i-1}^2} & 0 \\ -3n_{11} & (3n_{11} - n_{23}) & \frac{3\nabla_i \nabla_{i-1}}{\nabla_{i-1}^3 \nabla_{i-1}^2} & 0 \\ 3n_{11} & -(3n_{11} + n_{33}) & \frac{3(\nabla_i)^2}{\nabla_{i-1}^3 \nabla_{i-1}^2} & 0 \\ -n_{11} & (n_{11} - n_{43} - n_{44}) & n_{43} & \frac{(\nabla_i)^2}{\nabla_i^3 \nabla_i^2} \end{bmatrix}$$

$$n_{43} = -\{\frac{1}{3} n_{33} + n_{44} + (\nabla_i)^2 / (\nabla_i^2 \nabla_{i-1}^3)\},$$

$$\nabla_i^k = \nabla_i + \nabla_{i+1} + ... + \nabla_{i+k-1}.$$

Since a curve segment $\mathbf{r}^i(u)$ spans from P_i to P_{i+1}, the following relation holds at each data point ($for\ i = 0, ..., n-1$):

$$\mathbf{r}^i(0) = P_i \quad and$$

$$\mathbf{r}^i(1) = P_{i+1}.$$

By evaluating the above relations using the NUB-curve equation (4.30), the following set of linear equations are obtained:

$$f_i V_i + h_i V_{i+1} + g_i V_{i+2} = P_i \; ; \quad i = 0, 1, ..., n \tag{4.31}$$

$$\begin{aligned} where, \quad h_i &= (1 - f_i - g_i), \\ f_i &= (\nabla_i)^2 / (\nabla_{i-1}^2 \nabla_{i-2}^3), \\ g_i &= (\nabla_{i-1})^2 / (\nabla_{i-1}^2 \nabla_{i-1}^3). \end{aligned}$$

Two more linear equations are obtained by evaluating the curve derivatives at the very ends of the composite curve in Fig. 4.7. That is, the *end tangent* at the beginning of the composite curve is evaluated as

$$\begin{aligned} \hat{\mathbf{t}}_0 &\equiv \dot{\mathbf{r}}^0(0) \\ &= a_0 V_2 + (b_0 - a_0) V_1 - b_0 V_0 \end{aligned} \tag{4.32 - a}$$

$$\begin{aligned} where, \quad a_0 &= 3(\nabla_0 \nabla_{-1}) / (\nabla_{-1}^2 \nabla_{-1}^3), \\ b_0 &= 3(\nabla_0)^2 / (\nabla_{-1}^2 \nabla_{-2}^3), \end{aligned}$$

and the *end tangent* at the very end of the composite curve is given by

$$\begin{aligned} \hat{\mathbf{t}}_n &= \dot{\mathbf{r}}^{n-1}(1) \\ &= a_1 V_{n+2} + (b_1 - a_1) V_{n+2} - b_1 V_n \end{aligned} \tag{4.32 - b}$$

$$\begin{aligned} where, \quad a_1 &= 3(\nabla_{n-1})^2 / (\nabla_{n-1}^2 \nabla_{n-1}^3), \\ b_1 &= 3(\nabla_n \nabla_{n-1}) / (\nabla_{n-1}^2 \nabla_{n-2}^3). \end{aligned}$$

If the extended knot spans are determined from (4.29-b), namely, multiple knots are defined such that

$$\nabla_{-2} = \nabla_{-1} = \nabla_{n+1} = \nabla_n = 0,$$

then some of the coefficients in (4.31) and (4.32) are trivially evaluated as

$$\begin{aligned} f_0 &= 1 \; ; \quad g_0 = 0, \\ f_n &= 0 \; ; \quad g_n = 1, \\ a_0 &= 0 \; ; \quad b_0 = 3, \\ a_1 &= 3 \; ; \quad b_1 = 0. \end{aligned}$$

With these values of f_i, g_i, a_i, b_i, the following linear equation system for **NUB-curve fitting** is obtained from (4.31) and (4.32):

$$\begin{bmatrix} -3 & 3 & 0 & 0 & & & \\ 1 & 0 & 0 & 0 & & & \\ 0 & f_1 & h_1 & g_1 & & & \\ & & & & & & \\ & & & & & & \\ & & & & f_{n-1} & h_{n-1} & g_{n-1} \\ & & & & 0 & 0 & 1 \\ & & & & 0 & -3 & 3 \end{bmatrix} \begin{bmatrix} V_0 \\ V_1 \\ V_2 \\ \cdot \\ \cdot \\ \cdot \\ V_{n+1} \\ V_{n+2} \end{bmatrix} = \begin{bmatrix} \hat{\mathbf{t}}_0 \\ P_0 \\ P_1 \\ \cdot \\ \cdot \\ \cdot \\ P_n \\ \hat{\mathbf{t}}_n \end{bmatrix} \tag{4.33}$$

$$\begin{aligned} where,\quad h_i &= (1 - f_i - g_i), \\ f_i &= (\nabla_i)^2/(\nabla_{i-1}^2 \nabla_{i-2}^3), \\ g_i &= (\nabla_{i-1})^2/(\nabla_{i-1}^2 \nabla_{i-1}^3), \\ \nabla_i^k &= \nabla_i + \ldots + \nabla_{i+k-1}. \end{aligned}$$

The system of linear equations given by (4.33) is also easily solved by using the substitution method of Appendix A. In applying the procedure for solving *tridiagonal matrices*, a care must be taken because the coefficient matrix of (4.33) has some diagonal elements whose values are zero. That is, row operations are needed to make the (zero) diagonal elements non-zero.

When all the "supporting knot spans" in (4.29-a) are identical, the linear equation (4.31) would become the uniform B-spline linear equation (4.14). Further, the two linear equations in (4.32) would become those in (4.15) if the "extended knot spans" are determined from (4.29-c). Thus, the uniform B-spline curve fitting procedure of §4.2.3 is a special case of the NUB-curve fitting. In other words, the NUB-curve fitting procedure discussed in this section degenerates to the uniform B-spline case when all the knot spans are identical. This is an expected result.

4.4 SPECIAL TECHNIQUES FOR 2D CURVE MODELING

Presented in this section are special techniques for *2D curve fitting* and methods of parametrizing conic section curves. Topics to be discussed are

- biarc curve fitting,
- conic section curve fitting,
- rational parametric representation of conic sections, and
- Bezier curve approximation of circular arcs.

It should be noted that 2D curve fitting problem is a subset of 3D curve fitting problem, and all the curve fitting methods introduced earlier in this chapter are equally applicable to 2D curve fitting.

4.4.1 Biarc Curve Fitting

A sequence of 2D points $\{P_i\}$ can be interpolated by a smooth composite curve composed of circular arcs. Each curve segment *spanning* $[P_i,\ P_i+1]$ is represented by two circular arcs, so it is called a **biarc curve**. This method has long been used in NC part programming languages, and Bolton (1975) published a version of it in the open literature. Biarc fitting is very simple and the resulting composite curve is conveniently machined on NC machines because most NC controllers support circular interpolations. A biarc fitting consists of two separate steps:

- *estimation of the gradient at each data point, and*
- *construction of a biarc at each span.*

There are two types of biarc, namely, *unimodal biarc and inflection biarc.*

1) Gradient Estimation at Data Point:

For a given sequence of data points $\{P_i : i = 0, 1, ..., n\}$, many choices are possible for the estimation of a **gradient** (ie, unit vector) $\mathbf{T}_i$ at each data point P_i. For example, a low degree polynomial may be fitted through a few points near P_i, or any one of the curve fitting methods presented in this chapter may be used. The cubic spline fitting method of §4.2.2 was employed in Bolton (1975). The simplest choice for *the gradients at interior data points* is to use the following linear estimator:

$$\mathbf{T}_i = (P_{i+1} - P_{i-1})/|P_{i+1} - P_{i-1}|\ ; \quad i = 1, 2, ..., n-1 \qquad (4.34)$$

The *end gradients* $\mathbf{T_0}, \mathbf{T_n}$ may be obtained by applying any one of the "end tangent conditions" introduced in §4.2.2. Having determined the gradient at each data point P_i, the next step is to construct a biarc curve at each span $[P_i, P_{i+1}]$.

2) Construction of Unimodal Biarc:

Consider a *unimodal curve* segment $[P_0, P_1]$ shown in Fig. 4.8. As shown in the figure, $\mathbf{T_0}$ and $\mathbf{T_1}$ are the gradients at P_0 and P_1, respectively. We want to construct two circular arcs having radii r_0 and r_1. Let $\mathbf{d}$ denote a unit vector from P_0 to P_1 such that

$$\mathbf{d} = (P_1 - P_0)/|P_1 - P_0| \tag{4.35}$$

then, by the definition of *scalar product* (See Eqn. (2.3)), the angles θ_0, θ_1 between $\mathbf{d}$ and the gradients $\mathbf{T_0}, \mathbf{T_1}$ can be obtained from

$$cos\,\theta_0 = \mathbf{d} \cdot \mathbf{T}_0 \qquad (4.36-a)$$

$$cos\,\theta_1 = \mathbf{d} \cdot \mathbf{T}_1. \qquad (4.36-b)$$

In Fig. 4.8, AB is a line parallel with the chord P_0P_1 and C is a point on the line AB.

We want to construct two circular arcs each of them in tangential contact with the line AB at the same point C. From the trigonometric relations in the figure, the following relations are obtained:

$$r_0 \;\; sin\,\theta_0 + r_1 \;\; sin\,\theta_1 = c \quad and$$

$$r_1 \;\; cos\,\theta_1 - r_0 \;\; cos\,\theta_0 = r_1 - r_0$$

which are easily solved for the unknown radii. That is, the radii of the **unimodal biarc** are expressed as

$$r_0 = c(1 - cos\,\theta_1)/(sin\,\theta_0 + sin\,\theta_1 - sin(\theta_0 + \theta_1)) \qquad (4.37-a)$$

$$r_1 = c(1 - cos\,\theta_0)/(sin\,\theta_0 + sin\,\theta_1 - sin(\theta_0 + \theta_1)) \qquad (4.37-b)$$

where $c = |P_1 - P_0|$ is the chord-length. Using these radius values, the starting and ending points of each circular arc are easily obtained.

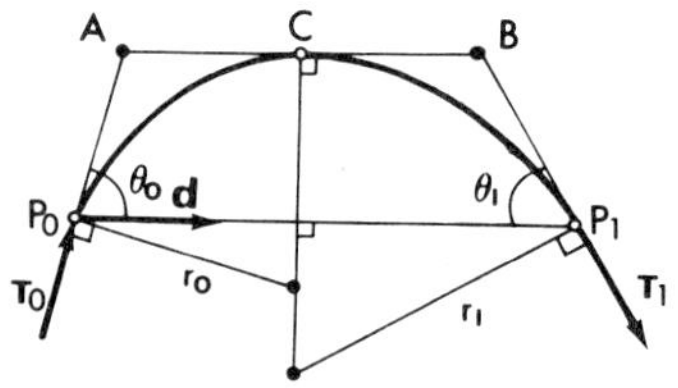

Figure 4.8 Unimodal Biarc Construction

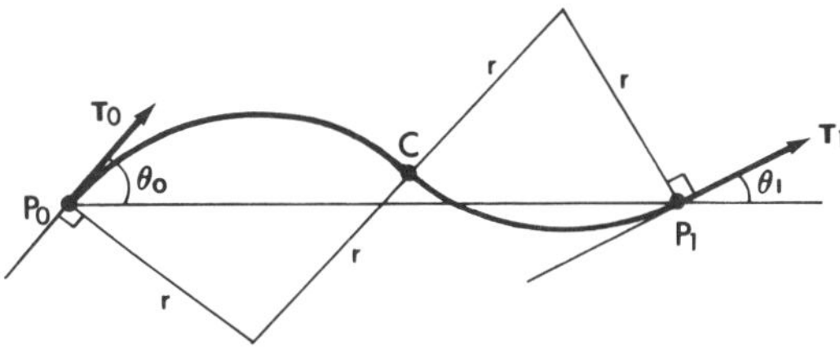

Figure 4.9 Inflection Biarc Construction

3) Construction of Inflection Biarc:

Another case of end tangent configuration is shown in Fig. 4.9 where an *inflection point* is allowed inside the curve segment. In this case, it is always possible to construct two circular arcs having the same radius r. From the trigonometric relations in the figure, the radius value is given by

$$r = c/[sin\,\theta_0 + sin\,\theta_1 + \{4 - (cos\,\theta_0 + cos\,\theta_1)^2\}^{1/2}], \tag{4.38}$$

where c is the chord-length and the angles θ_0, θ_1 are as given in (4.36).

The biarc construction is not unique. One may introduce some optimality measures in determining the radius values, such as minimum radii difference $|r_1 - r_0|$ or minimum curvature difference $|1/r_1 - 1/r_0|$. However, tests showed that the methods introduced here give satisfactory results. In practical implementation of the biarc fitting method, special cases may arise. For example, when more than two points are collinear, the chords belonging to this straight line region may be fitted with straight lines.

4.4.2 Conic Section Curve Fitting

This section presents another method of constructing a curve segment for the unimodal span shown in Fig. 4.8. This time, the same curve span is represented as a conic section instead of a biarc. A **conic section curve** is expressed as an *implicit quadratic equation:*

$$S(x, y) = ax^2 + by^2 + c + 2fx + 2gy + 2hxy = 0. \tag{4.39}$$

The above quadratic equation becomes

- *an ellipse if* $h^2 < ab$,
- *parabola if* $h^2 = ab$, *or*
- *hyperbola if* $h^2 > ab$.

A conic section curve is obtained when a right cone is cut by a plane, and the resulting intersection curve becomes an ellipse, a parabola, or a hyperbola depending on where it is cut (see §3.2.1). If the cutting plane contains the axis of the cone, the resulting conic section becomes a pair of straight lines. In this degenerate case, the conic section equation (4.39) can be factored out as

$$S(x,y) = (a_1x + b_1y + c_1)(a_2x + b_2y + c_2).$$

An important property of conic section is that a blending of two conic sections is also a conic section. That is, if

$$S_1(x,y) = 0 \quad and \quad S_2(x,y) = 0$$

are conic sections, then

$$S_1(x,y) + \lambda S_2(x,y) = 0$$

is *also* a conic section.

We want to find a mathematical expression for the unimodal curve segment of Fig. 4.8 by using the conic section equation (4.39). Let

$$L_i(x,y) = a_ix + b_iy + c_i \quad (i = 1,2,3,4)$$

denote straight lines as shown in Fig. 4.10a. Then, for a given parameter $0 \leq \lambda \leq 1$, a quadratic curve passing through the intersection points of the above straight lines is expressed as:

$$S(x,y) = (1-\lambda)L_1L_2 + \lambda L_3L_4 = 0. \tag{4.40}$$

When the two lines L_3, L_4 become coincident, namely if

$$L_3 \equiv L_4,$$

then the lines L_1, L_2 become tangent lines to the conic section curve as shown in Fig. 4.10b. This property of quadratic equation of the form (4.40) is utilized in modeling a unimodal curve segment as a conic section curve.

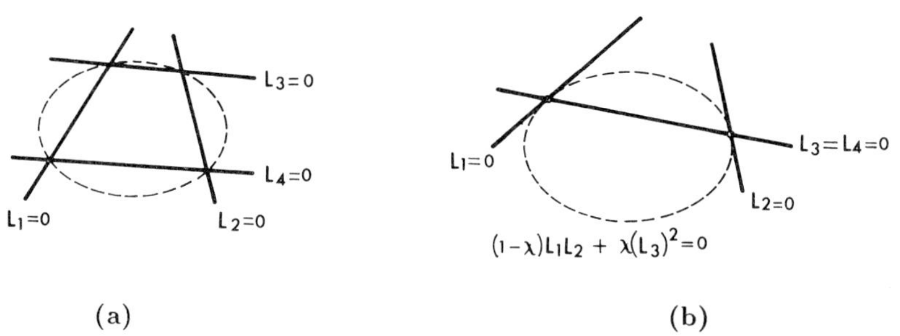

Figure 4.10 Conic Section Curve as a Family of Curves

A unimodal curve span similar to Fig. 4.8 is depicted in Fig. 4.11. In the figure, the following are defined:

P_0, P_1 : *end points of curve segment,*
$\mathbf{T_0}, \mathbf{T_1}$: *end gradients of the curve segment,*
Q : *intersection of the two end tangent lines,*
M : *middle point of P_0 and P_1, and*
F : *intersection between the line MQ and the curve segment.*

Further, let us define three line equations as

$L_1(x, y) = 0$ *equation of the line P_0Q,*
$L_2(x, y) = 0$ *equation of the line P_1Q, and*
$L_3(x, y) = 0$ *equation of the line P_0P_1.*

Then the conic section equation (4.40) can be rewritten as

$$S(x, y) = (1 - \lambda)L_1L_2 + \lambda(L_3)^2 = 0 \quad ; \quad 0 \leq \lambda \leq 1 \qquad (4.41)$$

The quadratic function (4.41) represents a family of curves, depending on the value of λ which has to be specified by some means. In the following, a method of determining λ is described.

The point $F(x, y)$ in Fig. 4.11 is called a *shoulder point* and the **fullness factor** ρ of the conic section curve is defined as

$$\rho = |F - M|/|Q - M|.$$

In practice, the fullness factor ρ is given and the **shoulder point** F is computed from the above relation. That is, we have

$$F = (x_s,\ y_s) = \rho Q + (1 - \rho)(P_0 + P_1)/2, \qquad (4.42)$$

where ρ is fullness factor, Q is the tangent intersection point, and P_i's are end points of the conic section curve.

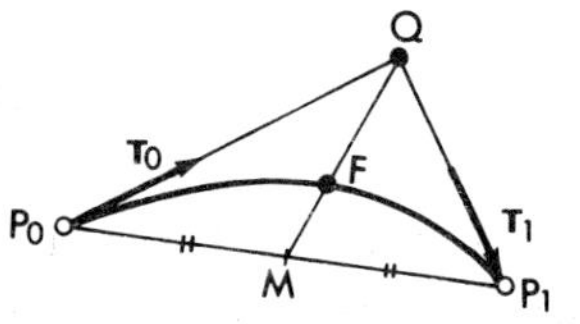

Figure 4.11 Geometric Construction of a Conic Section Curve

Once the shoulder point $F = (x_s,\ y_s)$ is determined from (4.42), the parameter λ of the conic section is determined by solving (4.41) for λ:

$$\lambda = \frac{L_1(x_s, y_s)L_2(x_s, y_s)}{L_1(x_s, y_s)L_2(x_s, y_s) - (L_3(x_s, y_s))^2}. \tag{4.43}$$

In defining the line equations $L_i(x, y) = 0$, care must be taken to make sure that $0 \leq \lambda \leq 1$.

Example 4.3: *Construct a conic section curve spanning from* $P_0 = (0, 1)$ *to* $P_1 = (1, 0)$ *with end tangent lines intersecting at* $Q = (1, 1)$. *Use a fullness factor* $\rho = 1/2$:

From the relation given in (4.42), the shoulder point F is obtained as

$$F = \frac{1}{2}(1, 1) + \frac{1}{4}\{(0, 1)\} + (1, 0) = (3/4, 3/4).$$

The line equations in (4.41) *are then expressed as*

$$L_1 \equiv (y - 1)\ ; \quad L_2 \equiv (1 - x)\ ; \quad L_3 \equiv (x + y - 1).$$

By evaluating (4.43) *at* $x = y = 3/4$, *we get* $\lambda = 1/5$. *Thus,* (4.41) *becomes*

$$S(x, y) = \frac{4}{5}(y - 1)(1 - x) + \frac{1}{5}(x_y - 1)^2 = 0$$

which simplifies to

$$x^2 + y^2 - 3 + 2x + 2y - 2xy = 0.$$

Comparing it with (4.39), *one may find that it is a parabola.* ◇

The above example shows that the resulting conic section curve is a parabola when the *fullness factor* ρ is 1/2. In general, a conic section represents

- a hyperbola when $\rho < 0.5$,
- an ellipse when $\rho > 0.5$, and
- a parabola when $\rho = 0.5$.

By varying the value of the fullness factor ρ, the conic section curve of Figure 4.11 can be made to pass any place inside the triangle P_0P_1Q. Thus, for a unimodal segment, the conic section model gives more flexibility than the biarc model. This approach of "conic section design" has long been used in the cross-sectional design of aircraft fuselages (Liming, 1944). However, the implicit form (4.41) of conic section curve is rarely used in modern design because the implicit form is difficult to evaluate in a sequential manner and it is not easily extended to a surface model. For this reason, a parametric form of conic section curve is widely used.

4.4.3 Parametrization of Conic Section Curves

A conic section curve is completely specified by its *end points, a tangent intersection point, and a fullness factor.* We want to represent the conic section curve as a parametric curve by using the rational quadratic Bezier curve model presented earlier in §3.5.2. Rewriting the rational quadratic Bezier curve equation (3.35), we have

$$\begin{aligned} \mathbf{r}(t) &= \frac{w_0(1-t)^2V_0 + 2w_1t(1-t)V_1 + w_2t^2V_2}{w_0(1-t)^2 + w_1t(1-t) + w_2t^2} \\ &= \sum_{i=0}^{2} w_iB_i^2(t)V_i \Big/ \Big(\sum_{i=0}^{2} w_iB_i^2(t) \Big) \\ &= \sum_{i=0}^{2} \phi_i(t)\ V_i \end{aligned} \tag{4.44}$$

$$\begin{aligned} \textit{where,}\quad & \phi_i(t) = w_iB_i^2(t)/h(t), \\ & h(t) = \sum_{i=0}^{2} w_iB_i^2(t), \\ & B_i^2(t) = \frac{2!}{(2-i)!\ i!} t^i(1-t)^i, \\ & V_i : \textit{Bezier control points}, \\ & \omega_i : \textit{weights}. \end{aligned}$$

Thus, the parametrization problem here is to determine

a) the Bezier points V_0, V_1, V_2, *and*

b) the weights $\omega_0, \omega_1, \omega_2$

from the following input data defining a conic section curve:

a) P_0, P_1*: end points,*

b) Q *: tangent intersection point, and*

c) ρ *: fullness factor.*

From the property of a quadratic Bezier curve (See §3.5.2), the Bezier control points are given by

$$V_0 = P_0\ ;\quad V_2 = P_1\ ;\quad V_1 = Q. \tag{4.45}$$

Before trying to determine the weights, we want to show that the rational curve (4.44) indeed represents a conic section curve. The rational Bezier form (4.44) can be directly derived from

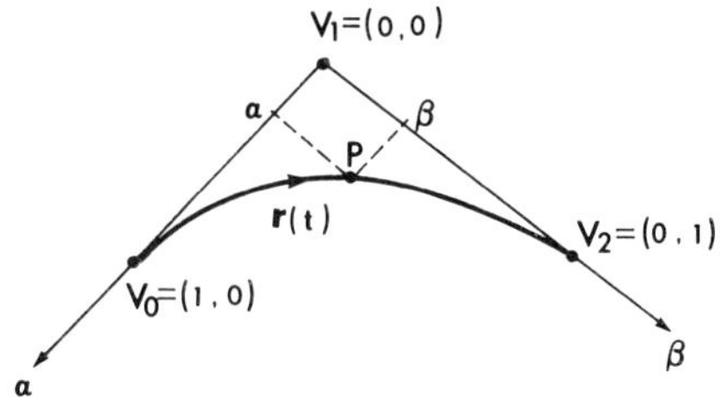

Figure 4.12 Oblique Coordinate System Defined on Control Vertices

the implicit form (4.41) as shown in Faux and Pratt (1980). It is also possible to *implicitize* (4.44) to obtain (4.41) as given in Lee (1987).

In order to show that the rational Bezier curve (4.44) is a conic section curve (ie, an implicit quadratic curve), an **oblique coordinate system** is defined with respect to the three control vertices V_0, V_1, V_2 such that they have the following coordinate values (α, β):

$$V_0 = (1,0)\ ;\quad V_1 = (0,0)\ ;\quad V_2 = (0,1).$$

The oblique coordinate system is depicted in Fig. 4.12. Let P be a point on the curve $\mathbf{r}(t)$, then it is expressed in terms of the oblique coordinates as

$$\begin{aligned} P &= V_1 + \alpha(V_0 - V_1) + \beta(V_2 - V_1) \\ &= \alpha V_0 + (1 - \alpha - \beta)V_1 + \beta V_2 \\ &= \alpha(t)V_0 + \{1 - \alpha(t) - \beta(t)\}V_1 + \beta(t)V_2 \\ &= \mathbf{r}(t). \end{aligned} \tag{4.46}$$

Since both equations, (4,44) and (4.46), represent the same curve $\mathbf{r}(t)$, the coefficients of V_i in both equations should be the same. Namely, we should have

$$\begin{aligned} \phi_0(t) &\equiv \alpha(t), \\ \phi_1(t) &\equiv (1 - \alpha(t) - \beta(t)), \\ \phi_2(t) &\equiv \beta(t). \end{aligned}$$

Here, we want to find a functional relationship between the two coordinate variables α and β. From the identity of Bernstein polynomial functions

$$B_0^2(t)B_2^2(t) \equiv \{B_1^2(t)\}^2/4,$$

we obtain the following quadratic function of α, β:

$$\begin{aligned}\alpha(t)\beta(t) &= \phi_0(t)\ \phi_2(t) \\ &= \kappa\{\phi_1(t)\}^2 \\ &= \kappa\{1-\alpha(t)-\beta(t)\}^2,\end{aligned}$$

where $\kappa = (w_0 w_2)/(2w_1)^2$. The above equation tells that the rational quadratic Bezier curve (4.44) is indeed an implicit quadratic curve, that is, a conic section curve.

Having verified that the rational curve (4.44) is a conic section, the next step is to determine the values of the weights ω_i. The procedure is similar to that of constructing an implicit conic section curve (see Fig. 4.11). Here, we want to find an expression for the weights ω_i with respect to the *fullness factor* ρ. In doing so, we need to use the concept of *barycentric coordinates*. In general, **barycentric coordinates** are defined as follows: Let **p**, **a**, **b** be three collinear points such that

$$\mathbf{p} = \alpha\mathbf{a} + \beta\mathbf{b} \quad with \quad \alpha + \beta = 1,$$

then α and β are called barycentric coordinates of **p** with respect to **a** and **b**. Similarly, let **p**, **a**, **b**, **c** be four co-planar points such that

$$\mathbf{p} = \alpha\mathbf{a} + \beta\mathbf{b} + \gamma\mathbf{c} \quad with \quad \alpha + \beta + \gamma = 1,$$

then α, β, γ become barycentric coordinates with respect to **a**,**b**,**c**. In this case the barycentric coordinates correspond to *area ratios*. If all the coordinates are positive, the point **p** is located inside the triangle formed by the three vertex points **a**,**b**,**c**.

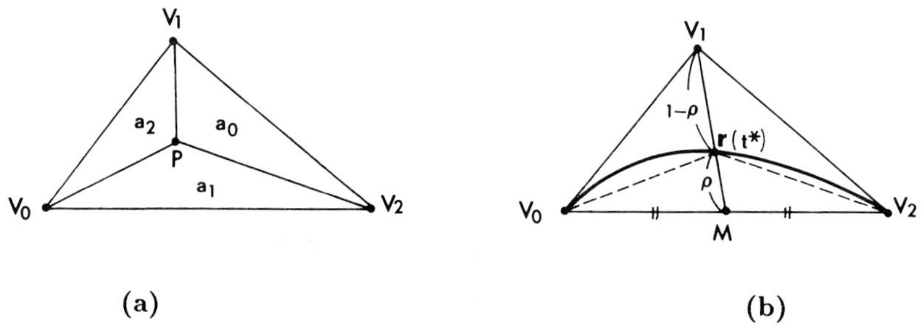

Figure 4.13 Area Ratios and Shoulder Point

A triangle formed by the three vertices V_0, V_2, V_1 is shown in Fig. 4.13a. The "big" triangle is decomposed into three "subtriangles" by a point P, and the *area ratios* of the subtriangles are denoted by a_0, a_1, a_2. Shown in Fig. 4.13b is a *shoulder point* $\mathbf{r}(t^*)$ on the conic section curve formed by the same vertices V_i. Let ρ be the fullness factor for the shoulder point, then the area ratios of the subtriangles in Fig. 4.13b are given by

$$a_0 \ : \ a_1 \ : \ a_2 \ = \ (1-\rho) \ : \ 2\rho \ : \ (1-\rho). \qquad (4.47)$$

Let us take the value of the parameter t in (4.44) at the shoulder point to be 1/2, that is $t = 1/2$, then we have

$$\begin{aligned} \mathbf{r}(0.5) &= \phi_0(0.5)V_0 + \phi_1(0.5)V_1 + \phi_2(0.5)V_2 \\ &= w_0 V_0 + 2w_1 V_1 + w_2 V_2. \end{aligned}$$

The weights (ie, coefficients) of V_i's in the above equation form a set of *barycentric coordinates*. On the other hand, the inside point P in Fig. 4.13a is also expressed as a barycentric point with respect to the three vertices:

$$P = a_0 V_0 + a_1 V_1 + a_2 V_2.$$

If we let $P \equiv \mathbf{r}(0.5)$, the above two relations lead to the following:

$$a_0 \ : \ a_1 \ : \ a_2 \ = \ w_0 \ : \ 2w_1 \ : \ w_2. \qquad (4.48)$$

Finally, from (4.47) and (4.48), the ratios of the *weights* w_0, w_1, w_2 are expressed, in terms of the fullness factor ρ, as:

$$w_0 \ : \ w_1 \ : \ w_2 \ = \ 1 \ : \rho/(1-\rho) \ : \ 1 \qquad (4.49-a)$$

which is equivalent to

$$w_0 = w_2 = 1 \quad \& \quad w_1 = \rho/(1-\rho). \qquad (4.49-b)$$

Example 4.4: *Construct a rational Bezier curve for the conic section curve given in Example 4.3:*

The three control vertices V_0, V_2, V_1 and fullness factor ρ are given as:

$$V_0 = (0,1) \ ; \quad V_1 = (1,1) \ ; \quad V_2 = (1,0) \ ; \quad \rho = 1/2.$$

From (4.49), the weights are $w_0 = w_1 = w_2 = 1$, which means that the denominator h(t) in (4.44) is 1. Thus, the rational curve (4.44) reduces to

$$\mathbf{r}(t) = (1-t)^2 V_0 + 2t(1-t)V_1 + t^2 V_2 \; ; \quad 0 \le t \le 1$$

which in fact is an ordinary quadratic Bezier curve. Written in component form, we get

$$x(t) = 2t(1-t) + t^2 = 2t - t^2 \; ; \quad y(t) = (1-t)^2 + 2t(1-t) = 1 - t^2.$$

If we remove the parameter t from the above equations, the implicit equation given in Example 4.3 may be obtained. ◇

The conic section curve in Example 4.3 is a parabola, so is the rational curve in Example 4.4. The above example tells that a parabola is parametrized as a (non-rational) quadratic Bezier curve. In general, the rational quadratic curve (4.44) becomes a *parabola* if

$$\begin{aligned} \kappa &= w_0 w_2/(2w_1)^2 \\ &= 1/4 \end{aligned} \tag{4.50}$$

Since $\kappa = (1-\rho)^2/(2\rho)^2$, (4.50) confirms the fact that a rational quadratic Bezier curve becomes a parabola when the fullness factor $\rho = 1/2$. Further, it becomes an ellipse if $\kappa > 1/4$ (*or* $\rho > 1/2$), and a hyperbola if $\kappa < 1/4$ (*or* $\rho < 1/2$). For more details on the subject of parametrizing conic sections, the reader is referred to Lee (1987).

4.4.4 Parametrization of Circular Arc

In engineering drawings, circular arcs are most commonly used in defining "free-formed" curves. A circular arc can be exactly modeled by a *rational quadratic* Bezier curve or be approximated by a (non-rational) Bezier curve.

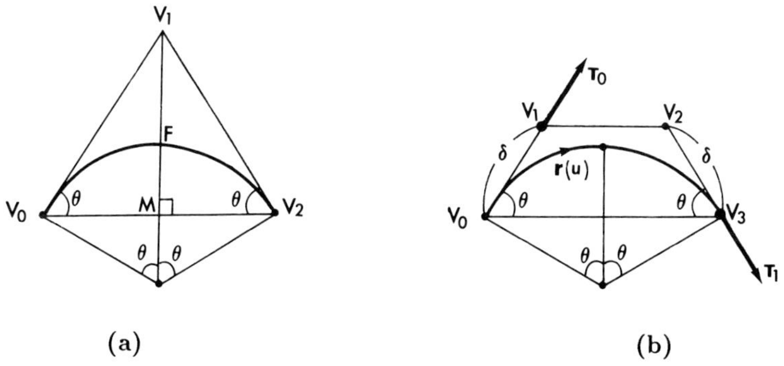

Figure 4.14 Parametrization of Circular Arcs

Figure 4.14a shows a circular arc with a subtending angle of 2θ. In the figure, V_0, V_1, V_2 are control vertices, M is the mid-point of V_0 and V_2, and F is a *shoulder point.* From the trigonometric relations in the figure, the *fullness factor* ρ is expressed in terms of the angle θ as

$$\rho = |F - M|/|V_1 - M|$$
$$= \cos\theta/(1 + \cos\theta)$$

which in turn gives, from (4.49), an expression for the weight w_1. That is,

$$w_1 = \rho/(1-\rho) = \cos\theta. \tag{4.51}$$

Thus, for the circular arc in Fig. 4.14, Eqn. (4.44) is expressed as $(w_0 = w_2 = 1)$:

$$\mathbf{r}(t) = \frac{(1-u)^2 V_0 + 2\cos\theta u(1-u)V_1 + u^2 V_2}{(1-u)^2 + 2\cos\theta u(1-u) + u^2} \tag{4.52}$$

The same circular arc can be approximated by a *cubic Bezier curve* given earlier in Eqn. (3.11). With the four control vertices defined as in Fig. 4.14b, the following distance has to be computed *(note $V_0 = P_0$ & $V_3 = P_1$)*

$$\delta = |V_1 - V_0| = |V_3 - V_2|$$

so that the approximation error is minimized. By enforcing $\mathbf{r}(0.5)$ to be the middle point of the circular arc, it is expressed as

$$\delta = \frac{2}{3}\frac{|V_0 - V_3|}{1 + \cos\theta}. \tag{4.53}$$

Then the middle control vertices are given by

$$V_1 = V_0 + \delta\, \mathbf{T}_0,$$
$$V_2 = V_3 - \delta\, \mathbf{T}_1,$$

where $\mathbf{T}_0$, $\mathbf{T}_1$ are end gradients. The approximation error is found to be about 0.13% when the *subtending angle* is 90°. This approximation error can be reduced by the factor of 1000 if a quintic Bezier curve is used (Lee, 1990).

4.5 DISCUSSIONS

Presented in this chapter were various methods for constructing *smooth* curve from a sequence of data points $\{P_i\}$. In general, a curve fitting problem is equivalent to solving a system of linear equations.

When the chord-lengths $|P_i - P_{i-1}|$ are identical, the NUB curve of §4.3.3 becomes the uniform B-spline curve of §4.2.3 and the chord-length spline becomes a (standard) cubic spline curve of §4.2.3. Both the cubic spline curve and the chord-length spline curve are composed of Ferguson curves which are easily converted to cubic Bezier curves. A B-spline curve can also be converted to a sequence of Bezier curves. The curve fitting methods introduced in this chapter can be directly extended to fitting composite surfaces as will be seen in Chapter 7.

When the point data are 2D points, more intuitive curve fitting methods can be used: Namely, the biarc fitting and the conic section curve fitting. The conic section method can not handle curve segments having an inflection point, but it gives more freedom than the biarc method does. Both the biarc curve and conic section curve can be converted to rational quadratic Bezier curves.

In practice, the input point data are in most cases obtained from physical models. In this case, measurement errors are inevitable. Thus, some form of adjustment of input point data becomes necessary in order to obtain a smoother curve. The process is called **fairing**. There have been diverse fairing methods proposed in the literature, but a relatively simple one by Kjellander (1983) will be introduced here. In case of the cubic spline fitting discussed in §4.2.2, a suspected point P_i may be "corrected" by using the following relation:

$$\tilde{P}_i = (P_{i-1} + P_{i+1})/2 + \alpha(\mathbf{t}_{i-1} - \mathbf{t}_{i+1})/4 \tag{4.54}$$

$$where, \quad \mathbf{t}_i : \ tangent\ vector\ at\ P_i,$$

$$\alpha \ : \ damping\ factor\ (usually\ \alpha = 0.4).$$

The user has to specify a maximum amount ϵ_i that is allowed for P_i to be corrected. That is, $|\tilde{P}_i - P_i| \le \epsilon_i$ should be maintained, where P_i is the original input value. In order to apply the fairing formula (4.54), a cubic spline curve has to be fitted first to obtain the tangents $\mathbf{t}_i$. And then, the fairing equation is applied to each point P_i. More details on fairing and smoothing may be found in Chapter 7.

CHAPTER 5

SURFACE PATCH MODELS

5.1 INTRODUCTION

This chapter presents various types of surface patch models that will be used as building blocks in constructing more complex surface shapes. The types of surface models to be introduced are:

- parametric polynomial surface patch models,
- boundary interpolating patch models,
- sweep surface patches, and
- quadric surface primitives.

There are practically unlimited types of surface models available in the literature, but we limit our discussions to simple and yet practical ones.

1) Parametric Polynomial Patch Models:

The *parametric polynomial patch* models to be discussed are *Ferguson patch, Bezier patch, and B-spline patch.* With the exception of *triangular Bezier patch*, they are rectangular patches defined as a *tensor product* of respective curve models introduced in Chapter 3. The parametric polynomial patch models are used mainly in constructing a *composite surface* interpolating over an array of point data. Bezier and B-spline patch models are used in *curve-net interpolation* as well.

2) Boundary Interpolating Patch Models:

The second type, the *boundary curve interpolating surface patch models*, is defined as a *blending* of boundary curves. Popular models in this category include *Coons patches and Gregory patches.*

3) Sweep Surface Patch Models:

The third type, *sweep surface models*, is expressed as a *transformation* of curves. This surface type has received less attention than the first two types. However, it plays a very important role in surface modeling as will be seen in later chapters (See Chapters 9 and 14).

4) Quadric Primitives:

The fourth type, *quadric surfaces*, has mainly been used in solid modeling. Its usefulness in surface modeling is expected to increase. Many of "free-formed" surfaces of practical importance can be modeled by using quadric surface primitives when suitable blending techniques are incorporated. They are very useful also in constructing compound surfaces.

5.2 PARAMETRIC POLYNOMIAL SURFACE PATCH MODELS

Presented in this section are (cubic) polynomial surface models that are widely used in constructing surfaces from a set of 3D points. The surface patch models to be introduced are

a) standard polynomial surface patches,

b) Ferguson surface patch,

c) Bezier surface patches (rectangular),

d) uniform B-spline surface patches,

e) non-uniform B-spline surface patches, and

f) triangular Bezier patch.

As a review, the <u>cubic polynomial curve models</u> introduced in Chapter 3 are summarized as follows:

$$\mathbf{r}(u) = \mathbf{U}\ \mathbf{A} \quad : \quad \textit{standard polynomial curve (3.5)} \qquad (5.1-a)$$

$$\mathbf{r}(u) = \mathbf{U}\ \mathbf{C}\ \mathbf{S} \quad : \quad \textit{Ferguson curve (3.8)} \qquad (5.1-b)$$

$$\mathbf{r}(u) = \mathbf{U}\ \mathbf{M}\ \mathbf{R} \quad : \quad \textit{Bezier curve (3.11)} \qquad (5.1-c)$$

$$\mathbf{r}(u) = \mathbf{U}\ \mathbf{N}\ \mathbf{R} \quad : \quad \textit{uniform B-spline curve (3.17)} \qquad (5.1-d)$$

$$\mathbf{r}(u) = \mathbf{U}\ \mathbf{N_c}\ \mathbf{R} \quad : \quad \textit{non-uniform B-spline curve (3.28)} \qquad (5.1-e)$$

where $\mathbf{U} = [1\ u\ u^2\ u^3]$,

$\mathbf{A} = [\mathbf{a}\ \mathbf{b}\ \mathbf{c}\ \mathbf{d}]^T$,

$\mathbf{S} = [P_0\ P_1\ \mathbf{t}_0\ \mathbf{t}_1]^T$: *end condition vector,*

$\mathbf{R} = [V_0\ V_1\ V_2\ V_3]^T$: *control point vector,*

$\mathbf{C}$: *Ferguson's coefficient matrix,*

$\mathbf{M}$: *Bezier coefficient matrix (cubic),*

$\mathbf{N}$: *uniform B-spline coefficient matrix (cubic),*

$\mathbf{N_c}$: *non-uniform B-spline coefficient matrix (cubic).*

5.2.1 <u>Standard Polynomial Surface Patches</u>

Consider a vector-valued polynomial function $\mathbf{r}(u, v)$ whose degrees are cubic in both u and v with coefficients $\mathbf{d}_{ij}$ (for $u^i v^j$). That is, a **bicubic (standard) polynomial patch** is defined as

$$\mathbf{r}(u, v) = \sum_{i=0}^{3} \sum_{j=0}^{3} \mathbf{d}_{ij} u^i v^j \qquad \textit{with} \quad 0 \le u, v \le 1 \qquad (5.2-a)$$

which is expressed in a matrix form as

$$\mathbf{r}(u,v) = \mathbf{U}\ \mathbf{D}\ \mathbf{V}^T \tag{5.2-b}$$

where, $\mathbf{U} = [1\ u\ u^2\ u^3]$,

$\mathbf{V} = [1\ v\ v^2\ v^3]$,

$$\mathbf{D} = \begin{bmatrix} \mathbf{d}_{00} & \mathbf{d}_{01} & \mathbf{d}_{02} & \mathbf{d}_{03} \\ \mathbf{d}_{10} & \mathbf{d}_{11} & \mathbf{d}_{12} & \mathbf{d}_{13} \\ \mathbf{d}_{20} & \mathbf{d}_{21} & \mathbf{d}_{22} & \mathbf{d}_{23} \\ \mathbf{d}_{30} & \mathbf{d}_{31} & \mathbf{d}_{32} & \mathbf{d}_{33} \end{bmatrix} \quad : \quad \textit{coefficients matrix.}$$

The bicubic polynomial surface patch (5.2) may be used in constructing a smooth surface interpolating to a 4×4 array of 3D points $\{P_{ij}\}$. Figure 5.1 shows a bicubic polynomial patch defined by 16 data points,

$$\{P_{ij} : i = 0, ..., 3;\ j = 0, ..., 3\}.$$

The parameter values at the corners may be assigned as follows:

$$u = v = 0\ at\ P_{00}; \quad u = 0,\ v = 1\ at\ P_{03}; \quad u = 1,\ v = 0\ at\ P_{30}; \quad u = v = 1\ at\ P_{33}.$$

Parameter values at the remaining data points may be determined based on chord-lengths. For example, the value of u at P_{11} can be determined as

$$u = |P_{11} - P_{01}| / \{|P_{11} - P_{01}| + |P_{21} - P_{11}| + |P_{31} - P_{21}|\}.$$

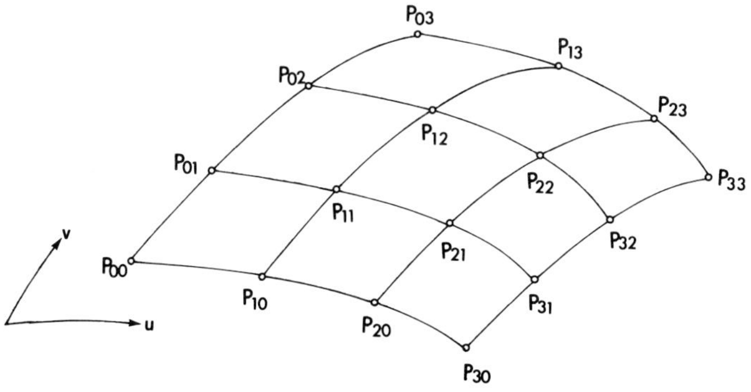

Figure 5.1 Standard Bicubic Polynomial Surface Patch

The degrees of u, v may be increased to m, n so that the surface patch interpolates over an $(m+1)\times(n+1)$ array of data points. In an earlier version of a CAD/CAM system, $m = n = 15$ is used. A major drawback of the standard polynomial model is that it is difficult to maintain a desired level of continuity across individual patch boundaries when a composite surface of complex shape is to be constructed. If the degrees of the surface patch are raised, on the other hand, the resulting surface tends to show unwanted oscillations.

Example 5.1 : *Give the equation of a plane containing 3 points P_0, P_1, P_2.*

Let $\mathbf{r}(u,v) = \mathbf{a} + \mathbf{b}u + \mathbf{c}v$ be a polynomial model, then by assigning parameter values, for example, as $u = v = 0$ at P_0, $u = 1$, $v = 0$ at P_1, and $u = 0$, $v = 1$ at P_2, we have

$$P_0 = \mathbf{r}(0,0) = \mathbf{a}; \quad P_1 = \mathbf{r}(1,0) = \mathbf{a} + \mathbf{b}; \quad P_2 = \mathbf{r}(0,1) = \mathbf{a} + \mathbf{c}.$$

On solving the above equations for the coefficients and then substituting into the surface equation, we have $\mathbf{r}(u,v) = P_0 + (P_1 - P_0)u + (P_2 - P_0)v$. ◇

5.2.2 Ferguson Surface Patch Model

A more useful way of using the bicubic polynomial patch $\mathbf{r}(u,v)$ of (5.2) is to construct a surface patch interpolating over the four corner points $\{P_{ij} : i,j = 0,1\}$ as depicted in Fig. 5.2. Since there are 16 unknown coefficients in (5.2), it is necessary to specify the same number of relations (ie, constraints). The first four constraints are provided by the corner points as follows:

$$\mathbf{r}(i,j) = P_{ij} \quad for\ i,j = 0,1. \tag{5.3}$$

In order to provide additional constraints, the following **corner conditions** have to be specified (see Fig. 5.2):

$\mathbf{s}_{ij}$: *u-direction tangent vector at P_{ij},*

$\mathbf{t}_{ij}$: *v-direction tangent vector at P_{ij},*

$\mathbf{x}_{ij}$: **twist vector** *at P_{ij}.*

Assuming the above vectors are given, we can introduce the following relations at the four corners of the surface patch ($for\ i,j = 0,1$):

$$\mathbf{r}_u(i,j) = \mathbf{s}_{ij}; \quad \mathbf{r}_v(i,j) = \mathbf{t}_{ij}; \quad \mathbf{r}_{uv}(i,j) = \mathbf{x}_{ij} \tag{5.4}$$

where, $\mathbf{r}_u(u,v) = \partial\mathbf{r}(u,v)/\partial u$,

$\mathbf{r}_v(u,v) = \partial\mathbf{r}(u,v)/\partial v$,

$\mathbf{r}_{uv}(u,v) = \partial^2\mathbf{r}(u,v)/\partial u\partial v$.

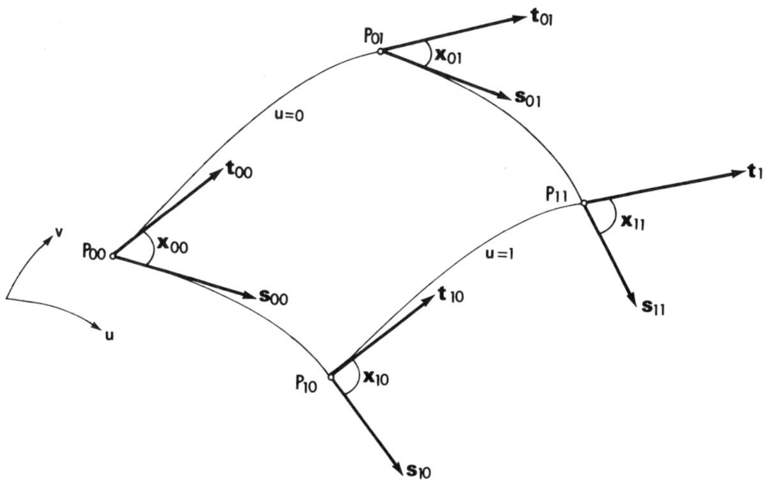

Figure 5.2 Ferguson Surface Patch

By solving the 16 linear equations in (5.3) & (5.4) for the unknowns coefficients $\mathbf{d}_{ij}$, the bicubic polynomial surface equation (5.2) can be converted to a **Ferguson patch equation**:

$$\begin{aligned}\mathbf{r}(u,v) &= \mathbf{U}\ \mathbf{D}\ \mathbf{V}^{\mathrm{T}} \\ &= \mathbf{U}\ \mathbf{C}\ \mathbf{Q}\ \mathbf{C}^{\mathrm{T}}\mathbf{V}^{\mathrm{T}} \ :\quad 0 \le u, v \le 1\end{aligned} \tag{5.5}$$

$$\begin{aligned}\textit{where,}\quad \mathbf{U} &= [1\ u\ u^2\ u^3], \\ \mathbf{V} &= [1\ v\ v^2\ v^3], \\ \mathbf{C} &= \begin{bmatrix} 1 & 0 & 0 & 0 \\ 0 & 0 & 1 & 0 \\ -3 & 3 & -2 & -1 \\ 2 & -2 & 1 & 1 \end{bmatrix} \ :\quad \textit{Ferguson coeff. matrix,} \\ \mathbf{Q} &= \begin{bmatrix} P_{00} & P_{01} & \mathbf{t}_{00} & \mathbf{t}_{01} \\ P_{10} & P_{11} & \mathbf{t}_{10} & \mathbf{t}_{11} \\ \mathbf{s}_{00} & \mathbf{s}_{01} & \mathbf{x}_{00} & \mathbf{x}_{01} \\ \mathbf{s}_{10} & \mathbf{s}_{11} & \mathbf{x}_{10} & \mathbf{x}_{11} \end{bmatrix} \ :\quad \textit{corner conditions.}\end{aligned}$$

Example 5.2: *Show that a boundary curve of a Ferguson patch is a Ferguson curve.*

Let's consider the $v = 0$ *boundary curve. With* $v = 0$, *the matrix multiplication* $\mathbf{Q}\ \mathbf{C}^{\mathbf{T}}\mathbf{V}^{\mathbf{T}}$ *in* (5.5) *is evaluated as:*

$$\mathbf{Q}\ \mathbf{C}^{\mathbf{T}}\mathbf{V}^{\mathbf{T}}|_{v=0} = \begin{bmatrix} P_{00} & P_{01} & \mathbf{t}_{00} & \mathbf{t}_{01} \\ P_{10} & P_{11} & \mathbf{t}_{10} & \mathbf{t}_{11} \\ \mathbf{s}_{00} & \mathbf{s}_{01} & \mathbf{x}_{00} & \mathbf{x}_{01} \\ \mathbf{s}_{10} & \mathbf{s}_{11} & \mathbf{x}_{10} & \mathbf{x}_{11} \end{bmatrix} \begin{bmatrix} 1 & 0 & -3 & 2 \\ 0 & 0 & 3 & -2 \\ 0 & 1 & -2 & 1 \\ 0 & 0 & -1 & 1 \end{bmatrix} \begin{bmatrix} 1 \\ 0 \\ 0 \\ 0 \end{bmatrix} = \begin{bmatrix} P_{00} \\ P_{10} \\ \mathbf{s}_{00} \\ \mathbf{s}_{10} \end{bmatrix} \equiv \mathbf{S}.$$

On substituting the above result into (5.5), *the Ferguson curve* $(5.1-b)$ *is obtained.* ◇

Recall from §3.3.2 that the Ferguson curve equation (5.1-b) is in fact a *Hermite blending of its end conditions.* That is,

$$\begin{aligned} \mathbf{r}(u) &= \mathbf{U}\ \mathbf{C}\ \mathbf{S} \\ &= H_0^3(u)P_0 + H_1^3(u)\mathbf{t}_0 + H_2^3(u)\mathbf{t}_1 + H_3^3(u)P_1 \quad : 0 \le u \le 1 \end{aligned}$$

$$\begin{aligned} \textit{where,}\quad H_0^3(u) &= (1 - 3u^2 + 2u^3), \\ H_1^3(u) &= (u - 2u^2 + u^3), \\ H_2^3(u) &= (-u^2 + u^3), \\ H_3^3(u) &= (3u^2 - 2u^3). \end{aligned}$$

By following the steps in Example 5.2 above, one may find that the boundary curves at $u = 0$ and $u = 1$ are expressed as Hermite blending of their end conditions. That is, for $i = 0, 1$

$$\mathbf{r}(i, v) = H_0^3(v)P_{i0} + H_1^3(v)\mathbf{t}_{i0} + H_2^3(v)\mathbf{t}_{i1} + H_3^3(v)P_{i1}. \tag{5.6}$$

Along the $u = 0, 1$ boundary curves, the *cross-boundary tangents* $\mathbf{r}_u(i, v)$ for $i = 0, 1$ are also expressed as Hermite blending of u-direction tangents and twist vectors at the ends:

$$\mathbf{r}_u(i, v) = H_0^3(v)\mathbf{s}_{i0} + H_1^3(v)\mathbf{x}_{i0} + H_2^3(v)\mathbf{x}_{i1} + H_3^3(v)\mathbf{s}_{i1}. \tag{5.7}$$

Now define $\mathbf{r}(u, v)$ as a Hermite blended function of the boundary curves (5.6) and the cross-boundary tangents (5.7) to obtain

$$\mathbf{r}(u, v) = H_0^3(u)\mathbf{r}(0, v) + H_1^3(u)\mathbf{r}_u(0, v) + H_2^3(u)\mathbf{r}_u(1, v) + H_3^3(u)\mathbf{r}(1, v).$$

The resulting equation is the Ferguson surface equation (5.5).

One would obtain the same result should he start with the $v = 0, 1$ boundary curves. This is an important property of a *tensor product* (or *Cartesian product*) surface. If a surface patch is completely defined in terms of the *corner conditions* (ie, P, s, t, x), it is called a **tensor product surface**. A tensor product patch has a rectangular topology and is expressed in a symmetric form (w.r.t. u and v) as can be seen in the standard bicubic polynomial patch (5.2) and the Ferguson patch (5.5).

5.2.3 Bezier Surface Patch Model

We may define a tensor product Bezier patch by generalizing the Bezier curve (5.1-c) the same way the Ferguson patch (5.5) is constructed from the Ferguson curve (5.1-b). Let us consider a 4×4 array of control vertices $\{V_{ij}\}$ as shown in Fig. 5.3. Then, by blending the control vertices with *Bernstein polynomials*, a **bicubic Bezier patch** is defined as follows (see Fig. 5.3):

$$\begin{aligned}\mathbf{r}(u,v) &= \sum_{i=0}^{3}\sum_{j=0}^{3} B_i^3(u)\ B_j^3(v) V_{ij} \\ &= \sum_{i=0}^{3}\sum_{j=0}^{3} \frac{3!}{(3-i)!\ i!} u^i (1-u)^{3-i} \frac{3!}{(3-j)!\ j!} v^j (1-v)^{3-j} V_{ij} \\ &= \mathrm{U\ M\ B\ M^T V^T}\end{aligned} \tag{5.8}$$

$$where,\quad \mathrm{U} = [1\ u\ u^2\ u^3]\ ;\quad \mathrm{V} = [1\ v\ v^2\ v^3],$$

$$\mathrm{M} = \begin{bmatrix} 1 & 0 & 0 & 0 \\ -3 & 3 & 0 & 0 \\ 3 & -6 & 3 & 0 \\ -1 & 3 & -3 & 1 \end{bmatrix}\ ;\quad \mathrm{B} = \begin{bmatrix} V_{00} & V_{01} & V_{02} & V_{03} \\ V_{10} & V_{11} & V_{12} & V_{13} \\ V_{20} & V_{21} & V_{22} & V_{23} \\ V_{30} & V_{31} & V_{32} & V_{33} \end{bmatrix}.$$

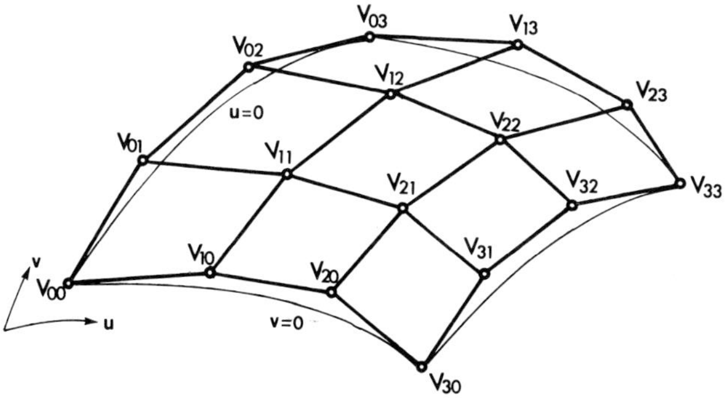

Figure 5.3 Bicubic Bezier Patch

In (5.8), the matrix M is called a (cubic) *Bezier coefficient matrix*, and **B** is called a *Bezier control point net* which forms a **characteristic polyhedron.** The above Bezier patch is also a *tensor product* surface patch.

As with Bezier curves of §3.2.2, the bicubic Bezier patch model can be generalized to a "degrees m, n" patch. Namely

$$\mathbf{r}(u,v) = \sum_{i=0}^{m} \sum_{j=0}^{n} B_i^m(u)\ B_j^n(v) V_{ij} \tag{5.9}$$

$$where \quad B_i^m(u) = \frac{m!}{(m-i)!\ i!} u^i (1-u)^{m-i},$$

$$B_j^n(v) = \frac{n!}{(n-j)!\ j!} v^j (1-v)^{n-j}.$$

In some commercial CAD/CAM systems, $m = n = 5$ or $m = n = 7$ are used. When $m = n = 5$, we need 36 control vertices to define a *biquintic* Bezier patch.

Example 5.3: *Show that a boundary of a bicubic Bezier patch is a cubic Bezier curve.*

Let's evaluate (5.8) at $u = 0$ *to obtain* $\mathbf{r}(0,v) = \mathbf{V\ M\ R}$, *where* **R** *is the control vertices at the* $u = 0$ *boundary curve (see Fig. 5.3). With* $u = 0$,

$$\mathbf{U\ M\ B}|_{u=0} = [1\ 0\ 0\ 0] \begin{bmatrix} 1 & 0 & 0 & 0 \\ -3 & 3 & 0 & 0 \\ 3 & -6 & 3 & 0 \\ -1 & 3 & -3 & 1 \end{bmatrix} \begin{bmatrix} V_{00} & V_{01} & V_{02} & V_{03} \\ V_{10} & V_{11} & V_{12} & V_{13} \\ V_{20} & V_{21} & V_{22} & V_{23} \\ V_{30} & V_{31} & V_{32} & V_{33} \end{bmatrix}$$

$$= [V_{00}\ V_{01}\ V_{02}\ V_{03}] \equiv \mathbf{R}^{\mathrm{T}}.$$

On substituting the above result into (5.8), *we have*

$$\mathbf{r}(0,v) = \mathbf{U\ M\ B\ M^T V^T}|_{u=0} = \mathbf{R^T M^T V^T} = \mathbf{V\ M\ R}. \qquad \diamond$$

One may easily express the corner conditions of the Ferguson patch (5.5) in terms of the control vertices of the bicubic Bezier patch (5.8) as both are the same bicubic polynomial. On equating the Ferguson patch equation (5.5) with the bicubic Bezier patch equation (5.8), we have

$$\mathbf{U\ C\ Q\ C^T V^T} = \mathbf{U\ M\ B\ M^T V^T}$$

which in turn leads to

$$\mathbf{C\ Q\ C^T} = \mathbf{M\ B\ M^T}.$$

Since the coefficient matrices **M, C** are non-singular, we have

$$\mathbf{Q} = (\mathbf{C}^{-1}\ \mathbf{M})\ \mathbf{B}\ (\mathbf{C}^{-1}\mathbf{M})^{\mathrm{T}}. \tag{5.10}$$

Since $\mathbf{C}^{-1}\mathbf{M}$ is given by

$$\mathbf{C}^{-1}\mathbf{M} = \begin{bmatrix} 1 & 0 & 0 & 0 \\ 0 & 0 & 0 & 1 \\ -3 & 3 & 0 & 0 \\ 0 & 0 & -3 & 3 \end{bmatrix},$$

the *corner conditions* at $u = v = 0$, for example, are expressed as

$$\begin{aligned} P_{00} &= V_{00}, \\ \mathbf{s}_{00} &= 3(V_{10} - V_{00}), \\ \mathbf{t}_{00} &= 3(V_{01} - V_{00}), \quad \textit{and} \\ \mathbf{x}_{00} &= 9(V_{00} - V_{01} - V_{10} + V_{11}). \end{aligned} \tag{5.11}$$

More detailed discussions on the properties of Bezier curves and surfaces will be provided in the next chapter.

5.2.4 Uniform B-spline Surface Patches

As with the bicubic Bezier patch (5.8), a bicubic B-spline patch is defined as a tensor product surface of the uniform cubic B-spline curve (5.1-d). For the same set of (Bezier) control vertices shown in Fig. 5.3, a bicubic uniform B-spline surface would look like the one shown in Fig. 5.4. The **bicubic uniform B-spline surface patch** is expressed as the following *tensor product* surface equation:

$$\begin{aligned} \mathbf{r}(u,v) &= \sum_{i=0}^{3}\sum_{j=0}^{3} N_i^3(u) N_j^3(v) V_{ij} \\ &= \mathbf{U\ N\ B\ N^T V^T} \quad \textit{with} \quad 0 \le u \le 1 \end{aligned} \tag{5.12}$$

$$\textit{where,} \quad \mathbf{U} = [1\ u\ u^2\ u^3]\ ; \quad \mathbf{V} = [1\ v\ v^2\ v^3],$$

$$\mathbf{B} = \begin{bmatrix} V_{00} & V_{01} & V_{02} & V_{03} \\ V_{10} & V_{11} & V_{12} & V_{13} \\ V_{20} & V_{21} & V_{22} & V_{23} \\ V_{30} & V_{31} & V_{32} & V_{33} \end{bmatrix}; \quad \mathbf{N} = \frac{1}{6}\begin{bmatrix} 1 & 4 & 1 & 0 \\ -3 & 0 & 3 & 0 \\ 3 & -6 & 3 & 0 \\ -1 & 3 & -3 & 1 \end{bmatrix},$$

$$\begin{aligned} N_0^3(u) &= (1 - 3u + 3u^2 - u^3)/6, \\ N_1^3(u) &= (4 - 6u^2 + 3u^3)/6, \\ N_2^3(u) &= (1 + 3u + 3u^2 - 3u^3)/6, \\ N_3^3(u) &= \frac{1}{6}u^3. \end{aligned}$$

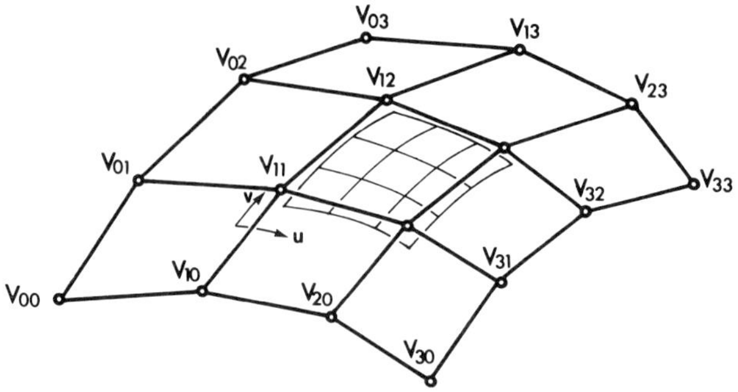

Figure 5.4 Uniform Bicubic B-spline Patch

A *biquadratic (uniform) B-spline patch* is defined similarly as a tensor product of the quadratic uniform B-spline curve (3.18). It is also possible to define a B-spline surface patch having different degrees in u- and v-directions. For example, a uniform B-spline surface patch of degrees 3 and 2 in u and v, respectively, may be defined as follows:

$$\mathbf{r}(u) = \mathbf{U}\ \mathbf{N_c}\ \mathbf{B}\ \mathbf{N_q^T V^T} \quad with \quad 0 \leq u \leq 1$$

$$where, \quad \mathbf{U} = [1\ u\ u^2\ u^3]$$

$$\mathbf{V} = [1\ v\ v^2],$$

$$\mathbf{B} = \begin{bmatrix} V_{00} & V_{01} & V_{02} \\ V_{10} & V_{11} & V_{12} \\ V_{20} & V_{21} & V_{22} \\ V_{30} & V_{31} & V_{32} \end{bmatrix}$$

$$\mathbf{N_c} = \frac{1}{6}\begin{bmatrix} 1 & 4 & 1 & 0 \\ -3 & 0 & 3 & 0 \\ 3 & -6 & 3 & 0 \\ -1 & 3 & -3 & 1 \end{bmatrix},$$

$$\mathbf{N_q} = \frac{1}{2}\begin{bmatrix} 1 & 1 & 0 \\ -2 & 2 & 0 \\ 1 & -2 & 1 \end{bmatrix} \quad (see\ Eqn.\ 3.18).$$

5.2.5 Non-uniform B-spline Surface Patches

A NUB (non-uniform B-spline) surface is defined as a tensor product of NUB curves. For example, with *knot span vectors* $\{\Delta_i\}$ and $\{\nabla_j\}$, a **biquadratic NUB surface** patch can be constructed from the quadratic NUB curve model (3.27) as follows:

$$\mathbf{r}(u,v) = \mathbf{U}\ \mathbf{N_s}\ \mathbf{B}\ \mathbf{N_t^T V^T} \qquad (5.13-a)$$

$$\textit{where,}\quad \mathbf{U} = [1\ u\ u^2],$$

$$\mathbf{V} = [1\ v\ v^2],$$

$$\mathbf{B} = \begin{bmatrix} V_{00} & V_{01} & V_{02} \\ V_{10} & V_{11} & V_{12} \\ V_{20} & V_{21} & V_{22} \end{bmatrix} \quad : \quad \textit{control vertices,}$$

$$\mathbf{N_s} = \begin{bmatrix} (\Delta_i/\Delta_{i-1}^2) & (\Delta_{i-1}/\Delta_{i-1}^2) & 0 \\ (-2\Delta_i/\Delta_{i-1}^2) & (2\Delta_i/\Delta_{i-1}^2) & 0 \\ (\Delta_i/\Delta_{i-1}^2) & \frac{\Delta_i}{\Delta_{i-1}}\{\frac{\Delta_i}{\Delta_{i-1}^2} - \frac{\Delta_{i-1}^3}{\Delta_i^2}\} & \frac{\Delta_i}{\Delta_i^2} \end{bmatrix}.$$

$$\mathbf{N_t} = \begin{bmatrix} (\nabla_j/\nabla_{j-1}^2) & (\nabla_{j-1}/\nabla_{j-1}^2) & 0 \\ (-2\nabla_j/\nabla_{j-1}^2) & (2\nabla_j/\nabla_{j-1}^2) & 0 \\ (\nabla_j/\nabla_{j-1}^2) & \frac{\nabla_j}{\nabla_{j-1}}\{\frac{\nabla_j}{\nabla_{j-1}^2} - \frac{\nabla_{j-1}^3}{\nabla_j^2}\} & \frac{\nabla_j}{\nabla_j^2} \end{bmatrix},$$

$\{\Delta_i\}$: *u-direction knot spans,*

$\{\nabla_j\}$: *v-direction knot spans.*

In the same way, a **bicubic NUB patch** is defined in terms of 16 control vertices $\{V_{ij} : j = 0,1,2,3\}$ and two knot span vectors $\{\Delta_i\}$ and $\{\nabla_j\}$.

$$\mathbf{r}(u,v) = \mathbf{U}\ \mathbf{N_s B}\ \mathbf{N_t^T}\ \mathbf{V^T} \qquad (5.13-b)$$

where $\mathbf{U} = [1\ u\ u^2\ u^3]$; $\mathbf{V} = [1\ v\ v^2\ v^3]$,

$\mathbf{B} = \{V_{ij}\}$: *control vertices as in* (5.12),

$\mathbf{N_s}$: *coefficient matrix as in* (3.28) *with knot spans* Δ_i,

$\mathbf{N_t}$: *coefficient matrix as in* (3.28) *with knot spans* ∇_j.

It should be noted in (5.13) that $\triangle_i$ and ∇_j are supporting know spans so that the surface patches are defined on the **knot area** $[\triangle_i \times \nabla_j]$. The supporting knot spans should be non-zero. If all the *extended knot spans* are zero (ie, *multiple knots*), namely, if

$$.. = \triangle_{i-1} = \triangle_{i+1} = .. = 0 \; and \; .. = \nabla_{j-1} = \nabla_{j+1} = .. = 0,$$

then the B-spline coefficient matrices N's becomes Bezier matrices M's of same degree. In other words, a NUB-patch degenerates to a Bezier patch when all the extended knot spans are zero. On the other hand, if all the knot spans (supporting and extended) are identical, that is, if

$$.. = \triangle_{i-1} = \triangle_i = \triangle_{i+1} = .. \; and \; .. = \nabla_{j-1} = \nabla_j = \nabla_{j+1} = ..,$$

the NUB patches in (5.13) reduce to uniform B-spline patches.

B-spline surfaces possess almost the same properties of Bezier patches (See Chapter 6): They have *affine invariance* and *convex hull* properties; they are subject to *degree elevation* and *subdivision*. After all, Bezier surface is a special case of NUB surface. For general discussions on the subject of B-spline surfaces, the reader is referred to Riesenfeld (1973) and Versprille (1975).

5.2.6 Triangular Bezier Patches

Presented in this subsection is a triangular polynomial patch model called a *triangular Bezier patch* which is expressed in terms of control vertices and bivariate Bernstein polynomials. Let us consider *barycentric* (*or* area ratio) *coordinates* (u, v, w) of the *domain triangle* $P_0P_1P_2$ shown in Fig. 5.5a. The triangular patch in Fig. 5.5b is defined in terms of the nine control vertices, which is a mapping from the domain triangle to 3D space.

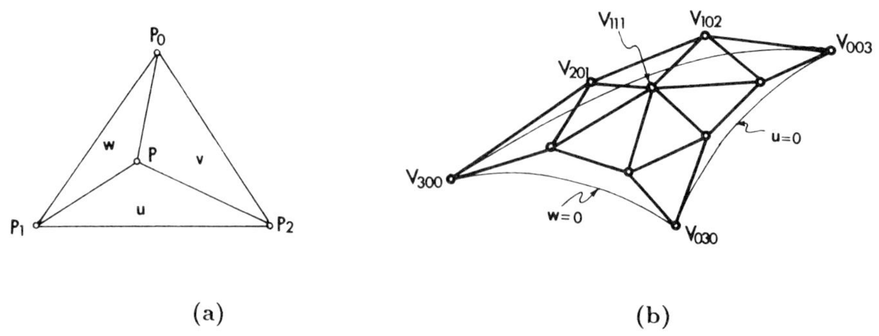

Figure 5.5 Triangular Cubic Bezier Patch

Let P be a point on the domain triangle, then from the property of barycentric coordinate (see §4.4.3), the following should hold:

$$P = u\ P_0 + v\ P_1 + w\ P_2 \qquad (5.14-a)$$

$$u + v + w = 1\ :\quad 0 \le u, v, w \le 1. \qquad (5.14-b)$$

Let us define **bivariate Bernstein polynomials** in terms of the barycentric coordinates (u, v, w):

$$B_{\mathbf{i}}^n(\mathbf{u}) = \frac{n!}{i!\ j!\ k!} u^i\ v^j\ w^k \quad for \quad 0 \le i, j, k \le n \qquad (5.15)$$

$$where, \quad \mathbf{u} = (u, v, w) : \ barycentric\ coordinates\ in (5.14),$$

$$\mathbf{i} = (i, j, k) \quad with \quad |\mathbf{i}| \equiv i + j + k = n.$$

Then, with the 10 control vertices $\{V_{ijk}\}$ in Fig. 5.5b, we may define a cubic polynomial surface by using the bivariate Bernstein polynomials (5.15) as blending functions. That is, when $n = 3$, a **cubic triangular Bezier patch** is defined as follows:

$$\mathbf{r}(\mathbf{u}) = \textstyle\sum_{|\mathbf{i}|=3} V_{\mathbf{i}}\ B_{\mathbf{i}}^3(\mathbf{u}) \qquad (5.16)$$

$$where, \quad \mathbf{i} = (i, j, k),$$

$$\mathbf{u} = (u, v, w),$$

$$V_{\mathbf{i}} : \ Bezier\ control\ points,$$

$$B_{\mathbf{i}}^3(\mathbf{u}) : \ bivariate\ cubic\ Bernstein\ polynomial.$$

Example 5.4: *Show that the* $w = 0$ *boundary curve of the triangular Bezier patch* (5.16) *is a cubic Bezier curve (See Fig.* 5.5 – *b).*

When $w = 0$, *the bivariate Bernstein polynomials in* (5.15) *are non-zero only for* $k = 0$. *Since* $v = 1 - u$ *and* $j = 3 - i$ *when* $w = 0$ *and* $k = 0$, (5.15) *becomes univariate Bernstein polynomials. That is,*

$$B_{\mathbf{i}}^3(\mathbf{u}) = \frac{3!}{i!\ j!\ 0!} u^i\ v^j\ 0^0 = \frac{3!}{i!(3-i)!} u^i (1-u)^{3-i} = B_i^3(u).$$

And the control vertices $V_{\mathbf{i}}$ *in* (5.16) *for* $i = 0, 1, 2, 3$ *are (see Fig.* 5.5 – *b)*

$$V_{030},\ V_{120},\ V_{210},\ and\ V_{300}.$$

Thus, for the $w = 0$ *boundary curve,* (5.16) *reduces to*

$$\mathbf{r}(\mathbf{u}) = \mathbf{r}(u, 1-u, 0) = B_0^3(u) V_{030} + B_1^3(u) V_{120} + B_2^3(u) V_{210} + B_3^3(u) V_{300}$$

which is exactly the equation of a cubic Bezier curve (See Eqn. 3.12) . ◇

5.3 CURVED BOUNDARY INTERPOLATING SURFACE PATCHES

In most conventional drawings, the descriptions of surfaces, which we call *descriptive shape models*, are provided in terms of boundary curves. Introduced in this section are methods of constructing a surface patch interpolating to a set of boundary curves. Surface patch models to be discussed in the section are *ruled surfaces*, *lofted surface*, *Coons surfaces*, and *Gregory patches*.

5.3.1 Ruled Surfaces

Let us consider two parametric curves, $\mathbf{r}_0(u)$ and $\mathbf{r}_1(u)$ with $0 \le u \le 1$, as shown in Fig. 5.6a. A linear blending of the two curves defines a surface patch called a *ruled surface*:

$$\mathbf{r}(u,v) = (1-v)\mathbf{r}_0(u) + v\ \mathbf{r}_1(u) \quad : \quad 0 \le u, v \le 1. \qquad (5.17-a)$$

It is the simplest possible surface that can be defined from boundary curves, and yet ruled surfaces are widely found in engineering products, such as the transition surface in CRT monitors (the surface between the CRT screen and the front face of a monitor) and a wing of an airplane.

The *blending functions* in (5.17) are not necessarily linear. In fact, the use of cubic *Hermite blending functions* is required in some application (see, for example, the *sweep surface* blending in §9.2.6). Let us rearrange the terms in the ruled surface (5.17) as follows

$$\mathbf{r}(u,v) = \mathbf{r}_0(u) + v(\mathbf{r}_1(u) - \mathbf{r}_0(u)) \quad : \quad 0 \le u, v \le 1. \qquad (5.17-b)$$

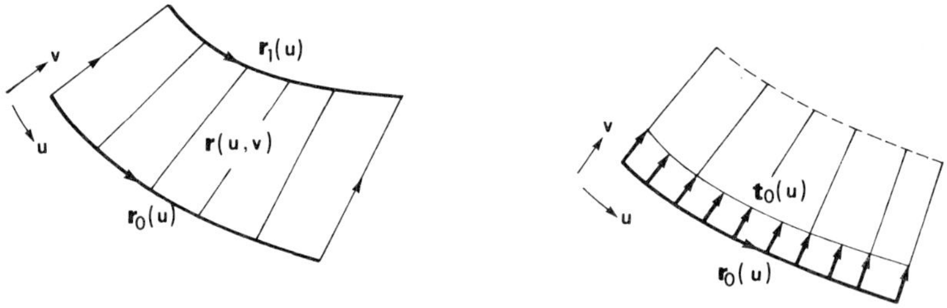

(a) (b)

Fiogure 5.6 Ruled Surface (a) and Linear Taylor Interpolant (b)

Then, the second term in the above equation represents a vector which is a function of u. A (unit) vector in the direction of $\mathbf{r}_1(u) - \mathbf{r}_0(u)$ in (5.17-b) is called a **ruling vector** $\mathbf{t}(u)$.

By using the ruling vector $\mathbf{t}_0(u)$ of a boundary curve $\mathbf{r}_0(u)$, we may define a surface equation similar to the ruled surface (5.17-b). That is,

$$\mathbf{r}(u,v) = \mathbf{r}_0(u) + v\ \mathbf{t}_0(u), \qquad (5.17-c)$$

where $\mathbf{t}_0(u)$ is a ruling vector. If the ruling vector in the above equation is the *cross-tangent vector* of the curve, the resulting surface equation becomes a **linear Taylor interpolant** as discussed in Gregory (1983). Figure 5.6b shows a linear Taylor interpolant defined by (5.17-c).

If the ruling vector $\mathbf{t}_0(u)$ in (5.17-c) is a constant vector $\mathbf{t}$ and the boundary curve is a plane curve, then a **TABCYL** (tabulated cylinder) surface is obtained. More details on ruled surfaces and TABCYL may be found in Faux and Pratt (1980).

5.3.2 Lofted Surfaces

An extension of the ruled surface models in (5.17) is the case where a pair of boundary curves together with their *cross-boundary tangents* are given:

$$\mathbf{r}_i(u)\ for\ i = 0,1 : \ boundary\ curves,\ and$$
$$\mathbf{t}_i(u)\ for\ i = 0,1 : \ cross\text{-}boundary\ tangents.$$

The situation is exactly the same as the Ferguson curve construction of §3.3.2, with the difference being that this time we have vector valued functions (instead of vectors). Thus, from the Ferguson curve equation (3.9), a **lofted surface** is constructed by blending the input data with (cubic) *Hermite blending functions* $H_i^3(v)$:

$$\mathbf{r}(u,v) = H_0^3(v)\mathbf{r}_0(u) + H_1^3(v)\mathbf{t}_0(u) + H_2^3(v)\mathbf{t}_1(u) + H_3^3(v)\mathbf{r}_1(u) \qquad (5.18)$$

$$\begin{aligned} where,\quad & H_0^3(v) = (1 - 3v^2 + 2v^3), \\ & H_1^3(v) = (v - 2v^2 + v^3), \\ & H_2^3(v) = (-v^2 + v^3), \\ & H_3^3(v) = (3v^2 - 2v^3), \\ & \mathbf{r}_i(u) : \ boundary\ curves(i = 0,1), \\ & \mathbf{t}_i(u) : \ cross\text{-}boundary\ tangents. \end{aligned}$$

The above surface patch is sometimes called a *cubic Hermite lofted surface*.

5.3.3 Rectangular Coons Surface Patches

Shown in Fig. 5.7a is a rectangular surface patch $\mathbf{r}(u, v)$ bounded by four *boundary curves*

$$\mathbf{a}_0(v),\ \mathbf{a}_1(v),\ \mathbf{b}_0(u),\quad and \quad \mathbf{b}_1(u)\ :\ 0 \le u, v \le 1.$$

We want to construct a surface patch $\mathbf{r}(u, v)$ which interpolates to the curve data at its patch boundaries. That is, the required **boundary conditions** are:

$$\mathbf{r}(i, v) = \mathbf{a}_i(v) \quad : \quad i = 0, 1 \qquad (5.19-a)$$

$$\mathbf{r}(u, j) = \mathbf{b}_j(u) \quad : \quad j = 0, 1 \qquad (5.19-b)$$

Also defined in Fig. 5.7a are *corner points* of the patch. They are

$$P_{ij} = \mathbf{r}(i, j) \quad : \quad i, j = 0, 1.$$

Let $\mathbf{r}_1(u, v)$ and $\mathbf{r}_2(u, v)$ be *ruled surfaces* satisfying the boundary conditions (5.19-a) and (5.19-b), respectively. Then, they are expressed as

$$\mathbf{r}_1(u, v) = (1 - u)\mathbf{a}_0(v) + u\ \mathbf{a}_1(v) \qquad (5.20-a)$$

$$\mathbf{r}_2(u, v) = (1 - v)\mathbf{b}_0(u) + v\ \mathbf{b}_1(u) \qquad (5.20-b)$$

These two ruled surfaces are depicted in Fig. 5.7b.

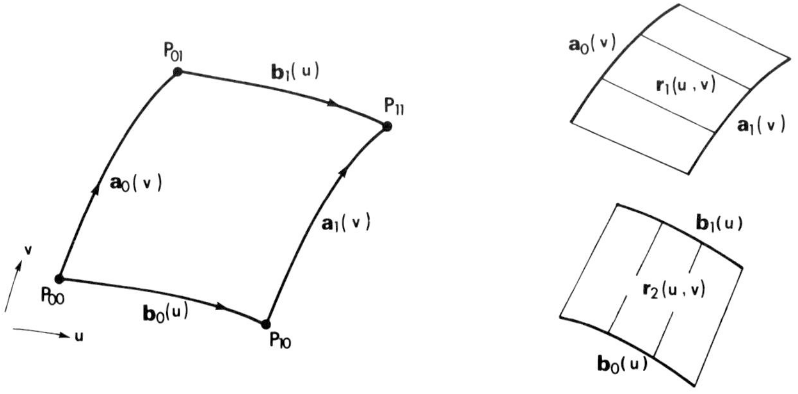

(a) (b)

Figure 5.7 Construction of a Coons Patch

Since the sum of the two ruled surfaces in (5.20) does not in general meet the boundary conditions, a *correction surface* $\mathbf{r}_3(u,v)$ has to be subtracted from the sum $\mathbf{r}_1(u,v)+\mathbf{r}_2(u,v)$. Then, the surface equation $\mathbf{r}(u,v)$ that meets the boundary condition (5.19) is given by

$$\mathbf{r}(u,v) = \mathbf{r}_1(u,v) + \mathbf{r}_2(u,v) - \mathbf{r}_3(u,v). \tag{5.21}$$

The above equation is said to be in a *Boolean sum form* (if $\mathbf{r}_3$ is an *intersection* of $\mathbf{r}_1$ and $\mathbf{r}_2$).

In order to determine the correction surface $\mathbf{r}_3(u,v)$, the surface equation (5.21) has to be evaluated at its boundaries. From the boundary condition (5.19-a) at $u = 0$, we have

$$\begin{aligned}\mathbf{a}_0(v) &\equiv \mathbf{r}(0,v)\\ &= \mathbf{r}_1(0,v) + \mathbf{r}_2(0,v) - \mathbf{r}_3(0,v)\\ &= \{\mathbf{a}_0(v)\} + \{(1-v)\mathbf{b}_0(0) + v\ \mathbf{b}_1(0)\} - \mathbf{r}_3(0,v),\end{aligned}$$

and the boundary condition at $u = 1$ becomes

$$\begin{aligned}\mathbf{a}_1(v) &\equiv \mathbf{r}(1,v)\\ &= \mathbf{r}_1(1,v) + \mathbf{r}_2(1,v) - \mathbf{r}_3(1,v)\\ &= \{\mathbf{a}_1(v)\} + \{(1-v)\mathbf{b}_0(1) + v\ \mathbf{b}_1(1)\} - \mathbf{r}_3(1,v).\end{aligned}$$

Rearranging the above expressions, boundary values of the *correction surface* at $u = 0, 1$ are obtained as

$$\begin{aligned}\mathbf{r}_3(0,v) &= (1-v)\mathbf{b}_0(0) + v\ \mathbf{b}_1(0)\\ &= (1-v)P_{00} + v\ P_{01}\end{aligned} \tag{5.22 – a}$$

$$\begin{aligned}\mathbf{r}_3(1,v) &= (1-v)\mathbf{b}_0(1) + v\ \mathbf{b}_1(1)\\ &= (1-v)P_{10} + v\ P_{11}.\end{aligned} \tag{5.22 – b}$$

Since the ruled surfaces in (5.21) are linearly blended surfaces, the correction surface $\mathbf{r}_3(u,v)$ may be defined as a linear blending of the two boundary curves in (5.22). Thus, the *correction surface* is expressed as

$$\begin{aligned}\mathbf{r}_3(u,v) &= (1-u)\mathbf{r}_3(0,v) + u\ \mathbf{r}_3(1,v)\\ &= (1-u)(1-v)P_{00} + (1-u)vP_{01} + u(1-v)P_{10} + uvP_{11}.\end{aligned} \tag{5.23}$$

Observe that the correction surface is defined as a *bilinear blending* of the four corner points.

On substituting the results of (5.20) and (5.23) into (5.21), the surface patch $\mathbf{r}(u,v)$ interpolating to the four boundary curves of Fig. 5.7a is expressed as follows:

$$
\begin{aligned}
\mathbf{r}(u,v) &= \mathbf{r}_1(u,v) + \mathbf{r}_2(u,v) - \mathbf{r}_3(u,v) \\
&= \{(1-u)\mathbf{a}_0(v) + u\ \mathbf{a}_1(v)\} + \{(1-v)\mathbf{b}_0(u) + v\ \mathbf{b}_1(u)\} \\
&\quad - \{(1-u)(1-v)P_{00} + (1-u)v\ P_{01} + u(1-v)\ P_{10} + uv\ P_{11}\} \\
&= [\alpha_0(u)\ \alpha_1(u)] \begin{bmatrix} \mathbf{a}_0(v) \\ \mathbf{a}_1(v) \end{bmatrix} + [\mathbf{b}_0(u)\ \mathbf{b}_1(u)] \begin{bmatrix} \alpha_0(v) \\ \alpha_1(v) \end{bmatrix} \\
&\quad - [\alpha_0(u)\ \alpha_1(u)] \begin{bmatrix} P_{00} & P_{01} \\ P_{10} & P_{11} \end{bmatrix} \begin{bmatrix} \alpha_0(v) \\ \alpha_1(v) \end{bmatrix} \quad : \quad 0 \le u, v \le 1
\end{aligned}
\tag{5.24}
$$

where, $\alpha_0(u) = (1-u)\ ; \quad \alpha_1(u) = u,$

$\alpha_0(v) = (1-v)\ ; \quad \alpha_1(v) = v.$

The surface equation (5.24) is called a **bilinear blended coons patch.**

One may easily verify that the above surface equation satisfies the boundary conditions. The patch (5.24), which is constructed solely in terms of its boundary curves, is the simplest one among the types of surfaces originally studied by Coons (1967).

The same way the lofted surface (5.18) is constructed, we can define a *cubic blended Coons* patch if the cross-boundary tangent of each boundary curve is available. As shown in Fig. 5.8, let $\mathbf{s}_i(v)$ and $\mathbf{t}_i(u)$ denote cross-boundary tangents along the boundary curves $\mathbf{a}_i(v)$ and $\mathbf{b}_i(u)$, respectively. Then, a lofted (ie, *Hermite blended*) surface of the form (5.18) can be constructed

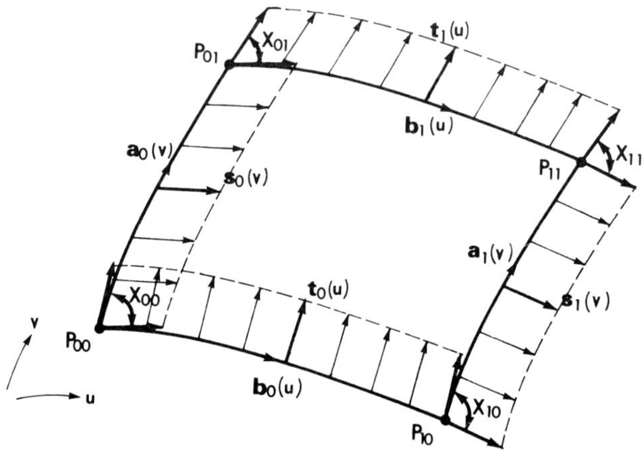

Figure 5.8 Hermite Blended Coons Patch

in terms of a pair of opposing boundary curves $\mathbf{a}_0(v), \mathbf{a}_1(v)$ and their cross-boundary tangents $\mathbf{s}_0(v), \mathbf{s}_1(v)$. That is, the *lofted surface* is expressed as follows:

$$\mathbf{r}_1(u,v) = \alpha_0(u)\mathbf{a}_0(v) + \alpha_1(u)\mathbf{a}_1(v) + \beta_0(u)\mathbf{s}_0(v) + \beta_1(u)\mathbf{s}_1(v), \qquad (5.25-a)$$

$$\begin{aligned} where, \quad \alpha_0(u) &= H_0^3(u) = (1 - 3u^2 + 2u^3), \\ \beta_0(u) &= H_1^3(u) = (u - 2u^2 + u^3), \\ \beta_1(u) &= H_2^3(u) = (-u^2 + u^3), \\ \alpha_1(u) &= H_3^3(u) = (3u^2 - 2u^3). \end{aligned}$$

Similarly, another *lofted surface* is obtained from the other pair of boundary curves $\mathbf{b}_0(u), \mathbf{b}_1(u)$ and their cross-boundary tangents $\mathbf{t}_0(u), \mathbf{t}_1(u)$:

$$\mathbf{r}_2(u,v) = \alpha_0(v)\mathbf{b}_0(u) + \alpha_1(v)\mathbf{b}_1(u) + \beta_0(v)\mathbf{t}_0(u) + \beta_1(v)\mathbf{t}_1(u), \qquad (5.25-b)$$

where $\alpha_i(v), \beta_i(v)$ are cubic Hermite blending functions given in (5.25-a).

As in the bilinear blending case, the interpolating surface $\mathbf{r}(u,v)$ can be expressed in a *Boolean sum form.* Namely,

$$\mathbf{r}(u,v) = \mathbf{r}_1(u,v) + \mathbf{r}_2(u,v) - \mathbf{r}_3(u,v), \qquad (5.26)$$

where $\mathbf{r}_3(u,v)$ is a correction surface. Let $\mathbf{x}_{ij}$ denote *twist vectors* at the patch corners such that (*for* $i,j = 0,1$)

$$\begin{aligned} \mathbf{x}_{ij} &= \partial^2\mathbf{r}(u,v)/\partial u\partial v|_{u=i,v=j} \\ &= \partial\mathbf{s}_i(v)/\partial v|_{v=j} \end{aligned} \qquad (5.27-a)$$

and

$$\begin{aligned} \mathbf{x}_{ij} &= \partial^2\mathbf{r}(u,v)/\partial v\partial u|_{u=i,v=j} \\ &= \partial\mathbf{t}_i(u)/\partial u|_{u=i}. \end{aligned} \qquad (5.27-b)$$

Then, by following the same steps for the linearly blended case, the "correction surface" $\mathbf{r}_3(u,v)$ for *Hermite blending* is obtained as:

$$\mathbf{r}_3(u,v) = [\alpha_0(u)\alpha_1(u)\beta_0(u)\beta_1(u)] \begin{bmatrix} P_{00} & P_{01} & \mathbf{t}_0(0) & \mathbf{t}_1(0) \\ P_{10} & P_{11} & \mathbf{t}_0(1) & \mathbf{t}_1(1) \\ \mathbf{s}_0(0) & \mathbf{s}_0(1) & \mathbf{x}_{00} & \mathbf{x}_{01} \\ \mathbf{s}_1(0) & \mathbf{s}_1(1) & \mathbf{x}_{10} & \mathbf{x}_{11} \end{bmatrix} \begin{bmatrix} \alpha_0(v) \\ \alpha_1(v) \\ \beta_0(v) \\ \beta_1(v) \end{bmatrix}. \qquad (5.28)$$

Finally, on substituting the results of (5.25) and (5.28) into the surface equation (5.26), the **cubic (Hermite) blended Coons patch** is obtained as

$$
\begin{aligned}
\mathbf{r}(u,v) &= \mathbf{r}_1(u,v) + \mathbf{r}_2(u,v) - \mathbf{r}_3(u,v) \\
&= [\alpha_0(u)\alpha_1(u)\beta_0(u)\beta_1(u)] \begin{bmatrix} \mathbf{a}_0(v) \\ \mathbf{a}_1(v) \\ \mathbf{s}_0(v) \\ \mathbf{s}_1(v) \end{bmatrix} + [\mathbf{b}_0(u)\mathbf{b}_1(u)\mathbf{t}_0(u)\mathbf{t}_1(u)] \begin{bmatrix} \alpha_0(v) \\ \alpha_1(v) \\ \beta_0(v) \\ \beta_1(v) \end{bmatrix} \\
&\quad - [\alpha_0(u)\alpha_1(u)\beta_0(u)\beta_1(u)] \begin{bmatrix} P_{00} & P_{01} & \mathbf{t}_0(0) & \mathbf{t}_1(0) \\ P_{10} & P_{11} & \mathbf{t}_0(1) & \mathbf{t}_1(1) \\ \mathbf{s}_0(0) & \mathbf{s}_0(1) & \mathbf{x}_{00} & \mathbf{x}_{01} \\ \mathbf{s}_1(0) & \mathbf{s}_1(1) & \mathbf{x}_{10} & \mathbf{x}_{11} \end{bmatrix} \begin{bmatrix} \alpha_0(v) \\ \alpha_1(v) \\ \beta_0(v) \\ \beta_1(v) \end{bmatrix}
\end{aligned} \tag{5.29}
$$

where, $\alpha_0(u) = (1 - 3u^2 + 2u^3)$,
$\alpha_1(u) = (3u^2 - 2u^3)$,
$\beta_0(u) = (u - 2u^2 + u^3)$,
$\beta_1(u) = (-u^2 + u^3)$,
$\mathbf{a}_i(v),\ \mathbf{b}_j(u)$: *boundary curves,*
$P_{ij},\ \mathbf{x}_{ij}$: *position and twist vectors at patch corners.*

The corner condition (5.27) is called the **twist compatibility condition**. If the twist compatibility condition does not hold, that is,

$$\partial \mathbf{s}_i(v)/\partial v|_{v=j} \neq \partial \mathbf{t}_j(u)/\partial u|_{v=i} \quad \textit{for some} \quad i,j = 0,1,$$

then the Coons patch (5.29) needs to be modified if we wish to maintain a geometric continuity across patch boundaries (when a composite surface is constructed from a mesh of curves). One such modification is called *Gregory square* as discussed in Gregory (1983).

If the boundary curves $\mathbf{a}_i(v), \mathbf{b}_j(u)$ and cross-boundary tangents $\mathbf{s}_i(v), \mathbf{t}_j(u)$ are all expressed in the form of cubic Hermite blending functions, they become Ferguson curves. In this case, each of the lofted surfaces $\mathbf{r}_1(u,v),\ \mathbf{r}_2(u,v)$ becomes identical to the correction surface $\mathbf{r}_3(u,v)$:

$$\mathbf{r}_1(u,v) = \mathbf{r}_2(u,v) = \mathbf{r}_3(u,v).$$

Under this condition, the *cubic Hermite blended* Coons patch (5.29) becomes the *tensor product* Ferguson patch (5.5).

5.3.4 Non-rectangular Gregory Patches

This subsection considers the problem of constructing a triangular surface patch specified by three boundary curves and their cross-boundary tangents. The boundary data depicted in Fig. 5.9 are:

$\mathbf{e}_i(s_i)$: i^{th} *boundary curve with* $0 \le s_i \le 1$ *for* $i = 1, 2, 3$, *and*

$\mathbf{t}_i(s_i)$: *cross-boundary tangent of* $\mathbf{e}_i(s_i)$.

We want to construct a triangular patch by combining three linear interpolants of the form (5.18), one for each boundary. Before proceeding with the surface construction scheme, an example from Gregory (1986) is reproduced here to help understand the nature of the problem.

Example 5.5: *Obtain parametric equations of the boundary curves and cross-boundary tangents of the surface given in Figure 5.9. The surface is the octant of a unit sphere centered at the origin.*

Since the circular arc, for example, on the xy-plane can be parameterized as $x = cos\theta$ *and* $y = sin\theta$, *the boundary curves are given by*

$$\mathbf{e}_1(s_1) = (0,\ cos(s_1\pi/2),\ sin(s_1\pi/2)) \ : \quad 0 \le s_1 \le 1$$

$$\mathbf{e}_2(s_2) = (sin(s_2\pi/2),\ 0,\ cos(s_2\pi/2)) \ : \quad 0 \le s_2 \le 1$$

$$\mathbf{e}_3(s_3) = (cos(s_3\pi/2),\ sin(s_3\pi/2),\ 0) \ : \quad 0 \le s_3 \le 1.$$

As the cross-boundary tangents are pointing inward parallel to the coordinate axes, they are expressed as (assuming their magnitudes are $\pi/2$*):*

$$\mathbf{t}_1 = (\pi/2,\ 0,\ 0),$$

$$\mathbf{t}_2 = (0,\ \pi/2,\ 0),$$

$$\mathbf{t}_3 = (0,\ 0,\ \pi/2). \qquad \diamond$$

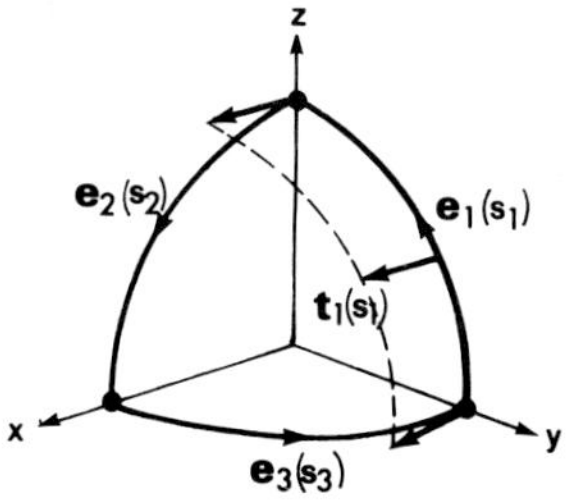

Figure 5.9 Boundary Data of a Triangular Patch

In order to construct a smooth surface patch from the boundary data of Fig. 5.9, we need to define a triangular parameter domain. Let us consider an equilateral triangle of height unity with vertices V_1, V_2, V_3. And let λ_i denote the perpendicular distance from an inside point $\mathbf{v}$ to the side facing V_i as shown in Fig. 5.10a. Then,

$$\mathbf{v} = (\lambda_1,\ \lambda_2,\ \lambda_3)$$

forms the *barycentric coordinates* of the domain triangle (See §4.4.3 for details). We use the convention that the directions of boundary curves are anti-clockwise as indicated in Fig. 5.10b. The barycentric coordinates of the vertices V_1, V_2, V_3 of the domain triangle are given by

$$V_1 = (1,\ 0,\ 0),$$
$$V_2 = (0,\ 1,\ 0), \quad \textit{and}$$
$$V_3 = (0,\ 0,\ 1)$$

and the vertices P_i of the triangular patch are given as

$$P_1 = \mathbf{e}_2(1) = \mathbf{e}_3(0),$$
$$P_2 = \mathbf{e}_3(1) = \mathbf{e}_1(0), \quad \textit{and}$$
$$P_3 = \mathbf{e}_1(1) = \mathbf{e}_2(0).$$

Further, the parameters s_i for the boundary curves can be defined in terms of λ_i as follows:

$$s_1 = \lambda_3/(\lambda_2 + \lambda_3)\ ; \quad s_2 = \lambda_1/(\lambda_3 + \lambda_1)\ ; \quad s_3 = \lambda_2/(\lambda_1 + \lambda_2) \tag{5.30}$$

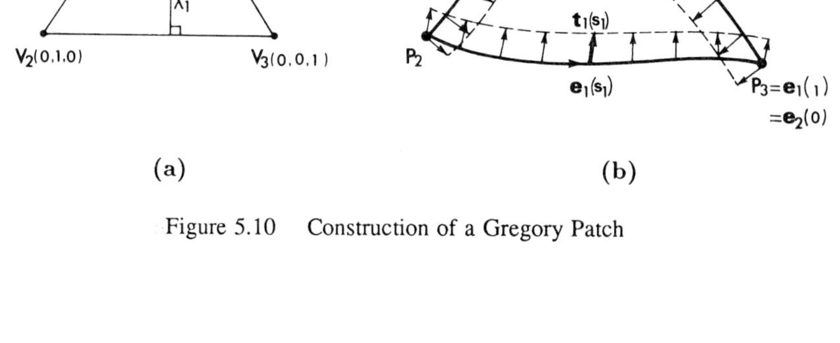

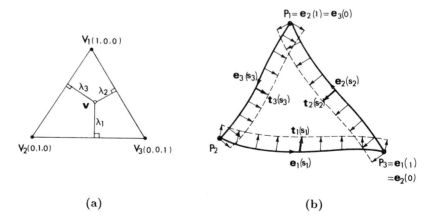

Figure 5.10 Construction of a Gregory Patch

Then we can define a **linear Taylor interpolant** $\mathbf{r}_i(s_i, \lambda_i)$ in terms of the boundary curve $\mathbf{e}_i(s_i)$ and its cross-boundary tangent $\mathbf{t}_i(s_i)$:

$$\mathbf{r}_i(s_i, \lambda_i) = \mathbf{e}_i(s_i) + \lambda_i \mathbf{t}_i(s_i) \quad for \quad i = 1, 2, 3 \tag{5.31}$$

where, $s_i \in [0,1]$: *parameter for boundary i as in* (5.30),

$\{\lambda_i\}$: *barycentric coordinates,*

$\mathbf{e}_i(s_i)$: *boundary curve i,*

$\mathbf{t}_i(s_i)$: *cross-boundary tangent for boundary i.*

Finally, a triangular surface patch filling in the inside of the three boundary curves of Fig. 5.10b is constructed as a *convex combination* of the three surfaces in (5.31). That is, a **triangular Gregory patch** is expressed as follows:

$$\mathbf{r}(\mathbf{v}) = \sum_{i=1}^{3} \gamma_i(\mathbf{v})\{\mathbf{e}_i(s_i) + \lambda_i \mathbf{t}_i(s_i)\} \tag{5.32}$$

where, $\mathbf{v} = (\lambda_1, \lambda_2, \lambda_2)$: *barycentric coordinates,*

s_i : *curve parameter as defined in*(5.30),

$$\gamma_i(\mathbf{v}) = \begin{cases} (1/\lambda_i)^2 / \sum_j (1/\lambda_j)^2 & if \quad \lambda_i \neq 0 \\ 1 \quad if \quad \lambda_i = 0. \end{cases}$$

The surface construction method of (5.32) is called **convex combination** because the resulting surface interpolates over the convex region formed by the three boundary curves. This method can easily be generalized to constructing an n-sided surface patch. For details, see Gregory (1986). The surface patch constructed in this way is called *Gregory patch* even though the convex combination scheme is due to Brown and Little (see Barnhill 1977).

It is also possible to constructed an n-sided patch based on the method of constructing the Coons patch (5.29) which is also known as *Boolean sum.* That is, a surface patch $\mathbf{r}(u, v)$ is expressed as a Boolean sum of participating surfaces:

$$\mathbf{r}(u,v) = \mathbf{r}_1(u,v) \oplus \mathbf{r}_2(u,v) = \mathbf{r}_1(u,v) + \mathbf{r}_2(u,v) - \mathbf{r}_3(u,v)$$

where $\mathbf{r}_3(u, v)$ is the "intersection" of $\mathbf{r}_1(u, v)$ and $\mathbf{r}_2(u, v)$. The surface construction method based on Boolean sum is due to Gregory (1983) and more details may be found in Gregory (1986).

5.4 SWEEP SURFACE PATCHES

This section presents methods of defining surface patches in terms of *cross-section curves* and *profile* curves. Here a surface patch is treated as a *swept trajectory* of one or more sectional curves. In engineering drawings, it is a common practice to describe surfaces in terms of *cross-section curves* (together with profile curves). As a matter of convenience, sweep surface patches are grouped into translational sweep patches, rotational sweep patches, and non-parametric sweep patches. The sweep surface patch models introduced in this section will be used in constructing general sweep-type surfaces later in Chapters 9 and 13.

5.4.1 Translational Sweep Surface Patches

Shown in Fig. 5.11 are two parametric space curves, $\mathbf{g}(u)$ and $\mathbf{d}(v)$. If the two space curves are regarded as rigid wires, then we can imagine a surface defined by the trajectory of the curve $\mathbf{g}(u)$ swept along the second curve $\mathbf{d}(v)$. The moving curve $\mathbf{g}(u)$ is called a **generator curve** and the guiding curve $\mathbf{d}(v)$ is called a **director curve**. The resulting sweep surface patch $\mathbf{r}(u, v)$ is then expressed as:

$$\mathbf{r}(u, v) = \mathbf{g}(u) + \mathbf{d}(v) - \mathbf{d}(0) \quad : \quad 0 \le u, v \le 1, \tag{5.33}$$

where $\mathbf{d}(0)$ denotes the starting point of director curve. The above surface is called a **translational sweep surface.**

A straightforward extension of the sweeping idea is the sweeping of a parametric curve defined by control vertices, such as Bezier curves and B-spline curves. In case of the cubic Bezier curve (5.1-c), we may move the control vertices V_0, V_1, V_2, V_3 along four director curves $\mathbf{d}_0(v)$, $\mathbf{d}_1(v)$, $\mathbf{d}_2(v)$, and $\mathbf{d}_3(v)$. Then, the resulting sweep surface $\mathbf{r}(u, v)$ is expressed as:

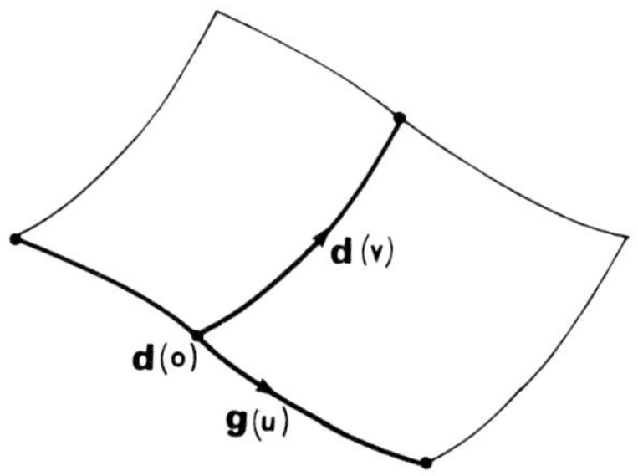

Figure 5.11 Sweep Surface Patch

$$\mathbf{r}(u,v) = \mathbf{U}\ \mathbf{M}\ \mathbf{R}(v) = \sum_{i=0}^{3} B_i^3(u)\ \mathbf{d}_i(v). \tag{5.34}$$

When the generator curve is a conic section curve and the director curves are cubic spline curve, the resulting sweep surface is sometimes called a *polyconic surface.* A variation of the polyconic surface is used in constructing "edge blend" surfaces in Chapter 11.

Example 5.6: *Construct a sweep surface (plane) containing three points P_0, P_1, P_2.*

We first define two straight lines P_0P_1 and P_0P_2, and treat them as generator and director curves. That is,

$$\mathbf{g}(u) = (1-u)P_0 + u\ P_1\ ;\ \mathbf{d}(v) = (1-v)P_0 + P_2.$$

Then, from (5.33), *the sweep surface is expressed as*

$$\mathbf{r}(u,v) = \mathbf{g}(u) + \mathbf{d}(v) - \mathbf{d}(0) = P_0 + (P_1 - P_0)u + (P_2 - P_0)v$$

which is identical to the plane equation given in Example 5.1. ◇

5.4.2 Rotational Sweep Surface Patches

Another simple yet widely used surface type is *rotational sweep surface*, also known as **surface of revolution**. Let us consider a section curve $\mathbf{s}(u)$ on the xz-plane as shown in Fig. 5.12a:

$$\mathbf{s}(u) = d(u)\mathbf{i} + z(u)\mathbf{k} = (d(u),\ 0,\ z(u)), \tag{5.35}$$

where $\mathbf{i} = (1,\ 0,\ 0)$ and $\mathbf{k} = (0,\ 0,\ 1)$. If we rotate the section curve (5.35) about the z-axis, the resulting sweep surface shown in Fig. 5.12b is expressed as a parameteric equation. That is, we have

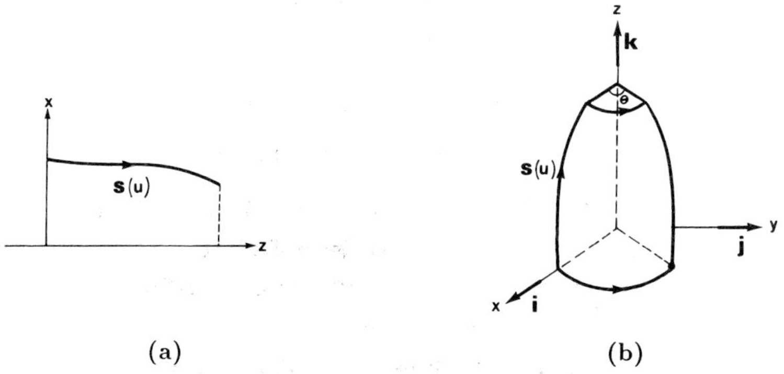

Figure 5.12 Surface of Revolution

$$\mathbf{r}(u,\theta) = (d(u)\cos\theta,\ d(u)\sin\theta,\ z(u)) \\ = d(u)\cos\theta\ \mathbf{i} + d(u)\sin\theta\ \mathbf{j} + z(u)\ \mathbf{k}, \tag{5.36}$$

where $d(u)$, $z(u)$ are section curves as given in (5.35).

Example 5.7: *Give an equation of a torus, with minor radius of* 1 *and major radius of* 4*, located at the origin as shown in Figure* 5.13.

Since the cross section curve of the torus on the xz-plane is a unit circle centered at x=4, z=0, its implicit equation is $(x-4)^2 + z^2 = 1$ and it has the parametric equation $\mathbf{s}(\alpha) = (4+\cos\alpha)\ \mathbf{i} + (\sin\alpha)\ \mathbf{k}$ with $0 \le \alpha \le 2\pi$. From (5.36)*, the torus surface is expressed as*

$$\mathbf{r}(\alpha,\beta) = (4+\cos\alpha)\cos\beta\ \mathbf{i} + (4+\cos\alpha)\sin\beta\ \mathbf{j} + \sin\alpha\ \mathbf{k} \\ = ((4+\cos\alpha)\cos\beta,\ (4+\cos\alpha)\cos\beta,\ \sin\alpha) : \ 0 \le \alpha,\beta \le 2\pi$$

◇

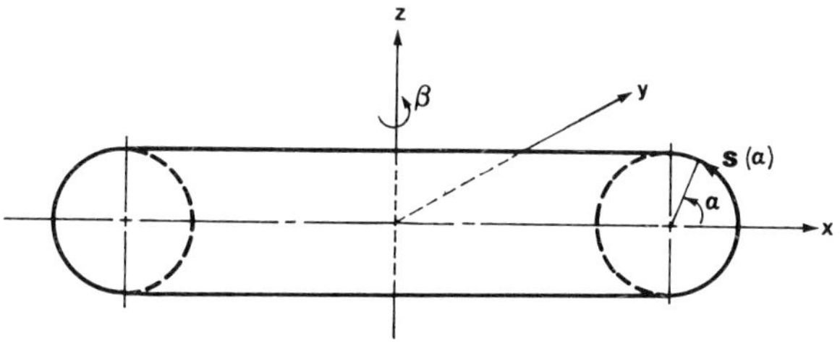

Figure 5.13 Torus as a Surface of Revolution

5.4.3 Non-parametric Sweep Surface

As discussed in Section 2.4.1, a parametric surface $\mathbf{r}(u,v)$ degenerates to a *nonparametric* surface when $x(u,v) \equiv u$ and $y(u,v) \equiv v$. That is,

$$\mathbf{r}(u,v) = (x(u,v),\ y(u,v),\ z(u,v)) \\ = (u, v, z(u,v)) \equiv (x, y, z(x,y)) \tag{5.37}$$

In effect, (5.37) is equivalent to an *explicit* equation $z = z(x,y)$. In this subsection, we present a method of constructing a *translational sweep surface* of the type $z = z(x,y)$. Depicted in Fig. 5.14a and b are *generator* and *director* curves, respectively, defined on xz- and yz-planes:

$$z = g(x)\ :\ x \in [x_0,\ x_1]\ ; \quad z = d(y)\ :\ y \in [y_0,\ y_1].$$

When the generator curve $z = g(x)$ is swept along the director curve $z = d(y)$, a *nonparametric sweep surface* will be obtained as depicted in Fig. 5.14c. From the definition of a translational

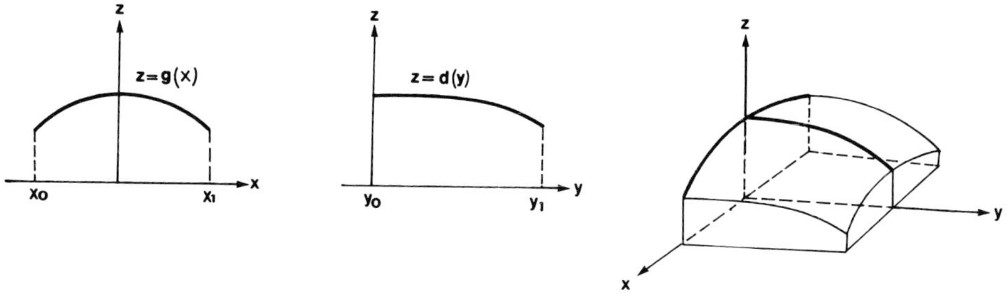

(a) (b) (c)

Figure 5.14 Non-parametric Sweep Surface Patch

sweep surface (5.33), the sweep surface is expressed as

$$z(x,y) = g(x) + d(y) - d(0) \ : \ x_0 \le x \le x_1 \ ; \ y_0 \le y \le y_1, \tag{5.38}$$

where $g(x)$ is a generator curve on xz-plane and $d(y)$ a director curve on yz-plane. Since the surface equation (5.38) can be rewritten as an implicit equation

$$f(x,y,z) \equiv -g(x) - d(y) + z + d(0) = 0$$

the surface normal N of the sweep surface (5.38) is easily obtained from Eqn. (2.25) in §2.4.2. That is,

$$N(x,y) = (-\partial z/\partial x, \ -\partial z/\partial y, \ 1) = (-\dot{g}(x), \ -\dot{d}(y), \ 1) \tag{5.39}$$

which may be normalized to obtained a unit normal vector.

Example 5.8: *Verify the surface normal expression* (5.39) *by using the parametric form given by* (5.37) *for the nonparametric sweep surface* (5.38).

The nonparametric sweep surface (5.38) *is expressed in parametric form, noting that* $x \equiv u$ *and* $y \equiv v$, *as*

$$\mathbf{r}(u,v) = (u, \ v, \ z(u,v)) = (u, \ v, \ g(u) + d(v) - d(0)).$$

Since the partial derivatives of $\mathbf{r}(u,v)$ *are expressed as*

$$\mathbf{r}_u = \partial \mathbf{r}/\partial u = (1, \ 0, \ \partial g(u)/\partial u) \ ; \ \mathbf{r}_v = \partial \mathbf{r}/\partial v = (0, \ 1, \ \partial d(v)/\partial v)$$

the surface nomral is obtained as

$$N(u,v) = \mathbf{r}_u \times \mathbf{r}_v = (1, \ 0, \ \partial g(u)/\partial u) \times (0, \ 1, \ \partial d(v)/\partial v)$$
$$= (-\partial g(u)/\partial u, \ -\partial d(v)/\partial v, \ 1) \equiv (-\dot{g}(u), \ -\dot{d}(v), \ 1).$$

◇

5.5 ANALYTIC SURFACE PATCHES

The term *analytic surface* is used to mean an implicit surface of the type $g(x, y, z) = 0$ in which the analytic function $g(*)$ is usually a polynomial in the coordinate variables x, y, z. If the degree of the polynomial $g(x, y, z)$ is 2, it is called a **conicoid** or a **quadric surface.** If the degree is 3, it is called a **cubicoid.** In practice, only quadric surfaces are usually used in representing physical shapes.

5.5.1 Quadric Surfaces

In general, an implicit quadratic polynomial function represents a quadric surface in 3D space. That is,

$$g(x, y, z) = \sum_{i=0}^{2} \sum_{j=0}^{2} \sum_{k=0}^{2} c_{ijk}\, x^i\, y^j\, z^k = 0, \tag{5.40}$$

where c_{ijk} are scalar coefficients. Since there are as many as 27 terms in (5.40), it does not give much geometric meaning in this form. In practice, the *standard quadric surfaces* shown in Fig. 5.15 are used as surface primitives, and general quadric surface are defined by transforming the standard ones. The standard quadric surfaces in Fig. 5.15 are *ellipsoid, hyperboloid of one sheet, hyperboloid of two sheets, elliptic paraboloid, hyperbolic paraboloid,* and *quadric cone.* Their implicit equations are:

$$(x/a)^2 + (y/b)^2 + (z/c)^2 = 1 \quad (\textit{ellipsoid}) \tag{5.41 - a}$$

$$(x/a)^2 + (y/b)^2 - (z/c)^2 = 1 \quad (\textit{hyperboloid of one sheet}) \tag{5.41 - b}$$

$$(x/a)^2 - (y/b)^2 - (z/c)^2 = 1 \quad (\textit{hyperboloid of two sheets}) \tag{5.41 - c}$$

$$(x/a)^2 + (y/b)^2 - z = 0 \quad (\textit{elliptic paraboloid}) \tag{5.41 - d}$$

$$(x/a)^2 - (y/b)^2 - z = 0 \quad (\textit{hyperbolic paraboloid}) \tag{5.41 - e}$$

$$(x/a)^2 + (y/b)^2 - (z/c)^2 = 0 \quad (\textit{quadric cone}). \tag{5.41 - f}$$

The geometric meaning of the "scaling constants" a, b, c in (5.41) may easily be identified by substituting, for example, $y = z = 0$ in each of the equations in (5.41) in order to see the effect of "a". If we want to define the quadric surfaces in a general 3D position, the standard equations are transformed by using a suitable (homogeneous) transformation matrix as will be discussed in Chapter 13.

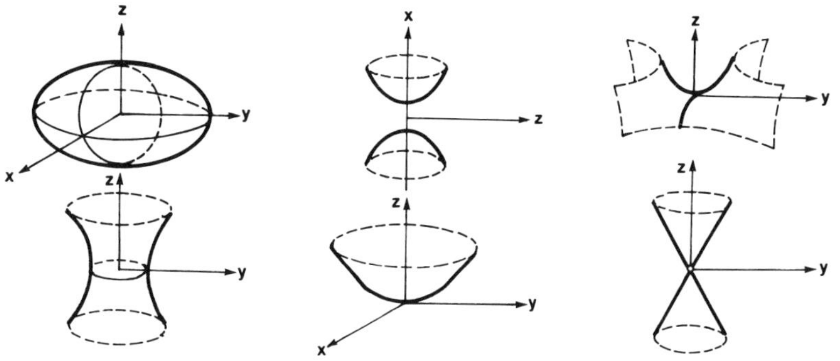

Figure 5.15 Standard Quadric Surfaces

5.5.2 Parametrization of Quadric Surfaces

The implicit functions in (5.41) are not a surface equation (in a differential geometry sense). They simply represent a boundry between two disjoint half spaces as discussed in Section 2.4. In most CAD/CAM applications, it is required to have parametric representations of the quadric surfaces. Methods of parametrizing quadric surfaces are briefly reviewed.

Let $\mathbf{r}(u,v) = (x(u,v),\ y(u,v),\ z(u,v))$ denote a parametric representation of a surface, then the standard quadric surfaces in Fig. 5.15 are expressed in parametric forms as:

$$\mathbf{r}(\alpha,\beta) = (a\ \cos\alpha\ \cos\beta,\ b\ \cos\alpha\ \sin\beta,\ c\ \sin\alpha)\ :\ \textit{ellipsoid} \qquad (5.42-a)$$

$$\mathbf{r}(\alpha,\beta) = (a\ \cos\beta/\cos\alpha,\ b\ \sin\beta/\cos\alpha,\ c\ \tan\alpha)\ :\ \textit{hyperboloid-1} \qquad (5.42-b)$$

$$\mathbf{r}(\alpha,\beta) = (a/\cos\alpha,\ b\ \tan\alpha\ \cos\beta,\ c\ \tan\alpha\ \sin\beta)\ :\ \textit{hyperboloid-2} \qquad (5.42-c)$$

$$\mathbf{r}(u,v) = (au,\ bv,\ u^2+v^2)\ :\ \textit{elliptic paraboloid} \qquad (5.42-d)$$

$$\mathbf{r}(u,v) = (au,\ bv,\ u^2-v^2)\ :\ \textit{hyperbolic paraboloid} \qquad (5.42-e)$$

$$\mathbf{r}(u,\beta) = (au\ \cos\beta,\ bu\ \sin\beta,\ cu)\ :\ \textit{elliptic cone} \qquad (5.42-f)$$

$where, -\pi/2 \le \alpha \le \pi/2\ ;\quad 0 \le \beta \le 2\pi\ ;\quad u, v\ :\ real.$

Parametrizations are not unique. A good parametrization is the one which gives more uniform *flow rates*. The trigonometric parametric forms in (5.42) may easily be converted to rational parametric forms through suitable substitutions. The following are homogeneous representations of the standard quadric surfaces (with the *scaling constants* set to unity: $a = b = c = 1$) as given in Mudur (1986):

(a) Ellipsoid (sphere):

$$\mathbf{R}(u,v) = ((1-u^2)(1-v^2),\ 2u(1-v^2),\ 2v(1+u^2),\ (1+u^2)(1+v^2)) \qquad (5.43-a)$$

(b) Hyperboloid of one sheet:

$$\mathbf{R}(u,v) = ((1-u^2)(4+v^2),\ 2u(4+v^2),\ 4v(1+u^2),\ (1+u^2)(4-v^2)) \qquad (5.43-b)$$

(c) Hyperboloid of two sheets:

$$\mathbf{R}(u,v) = ((4+u^2)(4+v^2),\ 4u(4+v^2),\ 4v(4-u^2),\ (4-u^2)(4-v^2)) \qquad (5.43-c)$$

(d) Elliptic paraboloid:

$$\mathbf{R}(u,v) = (u,\ v,\ u^2+v^2,\ 1) \qquad (5.43-d)$$

(e) Hyperbolic paraboloid:

$$\mathbf{R}(u,v) = (u,\ v,\ u^2-v^2,\ 1) \qquad (5.43-e)$$

(f) Elliptic cone:

$$\mathbf{R}(u,v) = (v(1-u^2),\ 2uv,\ v(1+u^2),\ (1+u^2)) \qquad (5.43-f)$$

Example 5.9: *Provide a rough sketch of the ellipsoid patch* $(5.43-a)$ *when the parameter domain is a unit square, that is,* $0 \le u, v \le 1$.

The ellipsoid equation $(5.43-a)$ *is expressed in a rational form as:*

$$\mathbf{r}(u,v) = \left(\frac{(1-u^2)(1-v^2)}{(1+u^2)(1+v^2)},\ \frac{2u(1-v^2)}{(1+u^2)(1+v^2)},\ \frac{2v(1+u^2)}{(1+u^2)(1+v^2)}\right).$$

Thus, patch corners are evaluated as

$$\mathbf{r}(0,0) = (1,0,0)\ ; \quad \mathbf{r}(0,1) = (0,1,0)\ ; \quad \mathbf{r}(1,0) = \mathbf{r}(1,1) = (0,0,1),$$

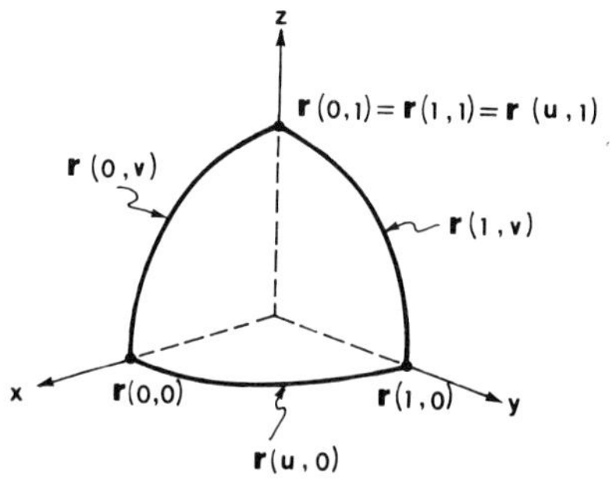

Figure 5.16 An Ellipsoid Patch

and, patch boundary curves are expressed as

$$\mathbf{r}(0,v) = ((1-v^2)/(1+v^2),\ 0,\ 2v/(1+v^2))\ ;$$
$$\mathbf{r}(1,v) = (0,(1-v^2)/(1+v^2),2v/(1+v^2))\ ;$$
$$\mathbf{r}(u,0) = ((1-u^2)/(1+u^2),\ 2u/(1+u^2),\ 0)\ ;$$
$$\mathbf{r}(u,1) = (0,\ 0,\ 1).$$

The ellipsoid patch is an octant of a unit sphere as sketched in Figure 5.16. *It is a degenerate triangular patch having a rectangular topology.* ◇

5.6 DISCUSSIONS

Four types of surface patch models have been introduced in this chapter. They are:

- point data interpolating patch models,
- curve data interpolating patch models,
- section-curve sweeping patch models, and
- quadric surface patch models.

For the construction of rectangular *interpolating* surface patches, cubic *blending functions* are employed. The blending functions employed are

- Hermite blending functions for Ferguson and Coons patches,
- Bernstein functions for Bezier patch, and
- B-spline functions for B-spline patches.

Surface patch models are represented in *power basis form* (ie, matrix form) where possible. Matrix representation is in general computationally more efficient than other representations. For example, when 100 points are to be evaluated from a bicubic NURB (non-uniform rational B-spline) patch, the matrix form would take about 9% of the execution time for evaluating the knot insertion algorithm (Choi *et al,* 1990b). In case of Bezier curves, however, the matrix form is found to be *numerically less stable* than the Bernstein polynomial form (Farin, 1988, p.302).

Among the (non-rational) rectangular patch models introduced in §5.2, NUB (non-uniform B-spline) model is the most general one in the sense that all the rest are special cases of NUB. On the other hand, Bezier model is the most adaptable one because all the rest can be converted to this type. For example, B-spline surfaces can be converted to Bezier surfaces by *inserting knots* (Boehm, 1981).

Perhaps, sweep surfaces are the most widely found surface types in conventional engineering drawings: Examples include ducts, ship hulls, fan blades, and injection-molded parts. *Sweep* surface patches are defined as coordinate transformations. This may be the main reason why sweep surfaces are the major surface type in most of commercial CAM systems. More details may be found in Chapter 14.

For the construction of non-rectangular surface patches, the concepts of

- *barycentric coordinates,*
- *convex combination, and*
- *Boolean sum*

are utilized. Quadric surfaces (except paraboloids) are parameterized in rational quadratic forms. The surface patch models introduced in this chapter will be used as building blocks in constructing more general surface shapes in later chapters.

CHAPTER 6

PROPERTIES OF BEZIER CURVES AND SURFACES

6.1 INTRODUCTION

Bezier surface model is probably the most versatile tool in surface modeling. In this chapter, we review some of the important characteristics of Bezier curves and rectangular Bezier surfaces. The topics to be discussed are

- *derivatives (tangent and normal vectors),*
- *subdivision (de Casteljau algorithm),*
- *degree elevation,*
- *basic properties, and*
- *conversion from B-spline to Bezier.*

Properties of *triangular Bezier patches* are to be discussed in Chapter 8. A Bezier curve can be expressed in a *Bezier form* or in a *matrix form.* The **Bezier form of a Bezier curve** is expressed as (*See* §3.3.3)

$$\mathbf{r}(u) = \sum_{i=0}^{n} B_i^n(u)\ V_i\ ;\ 0 \le u \le 1 \tag{6.1}$$

where $\{V_i : i = 0, ..., n\}$ are called *Bezier control vertices (or simply Bezier points)*, and n is the *degree* of the Bezier curve. The **Bernstein polynomial** in (6.1) is given by

$$B_i^n(u) = \frac{n!}{(n-i)!\ i!} u^i (1-u)^{n-i}. \tag{6.2}$$

The **matrix form of Bezier curve** is given by (Chang, 1982)

$$\mathbf{r}(u) = \mathbf{U\ M\ R} \quad ; \quad 0 \le u \le 1 \tag{6.3}$$

$$\textit{where,} \quad \mathbf{U} = [1\ u\ u^2 ... u^n], \quad \mathbf{R} = [V_0\ V_1\ V_2 ... V_n],$$

$$\mathbf{M} = \{m_{ij}\} : (n+1) \times (n+1) \quad \textit{coeff. matrix,}$$

$$m_{ij} = (-1)^{i-j} \binom{n}{i} \binom{i}{j}.$$

Sometimes, a "degree n" curve is denoted as $\mathbf{r}^n(u)$. A Bezier surface patch of degrees m, n is also expressed in *Bezier form* or in *matrix form.* The **Bezier form of a Bezier patch** of degrees m, n is given by (See §5.2.3)

$$\mathbf{r}^{m,n}(u, v) = \sum_{i=0}^{m} \sum_{j=0}^{n} B_i^m(u) B_j^n(v) V_{ij} \tag{6.4}$$

6.2 DERIVATIVES OF BEZIER CURVES AND SURFACES

The first derivative of the Bernstein polynomial (6.2) is evaluated as

$$\begin{aligned}\frac{d}{du}B_i^n(u) &= \frac{d}{du}\left(\frac{n!}{i!(n-i)!}u^i(1-u)^{n-i}\right)\\ &= \frac{i\,n!}{i!(n-i)!}u^{i-1}(1-u)^{n-i} - \frac{(n-i)n!}{i!(n-i)!}u^i(1-u)^{n-i-1} \qquad (6.5)\\ &= n\left(B_{i-1}^{n-1}(u) - B_i^{n-1}(u)\right).\end{aligned}$$

Thus, the **derivative of a Bezier curve** of degree n is determined as follows:

$$\begin{aligned}\frac{d}{du}\mathbf{r}^n(u) &= \frac{d}{du}\left(\sum_{i=0}^{n} B_i^n(u)V_i\right)\\ &= \sum_{i=0}^{n} n\left(B_{i-1}^{n-1}(u) - B_i^{n-1}(u)\right)V_i\\ &= n\sum_{i=1}^{n} V_i B_{i-1}^{n-1}(u) - n\sum_{i=0}^{n-1} V_i B_i^{n-1}(u) \quad ; \quad (B_{-1}^{n-1} = B_n^{n-1} = 0) \qquad (6.6)\\ &= n\sum_{j=0}^{n-1} V_{j+1} B_j^{n-1}(u) - n\sum_{i=0}^{n-1} V_i B_i^{n-1}(u) \quad ; \quad (j = i-1)\\ &= n\sum_{i=0}^{n-1}(V_{i+1} - V_i)B_i^{n-1}(u).\end{aligned}$$

The derivative (6.6) is also in a *Bezier form.* However, it is not a 3D curve because its *control points* $D_i = (V_{i+1} - V_i)$ are not 3D points (They are *vectors*). When the *control vectors* D_i are regarded as vectors starting from the origin 0, the derivative curve (6.6) is sometimes called a **hodograph.** The **second derivative of a Bezier curve** is also easily evaluated as

$$\begin{aligned}\frac{d^2}{du^2}\mathbf{r}^n(u) &= \frac{d}{du}\left(\frac{d}{du}\mathbf{r}^n(u)\right)\\ &= n\sum_{i=0}^{n-1}(V_{i+1} - V_i)\frac{d}{du}B_i^{n-1}(u) \qquad (6.7)\\ &= n(n-1)\sum_{i=0}^{n-2}(V_{i+2} - 2V_{i+1} + V_i)B_i^{n-2}(u).\end{aligned}$$

A partial derivative of a Bezier patch is evaluated as *(in u-direction)*

$$\begin{aligned}\frac{\partial}{\partial u}\mathbf{r}^{m,n}(u,v) &= \frac{\partial}{\partial u}\left(\sum_{i=0}^{m}\sum_{j=0}^{n} B_i^m(u)B_j^n(v)V_{ij}\right)\\ &= \sum_{j=0}^{n}\frac{\partial}{\partial u}\left(\sum_{i=0}^{m} B_i^m(u)V_{ij}\right)B_j^n(v)\\ &= \sum_{j=0}^{n}\left(m\sum_{i=0}^{m-1}(V_{i+1,j}-V_{i,j})B_i^{m-1}(u)\right)B_j^n(v)\\ &= m\sum_{i=0}^{m-1}\sum_{j=0}^{n}(V_{i+1,j}-V_{i,j})B_i^{m-1}(u)B_j^n(v)\end{aligned} \qquad (6.8-a)$$

The v-direction partial derivative is also expressed similarly as

$$\frac{\partial}{\partial v}\mathbf{r}^{m,n}(u,v) = n\sum_{i=0}^{m}\sum_{j=0}^{n-1}(V_{i,j+1}-V_{i,j})B_i^m(u)B_j^{n-1}(v) \qquad (6.8-b)$$

If the partial derivative (6.8-a) is evaluated at $u = 0$, it is sometimes called a **cross-boundary tangent** (along $u = 0$ boundary curve) which is given by

$$\frac{\partial}{\partial u}\mathbf{r}^{m,n}(0,v) = \frac{\partial}{\partial u}\mathbf{r}^{m,n}(u,v)|_{u=0} = m\sum_{j=0}^{n}(V_{1,j}-V_{0,j})B_j^n(v) \qquad (6.9)$$

The $u = 0$ boundary curve and its *control vector* $(V_{1,j} - V_{0,j})$ for a bicubic Bezier patch is shown in Fig. 6.1.

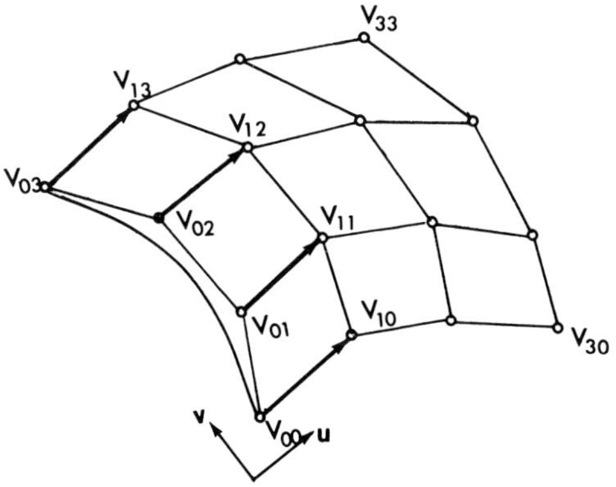

Figure 6.1 Control Vectors for a Cross-boundary Curve

The mixed partial derivative of a surface is called a *twist vector*. The *twist vector of a Bezier patch* is obtained as

$$\begin{aligned}\frac{\partial^2}{\partial u \partial v}\mathbf{r}^{m,n}(u,v) &= \frac{\partial}{\partial v}\left(\frac{\partial}{\partial u}\mathbf{r}^{m,n}(u,v)\right) \\ &= \frac{\partial}{\partial v}\left(m\sum_{i=0}^{m-1}\sum_{j=0}^{n}(V_{i+1,j}-V_{i,j})B_i^{m-1}(u)B_j^n(v)\right) \\ &= mn\sum_{i=0}^{m-1}\sum_{j=0}^{n-1}\Big((V_{i+1,j+1}-V_{i+1,j})-(V_{i,j+1}-V_{i,j})\Big)B_i^{m-1}(u)B_j^{n-1}(v)\end{aligned} \tag{6.10}$$

The twist vector at a patch corner is called a **corner twist vector** which is given by, at $u = v = 0$ for example,

$$\frac{\partial^2}{\partial u \partial v}\mathbf{r}^{m,n}(u,v)|_{u=v=0} = mn\{(V_{1,1}-V_{1,0})-(V_{0,1}-V_{0,0})\} \tag{6.11}$$

6.3 DE CASTELJAU ALGORITHM AND SUBDIVISION

Let $\{V_i : i = 0, 1, ..., n\}$ denote 3D *control vertices* defining a Bezier curve of degree n. Define the following recursive function

$$\mathbf{b}_i^r(u) = (1-u)\mathbf{b}_i^{r-1}(u) + u\mathbf{b}_{i+1}^{r-1}(u) \quad for \quad r = 1, ..., n\ ;\ i = 0, ..., n-r \tag{6.12}$$

with $\mathbf{b}_i^0(u) \equiv V_i$. Then it can be shown that $\mathbf{b}_0^n(u)$ in (6.12) represents the point on the Bezier curve (6.1) with parameter value u. That is,

$$\mathbf{b}_0^n(u) = \sum_{i=0}^{n} B_i^n(u)V_i \equiv \mathbf{r}^n(u) \tag{6.13}$$

The relation given by (6.12) is known as the **de Casteljau Algorithm**. Using this scheme, the **de Casteljau points** $\mathbf{b}_i^r$ are systematically obtained as follows (from left to right):

$$V_0 \equiv \mathbf{b}_0^0$$

$$V_1 \equiv \mathbf{b}_1^0 \to \mathbf{b}_0^1$$

$$V_2 \equiv \mathbf{b}_2^0 \to \mathbf{b}_1^1 \to \mathbf{b}_0^2$$

....

$$V_3 \equiv \mathbf{b}_n^0 \to \mathbf{b}_{n-1}^1 \to \mathbf{b}_{n-2}^2 \to \mathbf{b}_{n-3}^3 \cdots \to \mathbf{b}_0^n$$

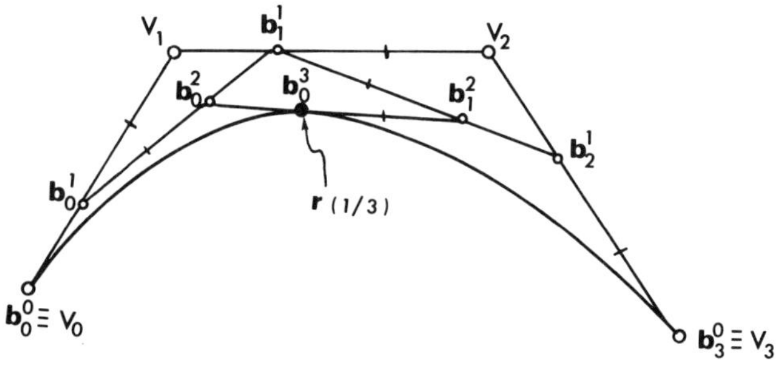

Figure 6.2 The de Casteljau Algorithm and Subdivision

An application of the de Casteljau algorithm to a cubic Bezier curve with $u = 1/3$ is shown in Fig. 6.2, in which the *de Casteljau points* are computed as

$$\begin{aligned}
\mathbf{b}_0^1 &= \frac{2}{3}\mathbf{b}_0^0 + \frac{1}{3}\mathbf{b}_1^0 \equiv \frac{2}{3}V_0 + \frac{1}{3}V_1 \\
\mathbf{b}_1^1 &= \frac{2}{3}\mathbf{b}_1^0 + \frac{1}{3}\mathbf{b}_2^0 \equiv \frac{2}{3}V_1 + \frac{1}{3}V_2 \\
\mathbf{b}_2^1 &= \frac{2}{3}\mathbf{b}_2^0 + \frac{1}{3}\mathbf{b}_3^0 \equiv \frac{2}{3}V_2 + \frac{1}{3}V_3 \\
\mathbf{b}_0^2 &= \frac{2}{3}\mathbf{b}_0^1 + \frac{1}{3}\mathbf{b}_1^1 \\
\mathbf{b}_1^2 &= \frac{2}{3}\mathbf{b}_1^1 + \frac{1}{3}\mathbf{b}_2^1 \\
\mathbf{b}_0^3 &= \frac{2}{3}\mathbf{b}_0^2 + \frac{1}{3}\mathbf{b}_1^2 = \mathbf{r}(1/3).
\end{aligned}$$

From the figure, one may observe that the tangent of the curve at $\mathbf{r}(1/3)$ is defined by the two de Casteljau points $\mathbf{b}_0^2, \mathbf{b}_1^2$. In fact, we have

$$\dot{\mathbf{r}}(1/3) = 3(\mathbf{b}_1^2 - \mathbf{b}_0^2).$$

In general, the following holds for a Bezier curve of degree n:

$$\frac{d}{du}\mathbf{r}(u) = n(\mathbf{b}_1^{n-1}(u) - \mathbf{b}_0^{n-1}(u)), \tag{6.14}$$

where $\mathbf{b}_i^{n-1}$ are de Casteljau points computed from the de Casteljau algorithm (6.12) for a degree n Bezier curve.

Let $\mathbf{r}^a(t)$ denote the first curve segment of $\mathbf{r}(u)$ in Fig. 6.2 corresponding to the parameter range $u \in [0, 1/3]$. Then it can be shown that $\mathbf{r}^a(t)$ for $t \in [0, 1]$ is also a cubic Bezier curve with control vertices

$$\{\mathbf{b}_0^0,\ \mathbf{b}_0^1,\ \mathbf{b}_0^2,\ \mathbf{b}_0^3\}.$$

Similarly, the second curve segment $\mathbf{r}^b(t)$ corresponding to $u \in [1/3, 1]$ is also a cubic Bezier curve with control vertices

$$\{\mathbf{b}_0^3,\ \mathbf{b}_1^2,\ \mathbf{b}_2^1,\ \mathbf{b}_3^0\}.$$

In other words, the cubic Bezier curve $\mathbf{r}(u)$ in Fig. 6.2 is *subdivided* into two cubic Bezier curves $\mathbf{r}^a(t)$, $\mathbf{r}^b(t)$ at the final de Casteljau point $\mathbf{b}_0^3(1/3)$.

The **subdivision of a Bezier curve** is achieved as a by-product of the de Casteljau algorithm. In general, the "diagonal" de Casteljau points $\{\mathbf{b}_0^i\}$ (See the iterative scheme right after Eqn. (6.13)) form the *Bezier polygon* of the first (subdivided) Bezier curve, and the second Bezier curve is defined by the de Casteljau points $\{\mathbf{b}_{n-1}^i\}$ at the bottom "row". That is, let

$\{V_i^a : i = 0, ..., n\}$ be Bezier points of the first segment $\mathbf{r}^a(t)$, and
$\{V_i^b : i = 0, ..., n\}$ be Bezier points of the second segment $\mathbf{r}^b(t)$,

then they are obtained from the de Casteljau points as

$$V_i^a = \mathbf{b}_0^i(u) \quad for\ i = 0, 1, ..., n, \tag{6.15 – a}$$

$$V_i^b = \mathbf{b}_i^{n-i}(u) \quad for\ i = 0, 1, ..., n. \tag{6.15 – b}$$

The **subdivision of a Bezier patch** may be carried out in two steps:

1) subdivide each "row" of Bezier control points array $\{V_{ij}\}$, and then
2) subdivide each "column" of the (subdivided) Bezier control points array.

6.4 DEGREE ELEVATION AND REDUCTION

Let $\mathbf{r}^n(u)$ denote a Bezier curve of degree n defined by $(n+1)$ control vertices $\{V_i : i = 0, ..., n\}$. If the Bezier curve can be exactly represented by another Bezier curve $\mathbf{r}^{n+1}(u)$ of degree $n+1$, the following should hold:

$$\begin{aligned}\mathbf{r}^n(u) &\equiv \sum_{i=0}^{n} V_i \binom{n}{i} u^i(1-u)^{n-1} \\ &= \sum_{i=0}^{n+1} V_i^+ \binom{n+1}{i} u^i(1-u)^{n+1-i} \equiv \mathbf{r}^{n+1}(u),\end{aligned} \tag{6.16}$$

where $\{V_i^+ : i = 0, ..., n+1\}$ are control vertices of the *degree-elevated* Bezier curve. On

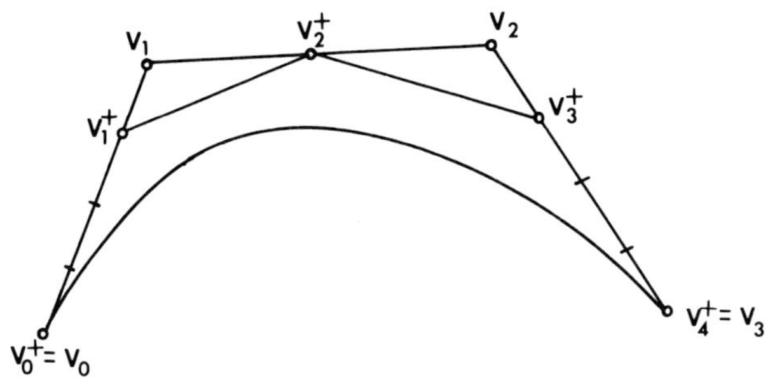

Figure 6.3 Degree Elevation of Cubic Bezier Curve

multiplying $1 = (u + (1 - u))$ to $\mathbf{r}^n(u)$ and then comparing the coefficients of the terms $u^i(1-u)^{n+1-i}$ on both sides of (6.16), the following **degree elevation formula** may be obtained:

$$V_i^+ = \frac{i}{n+1}V_{i-1} + (1 - \frac{i}{n+1})V_i \ ; \quad i = 0, 1, ..., n+1 \qquad (6.17)$$

where, V_i : *control vertices of a given Bezier curve of degree n,*

V_i^+ : *control vertices of the degree elevated Bezier curve.*

An example of degree elevation from *cubic* to *quartic* is shown in Fig. 6.3. If the degree elevation is continued, the *control polygon converges to the curve itself.* The degrees m, n of a Bezier patch $\mathbf{r}^{m,n}(u,v)$ can also be elevated the same way. Let us introduce

$\{V_{i,j}\}$: *control vertices of* $\mathbf{r}^{m,n}(u,v)$,
$\{V_{i,j}^{+0}\}$: *control vertices of* $\mathbf{r}^{m+1,n}(u,v)$,
$\{V_{i,j}^{0+}\}$: *control vertices of* $\mathbf{r}^{m,n+1}(u,v)$, and
$\{V_{i,j}^{++}\}$: *control vertices of* $\mathbf{r}^{m+1,n+1}(u,v)$.

If the u-direction degree m of $\mathbf{r}^{m,n}(u,v)$ is raised by one, the control vertices of the resulting Bezier patch $\mathbf{r}^{m+1,n}(u,v)$ is obtained from (6.17) as

$$V_{i,j}^{+0} = \frac{i}{m+1}V_{i-1,j} + (1 - \frac{i}{m+1})V_{i,j}; \quad i = 0, ..., m+1; \quad j = 0, ..., n \qquad (6.18-a)$$

Similarly, the control vertices of $\mathbf{r}^{m,n+1}(u,v)$ are given by

$$V_{i,j}^{0+} = \frac{j}{n+1}V_{i,j-1} + (1 - \frac{j}{n+1})V_{i,j}; \quad i = 0, ..., m; \quad j = 0, ..., n+1 \qquad (6.18-b)$$

By applying (6.18-a) and (6.18-b), the degree m, n of a Bezier patch are elevated by one in both u- and v-directions. The net result is expressed as

$$V_{i,j}^{++} = [\alpha_i \ (1-\alpha_i)] \begin{bmatrix} V_{i-1,j-1} & V_{i-1,j} \\ V_{i,j-1} & V_{i,j} \end{bmatrix} \begin{bmatrix} \beta_j \\ 1-\beta_j \end{bmatrix} \tag{6.19}$$

for $i = 0, ..., m+1; \quad j = 0, ..., n+1$, where $\alpha_i = i/(m+1)$ and $\beta_j = j/(n+1)$.

When a Bezier curve of degree $n-1$, $\mathbf{r}^{n-1}(u)$, is degree-elevated to a degree-n Bezier curve $\mathbf{r}^n(u)$, the degree elevation formula (6.17) is written as

$$V_i = \frac{i}{n} V_{i-1}^- + (1 - \frac{i}{n}) V_i^- \ ; \quad i = 0, 1, ..., n$$

where $\{V_i^- : i = 0, ..., n-1\}$ denote the control vertices of $\mathbf{r}^{n-1}(u)$. By inverting the above expression, two recursive **degree reduction formulae** may be obtained as

$$V_i^- = \frac{n}{n-i} V_i - \frac{i}{n-i} V_{i-1}^- \ ; \quad i = 0, 1, ..., n-1 \tag{6.20 - a}$$

$$V_{i-1}^- = \frac{n}{i} V_i - \frac{n-i}{i} V_i^- \ ; \quad i = n, n-1, ..., 1 \tag{6.20 - b}$$

If (6.20-a) and (6.20-b) give the same solution, the degree-n Bezier curve $\mathbf{r}^n(u)$ is "degree-reducible" meaning that it is in fact a drgree (n-1) curve. More exactly, a Bezier curve $\mathbf{r}^n(u)$ is **degree reducible** iff (Shin and Choi, 1990)

$$\sum_{i=0}^{n} V_i \binom{n}{i} (-1)^i = \mathbf{0}. \tag{6.21}$$

If a Bezier curve is not degree-reducible, degree reduction is an approximation process. In order to minimize the approximation error, V_i^- for $0, ..., [n/2]$ are obtained from (6.20-a) and the rest are obtained from (6.20-b). For example, a quartic Bezier curve may be degree reduced as follows ($n = 4$):

$$V_0^- = V_0 \quad \textit{from } (6.20 - a),$$

$$V_1^- = \frac{4}{3} V_1 - \frac{1}{3} V_0^- = \frac{4}{3} V_1 - \frac{1}{3} V_0 \quad \textit{from } (6.20 - a),$$

$$V_3^- = V_4 \quad \textit{from } (6.20 - b),$$

$$V_2^- = \frac{4}{3} V_3 - \frac{1}{3} V_3^- = \frac{4}{3} V_3 - \frac{1}{3} V_4 \quad \textit{from } (6.20 - b),$$

Degree reduction of a Bezier patch may be carried out by applying the degree reduction formulae (6.20) to each "row" of Bezier points $\{V_{i,j}\}$ and then to each "column" of the degree-reduced Bezier points, or the other way around.

6.5 INHERENT PROPERTIES OF BEZIER CURVE

Important properties inherent in Bezier curves are (Farin, 1988)

- *affine invariance property,*
- *convex hull property,*
- *invariance under affine parameter transformation, and*
- *variation diminishing property.*

1) Affine Invariance Property:

An important property of Bezier curves and surfaces is that they are *invariant* under affine maps. For a 3D point $\mathbf{r} = (x,\ y,\ z)$, an **affine map** Φ is defined as

$$\Phi\mathbf{r} = \mathbf{r}\ \mathbf{A} + \mathbf{t}, \tag{6.22}$$

where $\mathbf{A}$ is a 3×3 matrix and $\mathbf{t}$ is a 3D vector. On comparing (6.22) with the general 3D transformation expression (2.45) of §2.5.2, one may see that 3D coordinate transformations are in fact affine maps. The **affine invariance property** means that a Bezier curve can be transformed either 1) by transforming the evaluated point $\mathbf{r}(u)$ or 2) by transforming its *control points* $\{V_i\}$ and then evaluating the curve equation. Both schemes give the same result. The same is true for a Bezier surface patch.

2) Convex Hull Property:

The Bezier curve equation (6.1) is a *convex combination* of its control vertices $\{V_i\}$ because

$$\sum_{i=0}^{n} B_i^n(u) = 1 \quad and \quad B_i^n(u) \geq 0 \quad for\ u \in [0,1].$$

Thus, the curve $\mathbf{r}(u)$ is confined inside the *convex hull* of the *control polygon.* This property is called the *convex hull property.* The same property holds true for a Bezier patch since

$$\sum_{i=0}^{m}\sum_{j=0}^{n} B_i^m(u)B_j^n(v) = 1 \quad and \quad B_i^m(u)B_j^n(v) \geq 0.$$

This property is very useful, for example, in *detecting interferences* among Bezier patches because their *majorizing boxes* are easily computed from the control vertices alone.

3) Invariance under Affine Parameter Transformation:

An important implication of this property is that a Bezier curve $\mathbf{r}(u)$ for $u \in [0,1]$ is equivalent to $\mathbf{r}(t)$ with $t \in [a,b]$, where

$$t = a + (b-a)u \quad and \quad a < b.$$

The same invariance property holds for (tensor product) Bezier patches.

4) Variation Diminishing Property:

The *variation diminishing property* means that a Bezier curve $\mathbf{r}(u)$ has fewer intersections with a plane than its *control polygon* has. This property is useful in finding intersections between a 3D Bezier curve and a plane. For example, if the control polygon does not intersect with a plane then there is no intersection between the Bezier curve and the plane.

6.6 CONVERSION FROM B-SPLINE TO BEZIER

As discussed in §4.3.3, a (non-uniform) B-spline curve is defined by a sequence of *control vertices* $\{V_j\}$ and *knot spans* $\{\nabla_j\}$. A B-spline curve segment of degree 3, for example, is defined by four control vertices

$$\{V_i,\ V_{i+1},\ V_{i+2},\ V_{i+3}\} \quad and$$

five knot spans

$$\{\nabla_{i-2},\ \nabla_{i-1},\ \nabla_i,\ \nabla_{i+1},\ \nabla_{i+2}\}.$$

The following recursive relation is known as the **Boehm's knot insertion algorithm** (Boehm, 1980; Choi *et al*, 1990b):

$$\mathbf{b}_{j+d}^m(u) = \alpha_j^m(u)\mathbf{b}_{j+d}^{m-1}(u) + (1 - \alpha_j^m(u))\mathbf{b}_{j+d-1}^{m-1}(u) \ ; \quad 0 \le u \le 1 \tag{6.23}$$

$$where, \quad \alpha_j^m(u) = (u\nabla_i + \nabla_j^{i-j})/\nabla_j^{d-m+1} \quad for\ i - d + m \le j \le i,$$

$$\mathbf{b}_j^0 \equiv V_j,$$

$$d: \ degree\ of\ B\text{-}spline\ curve,$$

$$m: \ knot\ multiplicity\ (1 \le m \le d),$$

$$\nabla_j^k = \nabla_j + \nabla_{j+1} + \ldots + \nabla_{j+k-1}.$$

The above recursive relation can be used in computing *Bezier points* from the B-spline control points $\{V_i\}$. For a cubic B-spline curve (ie, $d = 3$), with $u = 0$ and $u = 1$, the knot insertion algorithm (6.23) is written as

$$\mathbf{b}_{j+3}^m(0) = \alpha_j^m(0)\mathbf{b}_{j+3}^{m-1}(0) + (1 - \alpha_j^m(0))\mathbf{b}_{j+2}^{m-1}(0) \tag{6.24 - a}$$

$$\mathbf{b}_{j+3}^m(1) = \alpha_j^m(1)\mathbf{b}_{j+3}^{m-1}(1) + (1 - \alpha_j^m(1))\mathbf{b}_{j+2}^{m-1}(1) \tag{6.24 - b}$$

$$where, \quad \alpha_j^m(0) = \nabla_j^{i-j}/\nabla_j^{4-m},$$

$$\alpha_j^m(1) = (\nabla_i + \nabla_j^{i-j})/\nabla_j^{4-m}, \quad for\ i - 3 + m \le j \le i.$$

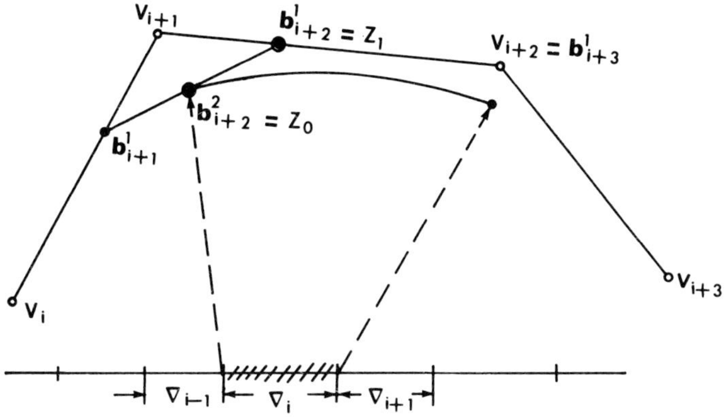

Figure 6.4 Knot Insertion at the Beginning of a Cubic B-spline Curve

By evaluating (6.24-a), the *intermediate* **de Boor points** $\mathbf{b}_i^m(0)$ depicted in Fig. 6.4 are obtained as follows:

a) $m = 1;\quad (j = i-2,\ i-1,\ i)$:

$$j = i-2: \quad \mathbf{b}^1_{i+1}(0) = (\nabla^2_{i-2}/\nabla^3_{i-2})\mathbf{b}^0_{i+1}(0) + (1 - \nabla^2_{i-2}/\nabla^3_{i-2})\mathbf{b}^0_i(0)$$

$$= ((\nabla_{i-2} + \nabla_{i-1})V_{i+1} + \nabla_i V_i)/\nabla^3_{i-2}.$$

$$j = i-1: \quad \mathbf{b}^1_{i+2}(0) = (\nabla_{i-1}V_{i+2} + (\nabla_i + \nabla_{i+1})V_{i+1})/\nabla^3_{i-1} \equiv Z_1.$$

$$j = i \quad : \quad \mathbf{b}^1_{i+3}(0) = V_{i+2}.$$

b) $m = 2;\quad (j = i-1,\ i)$:

$$j = i-1: \quad \mathbf{b}^2_{i+2}(0) = (\nabla_{i-1}\mathbf{b}^1_{i+2}(0) + \nabla_i \mathbf{b}^1_{i+1}(0))/\nabla^2_{i-1} \equiv Z_0.$$

$$j = i \quad : \quad \mathbf{b}^2_{i+3}(0) = \mathbf{b}^1_{i+2}(0).$$

Let the Bezier points of the cubic curve segment shown in Fig. 6.4 be denoted by $\{Z_i : i = 0, ..., 1\}$. Then from the above evaluation, we have obtained

$$Z_0 = \mathbf{b}^2_{i+2}(0), \quad \text{and} \quad Z_1 = \mathbf{b}^1_{i+2}(0).$$

By evaluating (6.24-b), the remaining Bezier points are obtained as

$$Z_2 = \mathbf{b}^1_{i+2}(1), \quad \text{and} \quad Z_3 = \mathbf{b}^2_{i+3}(1).$$

The sequence of evaluating the knot insertion algorithm (6.24) may be summarized as follows (from left to right):

1) Using (6.24 – a) :

$$V_i \equiv \mathbf{b}_i^0(0)$$
$$V_{i+1} \equiv \mathbf{b}_{i+1}^0(0) \rightarrow \mathbf{b}_{i+1}^1(0)$$
$$V_{i+2} \equiv \mathbf{b}_{i+2}^0(0) \rightarrow \mathbf{b}_{i+2}^1(0) \rightarrow \mathbf{b}_{i+2}^2(0)$$

2) Using (6.24 – b) :

$$V_{i+1} \equiv \mathbf{b}_{i+1}^0(1)$$
$$V_{i+2} \equiv \mathbf{b}_{i+2}^0(1) \rightarrow \mathbf{b}_{i+2}^1(1)$$
$$V_{i+3} \equiv \mathbf{b}_{i+3}^0(1) \rightarrow \mathbf{b}_{i+3}^1(1) \rightarrow \mathbf{b}_{i+3}^2(1)$$

The algebraic evaluation may seem to be complicated, but it's geometric interpretation is quite simple as depicted in Fig. 6.5. The Bezier points $\{Z_i : i = 0, 1, 2, 3\}$ are obtained by *subdividing the B-spline control polygon according to the ratios of the knot spans.* The Bezier points of a tensor product B-spline surface can be obtained by the same algorithm: It is first applied to each *column* of the B-spline polyhedron net and then to each row of the intermediate net thus obtained (Boehm, 1981).

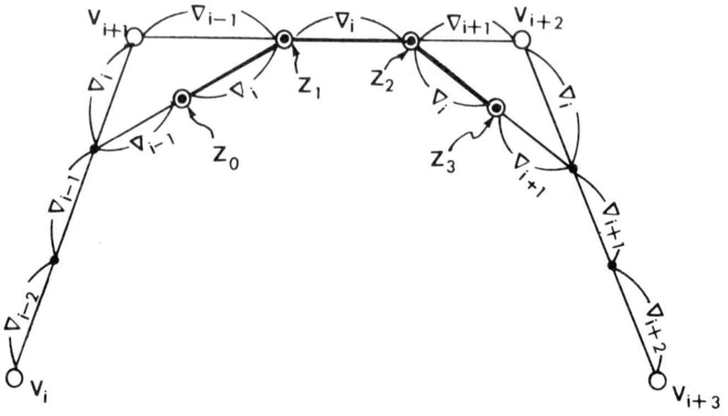

Figure 6.5 Construction of Bezier Points $\{Z_i\}$ from cubic B-spline Polygon

CHAPTER 7

SURFACE CONSTRUCTION FROM 3D DATA ARRAY

7.1 INTRODUCTION

Presented in this chapter are various methods of constructing a parametric surface $\mathbf{r}(u,v)$ interpolating over an array of 3D data points $\{P_{ij} : i = 0,1,...,m;\ j = 0,1,...,n\}$ which are topologically connected by rows and columns to form a rectangular mesh. The *parameters* u, v of the surface are associated with the *indices* i, j of the input data points as depicted in Fig. 7.1. One way of fitting the point data in Fig. 7.1 is to use the *standard polynomial model* of the form

$$\mathbf{r}(u,v) = \sum_{i=0}^{m}\sum_{j=0}^{n} \mathbf{a}_{ij} u^i v^j,$$

where $\mathbf{a}_{ij}$ are vector coefficients. However, the resulting surface may possess undesirable properties when the degrees m, n become large.

Thus, it is desired to use a low degree (usually cubic) polynomial patch model to form a *composite surface*. For this, we may use the Ferguson patch model, Bezier patch models, or a B-spline patch model. The surface fitting methods to be discussed in this chapter are:

a) the FMILL method,

b) composite Ferguson fitting method,

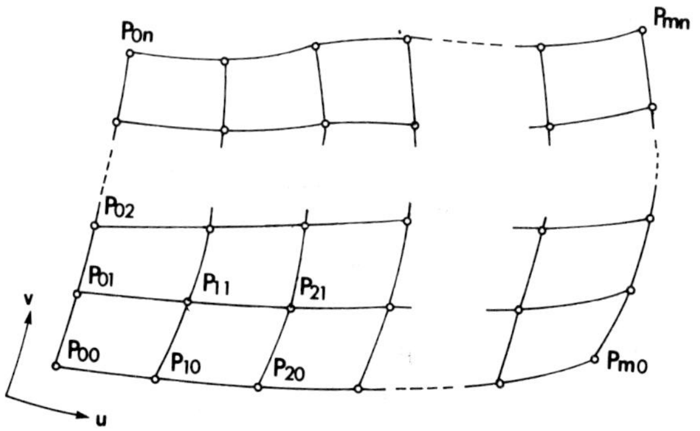

Figure 7.1 Input data Point $\{P_{ij}\}$ to be Interpolated

$\hat{\mathbf{x}}_{00}$	$\hat{\mathbf{s}}_{00}$	$\hat{\mathbf{s}}_{01}$	$\hat{\mathbf{s}}_{02}$	..	$\hat{\mathbf{s}}_{0j}$	...	$\hat{\mathbf{s}}_{0n}$	$\hat{\mathbf{x}}_{0n}$
$\hat{\mathbf{t}}_{00}$	P_{00}	P_{01}	P_{02}	..	P_{0j}	...	P_{0n}	$\hat{\mathbf{t}}_{0n}$
$\hat{\mathbf{t}}_{10}$	P_{10}	P_{11}	P_{12}	..	P_{1j}	...	P_{1n}	$\hat{\mathbf{t}}_{1n}$
.	.	.	.		.		.	.
.	.	.	.		.		.	.
$\hat{\mathbf{t}}_{i0}$	P_{i0}	P_{i1}	P_{i2}	..	P_{ij}	...	P_{in}	$\hat{\mathbf{t}}_{in}$
.	.	.	.		.		.	.
.	.	.	.		.		.	.
$\hat{\mathbf{t}}_{m0}$	P_{m0}	P_{m1}	P_{m2}	..	P_{mj}	...	P_{mn}	$\hat{\mathbf{t}}_{mn}$
$\hat{\mathbf{x}}_{m0}$	$\hat{\mathbf{s}}_{m0}$	$\hat{\mathbf{s}}_{m1}$	$\hat{\mathbf{s}}_{m2}$	..	$\hat{\mathbf{s}}_{mj}$	...	$\hat{\mathbf{s}}_{mn}$	$\hat{\mathbf{x}}_{mn}$

Figure 7.2 Input Data for Composite Surface Construction

c) uniform B-spline fitting method,

d) NUB fitting method, and

e) G^1 composite Bezier surface fitting method.

In all of the five methods, it is necessary to know the *cross-boundary tangents* at the *boundary mesh points* and *twist vectors* at the *corner mesh points* in addition to the *mesh points* as depicted in Fig. 7.2. That is the following boundary data are assumed to be given:

$\{\hat{\mathbf{s}}_{0j}, \hat{\mathbf{s}}_{mj}\}$: *u-direction cross-boundary tangents at boundary mesh points,*

$\{\hat{\mathbf{t}}_{i0}, \hat{\mathbf{t}}_{in}\}$: *v-direction cross-boundary tangents at boundary mesh points,*

$\{\hat{\mathbf{x}}_{00},\ \hat{\mathbf{x}}_{m0},\ \hat{\mathbf{x}}_{0n},\ \hat{\mathbf{x}}_{mn}\}$: *twist vectors at corner mesh points.*

The *FMILL method*, the *Ferguson* method, and *uniform B-spline fitting* method are applicable only to evenly spaced data points because they tend to produce surfaces having local flatness or bulges for unevenly spaced data points.

The *NUB (non-uniform B-spline) surface fitting* method is applicable to the data array whose physical spacings are semi-even, meaning that rows and columns of input data are parallel but uneven with each other (see Fig. 7.13a). When the physical spacing is uneven, the G^1 *Bezier method* is the only choice (among the five) in order to obtain a *smooth* surface which is free from local flatness or bulges. The Ferguson method and the NUB method result in a C^2 composite surface, while the FMILL surface and the composite Bezier surface are G^1 surfaces.

7.2 FMILL SURFACE FITTING

As discussed in §5.2.2, a Ferguson surface patch is completely specified by *position vectors, tangent vectors* in u- and v-directions, and *twist vectors* at the four "corner points". In the **FMILL** module, which is a surface-fitting routine in the APT system, the **twist vectors** $\mathbf{x}_{ij}$ at the data points are set to zero. The *u-direction tangents* $\mathbf{s}_{ij}$ are estimated from the following expression *(for i=1,...,m-1 and j=0,...,n):*

$$\mathbf{s}_{ij} = c_i(P_{i+1,j} - P_{i-1,j})/|P_{i+1,j} - P_{i-1,j}| \tag{7.1}$$

$$where,\ c_i = min\{|P_{i,j} - P_{i-1,j}|,\ |P_{i+1,j} - P_{i,j}|\}.$$

The *v-direction tangents* $\mathbf{t}_{ij}$ are similarly determined. Then, the **corner condition matrix Q** of the Ferguson patch (5.5) over the four data points $\{P_{ij},\ P_{i+1,j},\ P_{i,j+1},\ P_{i+1,j+1}\}$ is given by (see Fig. 7.3)

$$\mathbf{Q}^{ij} = \begin{bmatrix} P_{i,j} & P_{i,j+1} & \mathbf{t}_{i,j} & \mathbf{t}_{i,j+1} \\ P_{i+1,j} & P_{i+1,j+1} & \mathbf{t}_{i+1,j} & \mathbf{t}_{i+1,j+1} \\ \mathbf{s}_{i,j} & \mathbf{s}_{i,j+1} & 0 & 0 \\ \mathbf{s}_{i+1,j} & \mathbf{s}_{i+1,j+1} & 0 & 0 \end{bmatrix}.$$

Since the *corner condition matrix* $\mathbf{Q}^{ij}$ has been fixed, the rectangular region is represented as a Ferguson patch. In this way, a composite surface composed of $n \times m$ Ferguson patches is obtained. The FMILL method is very easy to implement and the resulting surface is visually smooth (it is a G^1 surface). It is also *aesthetically smooth* as long as the data points are evenly spaced. But it may produce undesirable surface quality (eg, *local flatness or unnatural surface normal at mesh points*) due to the assumption of *zero twist* vectors.

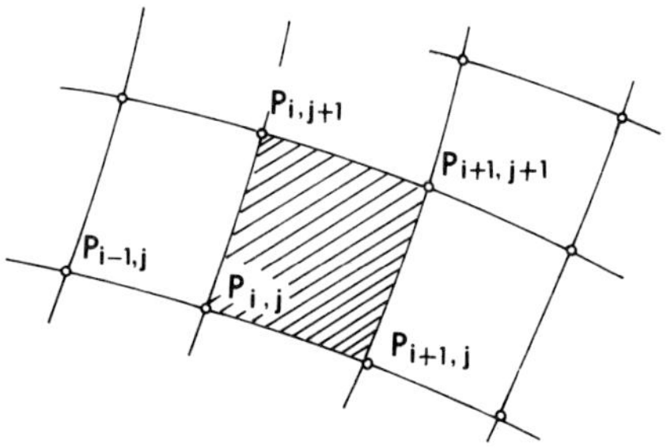

Figure 7.3 A Patch in the Mesh of Data Points

7.3 COMPOSITE FERGUSON SURFACE FITTING

This section presents a surface construction method introduced in Faux and Pratt (1980). If we remove the assumption of "zero twist vector" from the FMILL method, then the *twist vectors* $\mathbf{x}_{ij}$ at mesh points have to be determined. One way of determining $\mathbf{x}_{ij}$ is to impose the *parametric* C^2-condition (4.1) across patch boundaries.

Consider three Ferguson patches, $\mathbf{r}^1(u,v)$, $\mathbf{r}^2(u,v)$, and $\mathbf{r}^3(u,v)$, sharing the mesh point P_{ij} as depicted in Fig. 7.4. Recall from §5.2.2 that the Ferguson patch $\mathbf{r}^1(u,v)$ is expressed as (from Eqn. (5.5))

$$\mathbf{r}^1(u,v) = \mathbf{U}\ \mathbf{C}\ \mathbf{Q}^1\mathbf{C}^T\mathbf{V}^T : 0 \le u,v \le 1 \tag{7.2}$$

$$where,\ \mathbf{C} = \begin{bmatrix} 1 & 0 & 0 & 0 \\ 0 & 0 & 1 & 0 \\ -3 & 3 & -2 & -1 \\ 2 & -2 & 1 & 1 \end{bmatrix},$$

$$\mathbf{Q}^1 = \begin{bmatrix} P_{i-1,j-1} & P_{i-1,j} & \mathbf{t}_{i-1,j-1} & \mathbf{t}_{i-1,j} \\ P_{i,j-1} & P_{i,j} & \mathbf{t}_{i,j-1} & \mathbf{t}_{i,j} \\ \mathbf{s}_{i-1,j-1} & \mathbf{s}_{i-1,j} & \mathbf{x}_{i-1,j-1} & \mathbf{x}_{i-1,j} \\ \mathbf{s}_{i,j-1} & \mathbf{s}_{i,j} & \mathbf{x}_{i,j-1} & \mathbf{x}_{i,j} \end{bmatrix}.$$

An application of the C^2-condition (4.1) to the commmon boundary of $\mathbf{r}^1(u,v)$ and $\mathbf{r}^2(u,v)$ and to that of $\mathbf{r}^1(u,v)$ and $\mathbf{r}^3(u,v)$ would give

$$\mathbf{r}^1_{uu}(1,v) = \mathbf{r}^2_{uu}(0,v) \tag{7.3 – a}$$

$$\mathbf{r}^1_{vv}(u,1) = \mathbf{r}^3_{vv}(u,0) \tag{7.3 – b}$$

Since the u-direction second derivative of the Ferguson patch equation is expressed as

$$\begin{aligned} \mathbf{r}_{uu}(u,v) &= \ddot{\mathbf{U}}\ \mathbf{C}\ \mathbf{Q}\ \mathbf{C}^T\ \mathbf{V}^T \\ &= [0\ 0\ 2\ 6u]\ \mathbf{C}\ \mathbf{Q}\ \mathbf{C}^T\ \mathbf{V}^T, \end{aligned}$$

the left-hand side of (7.3-a) is evaluated as

$$\begin{aligned} \mathbf{r}^1_{uu}(1,v) &= [0\ 0\ 2\ 6]\ \mathbf{C}\ \mathbf{Q}^1\ \mathbf{C}^T\ \mathbf{V}^T \\ &= [6\ -6\ 2\ 4]\ \mathbf{Q}^1\ \mathbf{C}^T\ \mathbf{V}^T, \end{aligned}$$

and the right-hand side would become

$$\begin{aligned} \mathbf{r}^2_{uu}(0,v) &= [0\ 0\ 2\ 0]\ \mathbf{C}\ \mathbf{Q}^2\ \mathbf{C}^T\ \mathbf{V}^T \\ &= [-6\ \ 6\ -4\ -2]\ \mathbf{Q}^2\ \mathbf{C}^T\ \mathbf{V}^T. \end{aligned}$$

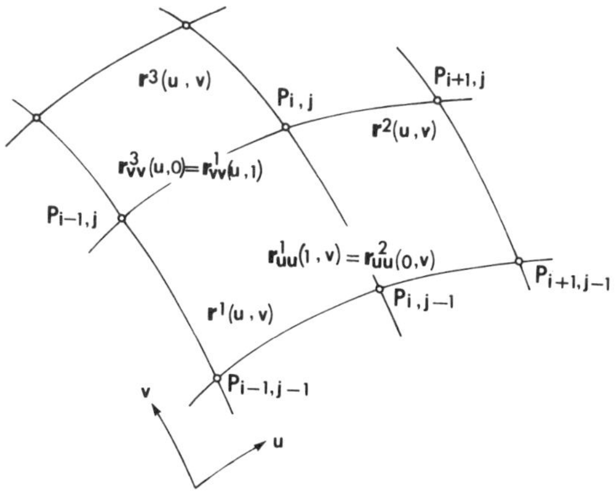

Figure 7.4 C^2-Conditions among Ferguson Patches

Thus, the u-direction C^2-condition (7.3-a) is rewritten as follows:

$$[6\ -6\ 2\ 4]\ \mathbf{Q^1} = [-6\ \ 6\ -4\ -2]\ \mathbf{Q^2} \tag{7.4}$$

$$where,\quad \mathbf{Q^1} = \begin{bmatrix} P_{i-1,j-1} & P_{i-1,j} & \mathbf{t}_{i-1,j-1} & \mathbf{t}_{i-1,j} \\ P_{i,j-1} & P_{i,j} & \mathbf{t}_{i,j-1} & \mathbf{t}_{i,j} \\ \mathbf{s}_{i-1,j-1} & \mathbf{s}_{i-1,j} & \mathbf{x}_{i-1,j-1} & \mathbf{x}_{i-1,j} \\ \mathbf{s}_{i,j-1} & \mathbf{s}_{i,j} & \mathbf{x}_{i,j-1} & \mathbf{x}_{i,j} \end{bmatrix}$$

$$\mathbf{Q}^2 = \begin{bmatrix} P_{i,j-1} & P_{i,j} & \mathbf{t}_{i,j-1} & \mathbf{t}_{i,j} \\ P_{i+1,j-1} & P_{i+1,j} & \mathbf{t}_{i+1,j-1} & \mathbf{t}_{i+1,j} \\ \mathbf{s}_{i,j-1} & \mathbf{s}_{i,j} & \mathbf{x}_{i,j-1} & \mathbf{x}_{i,j} \\ \mathbf{s}_{i+1,j-1} & \mathbf{s}_{i+1,j} & \mathbf{x}_{i+1,j-1} & \mathbf{x}_{i+1,j} \end{bmatrix}.$$

Note that the relation (7.4) gives four linear (vector) equations, one from each column of the **Q**-matrix. The 2nd equation in (7.4) is rearranged as (*for* i=*1,...,m-1*)

$$\mathbf{s}_{i-1,j} + 4\mathbf{s}_{i,j} + \mathbf{s}_{i+1,j} = 3(P_{i+1,j} - P_{i-1,j}) \tag{7.5 – a}$$

and the last equation in (7.4) is expressed as (*for* i=*1,...,m-1*)

$$\mathbf{x}_{i-1,j} + 4\mathbf{x}_{i,j} + \mathbf{x}_{i+1,j} = 3(\mathbf{t}_{i+1,j} - \mathbf{t}_{i-1,j}) \tag{7.5 – b}$$

In the same way, the v-direction C^2-condition (7.3-b) would produce

$$\mathbf{t}_{i,j-1} + 4\mathbf{t}_{i,j} + \mathbf{t}_{i,j+1} = 3(P_{i,j+1} - P_{i,j-1}) \tag{7.6 – a}$$

$$\mathbf{x}_{i,j-1} + 4\mathbf{x}_{i,j} + \mathbf{x}_{i,j+1} = 3(\mathbf{s}_{i,j+1} - \mathbf{s}_{i,j-1}) \tag{7.6 – b}$$

for $j = 1, ..., n-1$.

In summary, the tangent vectors and twist vectors at the mesh points are determined by solving the linear equations in (7.5) and (7.6) as follows:

1) $\{\mathbf{s}_{ij} : j = 0, 1, ..., n\}$ *are determined by solving*
Eqn.$(7.5 - a)$ *for i=1,...,m-1 together with* $\mathbf{s}_{0j} = \hat{\mathbf{s}}_{0j};\ \mathbf{s}_{mj} = \hat{\mathbf{s}}_{mj}$ (7.7)

2) $\{\mathbf{t}_{ij} : i = 0, 1, ..., m\}$ *are determined by solving*
Eqn.$(7.6 - a)$ *for j=1,...,n-1 together with* $\mathbf{t}_{i0} = \hat{\mathbf{t}}_{i0};\ \mathbf{t}_{in} = \hat{\mathbf{t}}_{in}$ (7.8)

3) $\{\mathbf{x}_{ij} : j = 0 \& j = n\}$ *are determined by solving*
Eqn.$(7.5 - b)$ *for i=1,...,m-1 together with* $\mathbf{x}_{0j} = \hat{\mathbf{x}}_{0j};\ \mathbf{x}_{mj} = \hat{\mathbf{x}}_{mj}$ (7.9)

4) $\{\mathbf{x}_{ij} : i = 0, 1, ..., m\}$ *are determined by solving*
Eqn.$(7.6 - b)$ *for j=1,...,n-1 together with* $\mathbf{x}_{i0} = \hat{\mathbf{x}}_{i0};\ \mathbf{x}_{in} = \hat{\mathbf{x}}_{in}$ (7.10)

Since the linear equation systems (7.7) through (7.10) have the same matrix form as (4.6), they are easily solved (See Appendix A).

Having determined all the tangent vectors and twist vectors, the next step is to form a *corner condition matrix* $\mathbf{Q}^{ij}$ for each rectangular region. The resulting surface is a C^2 surface (see Faux and Pratt, 1980, pp.205-209) composed of $m \times n$ Ferguson patches which is called a **composite Ferguson surface**. The major drawback of the composite Ferguson surface is that it may not be *aesthetically smooth* when the data points are unevenly spaced (the same is true for the FMILL surface).

Examples of composite Ferguson surfaces are shown in Fig. 7.5 and Fig. 7.6. The left figures are *mesh polyhedra*, and right figure represent *isoparametric* plotting of resulting surfaces. The input data points $\{\mathrm{P}_{ij}\}$ have been sampled from smooth surfaces. As depicted in the left figures, there are $4 \times 4 = 16$ data points in Fig. 7.5 and $4 \times 5 = 20$ data points in Fig. 7.6. In both examples, the cross boundary tangents at boundary mesh points (ie, $\mathbf{s}_{0j}$, $\mathbf{s}_{mj}$, $\mathbf{t}_{i0}$, $\mathbf{t}_{in}$) have all been estimated by using the "circular end condition" introduced in §4.2.2. Namely, the "u-direction" end tangents are estimated from each "column" of input data and the "v-direction" end tangents from each "row" of input data. The corner twist vectors have been set to zero. As shown in the right figure of Fig. 9.5, the Ferguson fitting method produces a very "smooth" surface when the physical spacing of data points is even. On the other hand, the composite Ferguson surface in Fig. 9.6 clearly shows that the surface fitting method cannot handle unevenly spaced input data.

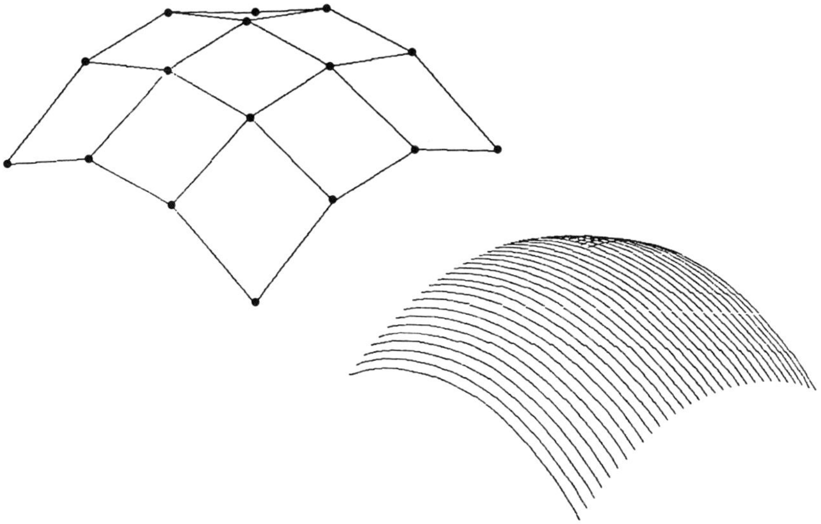

Figure 7.5 Composite Ferguson Surface for Evenly Spaced Data Points

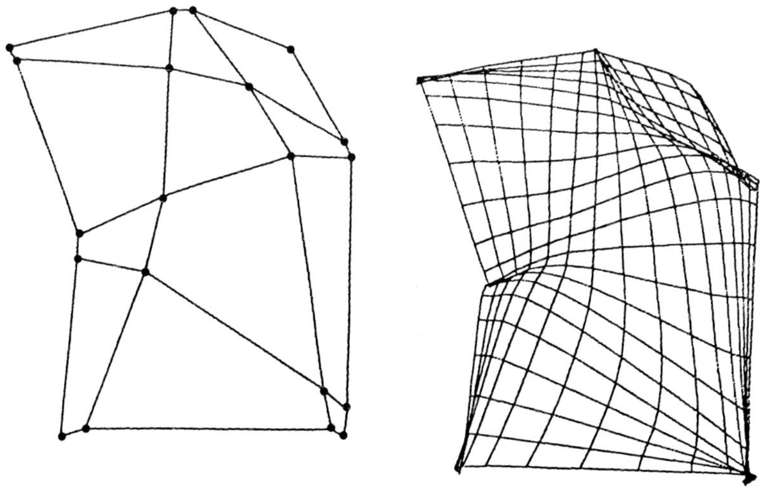

Figure 7.6 Composite Ferguson Surface for Unevenly Spaced Data Points

7.4 UNIFORM B-SPLINE SURFACE FITTING

This section presents another method of constructing a smooth composite surface from the input data of Fig. 7.2. This time, the uniform B-spline surface patch model (5.12) is employed. The properties of the resulting surface are the same as those of the composite Ferguson surface of §7.3.

7.4.1 Some Properties of Uniform B-spline Surface Patch

A bicubic uniform B-spline patch $\mathbf{r}(u,v)$ defined by 16 control vertices $\{V_{ij}\}$ is expressed in a *matrix form* (see Eqn. (5.12)) as

$$\mathbf{r}(u,v) = \mathbf{U}\ \mathbf{N}\ \mathbf{B}\ \mathbf{N}^{\mathrm{T}}\ \mathbf{V}^{\mathrm{T}} \qquad (7.11-a)$$

$$where,\quad \mathbf{U} = [1\ u\ u^2\ u^3],\quad \mathbf{V} = [1\ v\ v^2\ v^3],$$

$$\mathbf{B} = \begin{bmatrix} V_{00} & V_{01} & V_{02} & V_{03} \\ V_{10} & V_{11} & V_{12} & V_{13} \\ V_{20} & V_{21} & V_{22} & V_{23} \\ V_{30} & V_{31} & V_{32} & V_{33} \end{bmatrix},$$

$$\mathbf{N} = \frac{1}{6}\begin{bmatrix} 1 & 4 & 1 & 0 \\ -3 & 0 & 3 & 0 \\ 3 & -6 & 3 & 0 \\ -1 & 3 & -3 & 1 \end{bmatrix}.$$

Alternatively, it is expressed in a *B-spline basis form* as

$$\begin{aligned}\mathbf{r}(u,v) &= \sum_{j=0}^{3} N_j^3(v)\left(\sum_{i=0}^{3} N_i^3(u)V_{ij}\right) \\ &= \sum_{j=0}^{3} N_j^3(v)\mathbf{r}_j(u),\end{aligned} \qquad (7.11-b)$$

where $N_i^3(u)$, and $N_j^3(v)$ are as defined in (5.12). From (7.11-b), one may see that a B-spline surface is defined as a "two stage curve blending":

1) Each column j of control vertices are blended to form a curve $\mathbf{r}_j(u)$, and
2) the curves $\mathbf{r}_j(u)$ are blended again to form a surface.

As depicted in Fig. 7.7, the four control vertices $\{V_{i0} : i = 0,1,2,3\}$ in the first column of the control polyhedron define a cubic B-spline curve $\mathbf{r}_0(u)$, and $\mathbf{r}_1(u)$ is constructed from the control vertices $\{V_{i1} : i = 0,1,2,3\}$ in the second column, and so on. And then the surface $\mathbf{r}(u,v)$ is formed by blending the four *control curves* $\{\mathbf{r}_j : j = 0,1,2,3\}$.

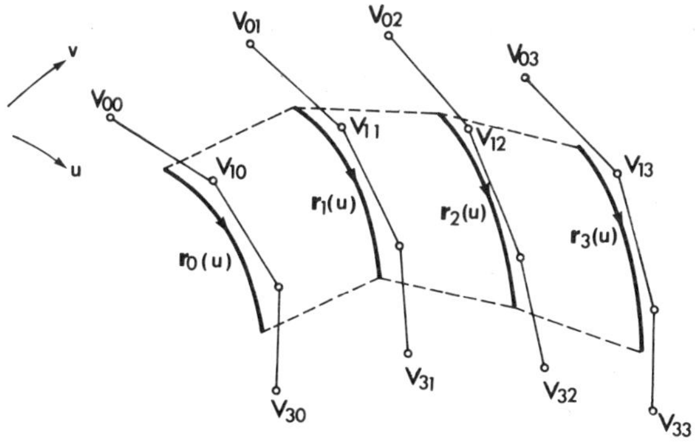

Figure 7.7 B-spline Patch as a Blending of B-spline Curves

The v-direction *derivatives* of the B-spline patch (7.11-a) at the patch corners are evaluated as follows:

$$\begin{aligned}\mathbf{r}_v(0,0) &= \mathbf{U\ N\ B\ N^T\ \dot{V}^T}|_{u=0,v=0} \\ &= [1\ 0\ 0\ 0]\ \mathbf{N\ B\ N^T}[0\ 1\ 0\ 0]^T \\ &= (V_{02}-V_{00})/12 + (V_{12}-V_{10})/3 + (V_{22}-V_{20})/12\end{aligned} \tag{7.12-a}$$

$$\begin{aligned}\mathbf{r}_v(0,1) &= [1\ 0\ 0\ 0]\ \mathbf{N\ B\ N^T}[0\ 1\ 2\ 3]^T \\ &= (V_{03}-V_{01})/12 + (V_{13}-V_{11})/3 + (V_{23}-V_{21})/12.\end{aligned} \tag{7.12-b}$$

Twist vectors at the corners are given by

$$\begin{aligned}\mathbf{r}_{uv}(0,0) &= \mathbf{\dot{U}\ N\ B\ N^T\ \dot{V}^T}|_{u=0,v=0} \\ &= [0\ 1\ 0\ 0]\ \mathbf{N\ B\ N^T}[0\ 1\ 0\ 0]^T \\ &= -(V_{02}-V_{00})/4 + (V_{22}-V_{20})/4\end{aligned} \tag{7.13-a}$$

$$\mathbf{r}_{uv}(0,1) = -(V_{03}-V_{01})/4 + (V_{23}-V_{21})/4. \tag{7.13-b}$$

7.4.2 Uniform B-spline Surface Fitting

Now we present a method of constructing a composite B-spline surface composed of $m \times n$ bicubic B-spline patches of the form (7.11). The input data are as given in Fig. 7.2. That is,

- $(m+1) \times (n+1)$ mesh points $\{P_{ij}\}$,
- u-direction cross-boundary tangents $\hat{\mathbf{s}}_{0j}, \hat{\mathbf{s}}_{mj} (j = 0, ..., n)$,
- v-direction cross-boundary tangents $\hat{\mathbf{t}}_{i0}, \hat{\mathbf{t}}_{in} (i = 0, ..., m)$, and
- corner twist vectors $\hat{\mathbf{x}}_{00}, \hat{\mathbf{x}}_{0n}, \hat{\mathbf{x}}_{m0}, \hat{\mathbf{x}}_{mn}$.

In interpolating the input data, the two-stage view of the B-spline surface definition (7.11-b) is utilized. That is, a set of *intermediate control vertices* is obtained for each "column" of input data, and then a set of B-spline control vertices is obtained for each "row" of intermediate control vertices.

1) Determination of Intermediate Control Vertices:

The first step is to construct an *intermediate* B-spline curve from each column "j" of the input data as depicted in Fig. 7.8. In the figure, the second subscript "j is omitted for brevity. The problem here is:

Given: $m + 1$ data points $\{P_{i(j)} : i = 0, ..., m\}$ and the two end tangents $\hat{\mathbf{s}}_{0(j)}, \hat{\mathbf{s}}_{m(j)}$.

Find: $m + 3$ **intermediate control vertices** $\{C_{i(j)} : i = 0, ..., m + 2\}$.

The situation is exactly the one for the curve fitting case discussed in §4.2.3. Thus, the *intermediate control vertices* $\{C_{ij}\}$ can easily be obtained by solving a linear system of the type given in Eqn. (4.17-a). That is, we have:

$$\begin{bmatrix} -1 & 0 & 1 & 0 & & & \\ 1 & 4 & 1 & 0 & & & \\ 0 & 1 & 4 & 1 & & & \\ & & & \cdot & & & \\ & & & & \cdot & & \\ & & & & 1 & 4 & 1 \\ & & & & -1 & 0 & 1 \end{bmatrix} \begin{bmatrix} C_{0,j} \\ C_{1,j} \\ C_{2,j} \\ \cdot \\ \cdot \\ C_{m+1,j} \\ C_{m+2,j} \end{bmatrix} = \begin{bmatrix} 2\hat{\mathbf{s}}_{0j} \\ 6P_{0j} \\ 6P_{1j} \\ \cdot \\ \cdot \\ 6P_{mj} \\ 2\hat{\mathbf{s}}_{mj} \end{bmatrix} \quad for\ j = 0, 1, ..., n \tag{7.14}$$

2) Determination of Boundary Vectors:

Having determined the $(m+3) \times (n+1)$ *intermediate control vertices* $\{C_{ij}\}$ using (7.14), the next step is to determine $(m + 3) \times (n + 3)$ *control vertices* $\{V_{ij}\}$ of the B-spline surface. As depicted in Fig. 7.9, each row "i" of the intermediate control vertices $\{C_{(i)j} : j = 0, ..., n\}$ is treated as a sequence of data points to be interpolated. Then from the result of §4.2.3, V_{ij} and C_{ij} are related by the following set of linear equations (see Eqn. (4.14)):

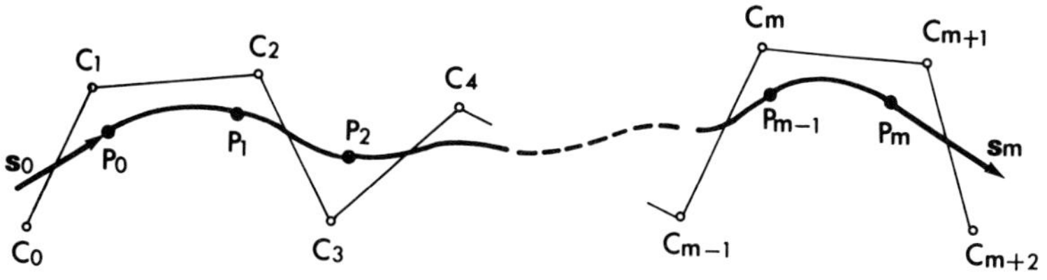

Figure 7.8 B-spline Curve Fitting of a Column of Input Data

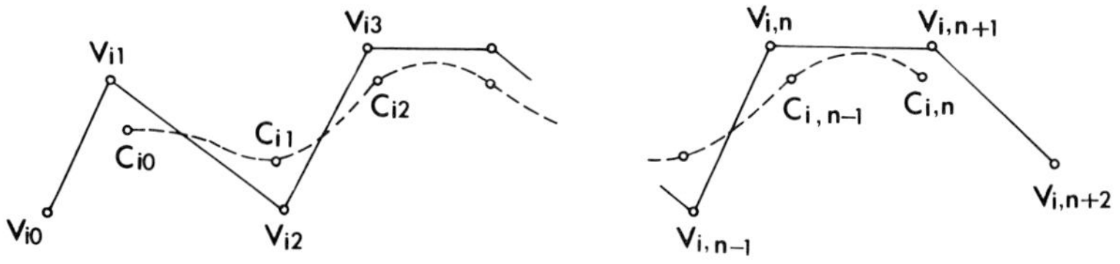

Figure 7.9 B-spline Curve Fitting of a Row of Intermediate Control Points

$$V_{i,j} + 4V_{i,j+1} + V_{i,j+2} = 6C_{i,j} \ ; \quad j = 0, 1, ..., n \qquad (7.15)$$

The above relation holds for all rows, namely, for $i = 0, 1, ..., m+2$.

Since there are two more unknowns than the number of equations in (7.15), we need two more linear equations for each "row i". The additional equations can be provided from the v-direction derivative equations (7.12) and the twist vector equations (7.13). Let us define **boundary vectors** *in v-direction* so that

$$\mathbf{d}_i = (V_{i2} - V_{i0})/2 \quad for\ i = 0, 1, ..., m+2$$

then, the v-direction cross-boundary vector, for example, at $u = v = 0$ with $i = 0$ can be expressed in terms of $\mathbf{d}_i$'s as follows (from Eqn. (7.12)):

$$\hat{\mathbf{t}}_{00} = \mathbf{r}_v(0,0) = \mathbf{d}_0/6 + 2\mathbf{d}_1/3 + \mathbf{d}_2/6.$$

On applying the above relation to each "left" *boundary mesh point* $\{P_{i0}\}$, the following set of linear equations is obtained:

$$\mathbf{d}_i + 4\mathbf{d}_{i+1} + \mathbf{d}_{i+2} = 6\hat{\mathbf{t}}_{i0} \ ; \quad i = 0, 1, ..., m, \qquad (7.16-a)$$

By evaluating (7.13) at "$u = v = 0$ with $i = 0, m$", we have

$$\hat{\mathbf{x}}_{00} = \mathbf{r}^{00}_{uv}(0,0) = -\mathbf{d}_0/2 + \mathbf{d}_2/2, \quad and \quad \hat{\mathbf{x}}_{m0} = \mathbf{r}^{m0}_{uv}(0,0) = -\mathbf{d}_m/2 + \mathbf{d}_{m+2}/2.$$

These *twist vector relations* at the "left" corners are rearranged as

$$\mathbf{d}_2 - \mathbf{d}_0 = 2\hat{\mathbf{x}}_{00} \quad ; \quad \mathbf{d}_{m+2} - \mathbf{d}_m = 2\hat{\mathbf{x}}_{m0}. \tag{7.16 – b}$$

A mesh point net and a B-spline control point net are shown in Fig. 7.10. Shown in Fig. 7.10a are cross boundary tangents $\mathbf{t}_{i0}$ and corner twist vectors $\mathbf{x}_{00}, \mathbf{x}_{m0}$. The boundary vectors are depicted in Fig. 7.10b. Written in a matrix form, the linear equation system (7.16) for the left **boundary vectors** $\{\mathbf{d}_i : i = 0, 1, ..., m+2\}$ is expressed as

$$\begin{bmatrix} -1 & 0 & 1 & 0 & & & \\ 1 & 4 & 1 & 0 & & & \\ 0 & 1 & 4 & 1 & & & \\ & & & \cdot & & & \\ & & & & \cdot & & \\ & & & & 1 & 4 & 1 \\ & & & & -1 & 0 & 1 \end{bmatrix} \begin{bmatrix} \mathbf{d}_0 \\ \mathbf{d}_1 \\ \mathbf{d}_2 \\ \cdot \\ \cdot \\ \mathbf{d}_{m+1} \\ \mathbf{d}_{m+2} \end{bmatrix} = \begin{bmatrix} 2\hat{\mathbf{x}}_{00} \\ 6\hat{\mathbf{t}}_{00} \\ 6\hat{\mathbf{t}}_{i0} \\ \cdot \\ \cdot \\ 6\hat{\mathbf{t}}_{m0} \\ 2\hat{\mathbf{x}}_{m0} \end{bmatrix}. \tag{7.17}$$

Let $\{\mathbf{e}_i\}$ denote "right boundary vectors" so that

$$\mathbf{e}_i = (V_{i,n+2} - V_{i,n})/2 \quad for\ i = 0, 1, ..., m+2.$$

Then, relations similar to (7.16-a) are obtained from (7.12-b) as

$$\mathbf{e}_i + 4\mathbf{e}_{i+1} + \mathbf{e}_{i+2} = 6\hat{\mathbf{t}}_{in} \; ; \quad i = 0, 1, ..., m \tag{7.18 – a}$$

From (7.13-b), *twist vector relations* at the "right" corners are obtained as

$$\mathbf{e}_2 - \mathbf{e}_0 = 2\hat{\mathbf{x}}_{0n} \; ; \quad \mathbf{e}_{m+2} - \mathbf{e}_m = 2\hat{\mathbf{x}}_{mn} \tag{7.18 – b}$$

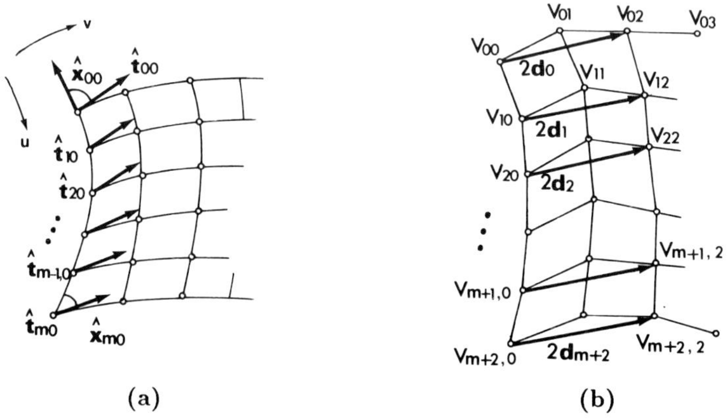

Figure 7.10 Boundary Conditions and Boundary Vectors $(\mathbf{d}_i)$

3) Determination of B-spline Control Vertices:

Now the "boundary vectors" $\mathbf{d}_i, \mathbf{e}_i$ are known values. Thus additional equations needed to solve (7.15) are given by

$$(V_{i,2} - V_{i,0}) = 2\mathbf{d}_i \quad and \quad (V_{i,n+2} - V_{i,n}) = 2\mathbf{e}_i$$

which are nothing but the definition of the "boundary vectors". The combined linear equation system (ie, Eqn. (7.15) and the above two equations) for **B-spline control vertices** is then expressed in a matrix form as :

$$\begin{bmatrix} -1 & 0 & 1 & 0 & & & \\ 1 & 4 & 1 & 0 & & & \\ 0 & 1 & 4 & 1 & & & \\ & & & \cdot & & & \\ & & & & \cdot & & \\ & & & & 1 & 4 & 1 \\ & & & & -1 & 0 & 1 \end{bmatrix} \begin{bmatrix} V_{i,0} \\ V_{i,1} \\ V_{i,2} \\ \cdot \\ \cdot \\ V_{i,n+1} \\ V_{i,n+2} \end{bmatrix} = \begin{bmatrix} 2\mathbf{d}_i \\ 6C_{i0} \\ 6C_{i1} \\ \cdot \\ \cdot \\ 6C_{in} \\ 2\mathbf{e}_i \end{bmatrix} \quad for\ i = 0, 1, ..., m+2 \qquad (7.19)$$

where $\{C_{ij}\}$ are intermediate control vertices given in (7.14) and $\{\mathbf{d}_i\}$ and $\{\mathbf{e}_i\}$ are boundary vectors in (7.17) and (7.18), respectively.

Once the control vertices are determined from (7.19), the *control vertex matrix* $\mathbf{B}^{ij}$ is constructed for the region $\{P_{i,j}, P_{i+1,j}, P_{i+1,j+1}, P_{i,j+1}\}$. Then the **bicubic uniform B-spline patch** $\mathbf{r}^{ij}(u,v)$ is expressed as:

$$\mathbf{r}^{ij}(u,v) = \mathbf{U}\ \mathbf{N}\ \mathbf{B}^{ij}\ \mathbf{N}^{\mathrm{T}}\ \mathbf{V}^{\mathrm{T}} \quad for\ i = 0, ..., m-1\ ;\ j = 0, ..., n-1 \qquad (7.20)$$

$$where, \quad \mathbf{U} = [1\ u\ u^2\ u^3],$$

$$\mathbf{V} = [1\ v\ v^2\ v^3],$$

$$\mathbf{N} = \begin{bmatrix} V_{00} & V_{01} & V_{02} & V_{03} \\ V_{10} & V_{11} & V_{12} & V_{13} \\ V_{20} & V_{21} & V_{22} & V_{23} \\ V_{30} & V_{31} & V_{32} & V_{33} \end{bmatrix},$$

$$\mathbf{N} = \frac{1}{6}\begin{bmatrix} 1 & 4 & 1 & 0 \\ -3 & 0 & 3 & 0 \\ 3 & -6 & 3 & 0 \\ -1 & 3 & -3 & 1 \end{bmatrix},$$

$$\mathbf{B}^{ij} = \begin{bmatrix} V_{i,j} & V_{i,j+1} & V_{i,j+2} & V_{i,j+3} \\ V_{i+1,j} & V_{i+1,j+1} & V_{i+1,j+2} & V_{i+1,j+3} \\ V_{i+2,j} & V_{i+2,j+1} & V_{i+2,j+2} & V_{i+2,j+3} \\ V_{i+3,j} & V_{i+3,j+1} & V_{i+3,j+2} & V_{i+3,j+3} \end{bmatrix}.$$

7.5 NON-UNIFORM B-SPLINE SURFACE FITTING

The steps for constructing a (composite) NUB surface are basically the same as those for uniform B-spline surface fitting. Figure 7.11 shows various geometric entities defining a *bicubic non-uniform B-spline patch* $\mathbf{r}^{ij}(u,v)$. They are

- $\{V_{i,j}$ *to* $V_{i+3,j+3}\}$: *16 control vertices,*
- $\{..,s_i,s_{i+1},..\}$: *u-direction knot vector,*
- $\{..,t_j,t_{j+1},..\}$: *v-direction knot vector.*

The domain of the NUB patch $\mathbf{r}^{ij}(u,v)$ is a rectangular **knot region** given by

$$(s,t) \in [s_i, s_{i+1}] \times [t_i, t_{i+1}].$$

In order to define the NUB patch in a *unit square domain* $(u,v) \in [0,1] \times [0,1]$, the parameters s, t are transformed as

$$u = (s - s_i)/\Delta_i, \quad and \quad v = (t - t_j)/\nabla_j,$$

where $\Delta_i = (s_{i+1} - s_i)$ and $\nabla_j = (t_{j+1} - t_j)$.

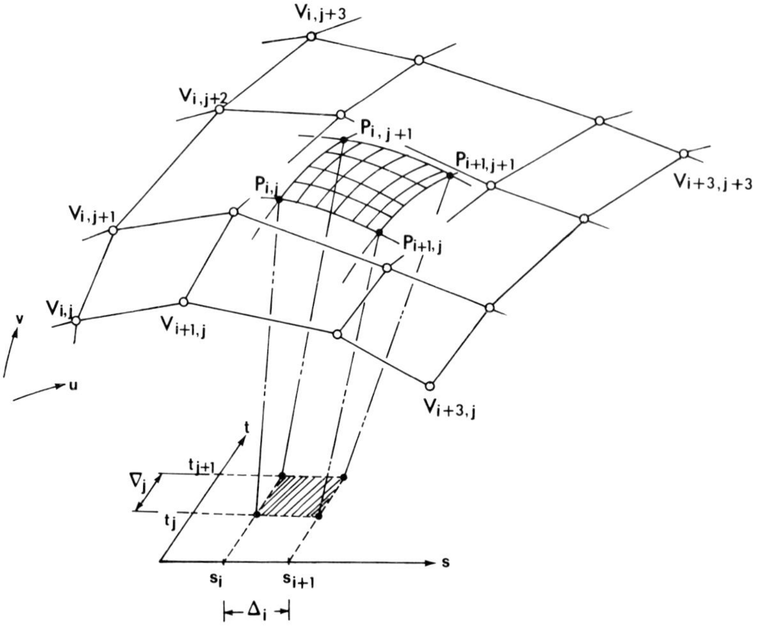

Figure 7.11 Geometric Handles Defining a Bicubic NUB Patch

With the above parameter transformation, a **bicubic NUB patch** is expressed in a matrix form as

$$\mathbf{r}^{ij}(u,v) = \mathbf{U}\ \mathbf{N_s}\ \mathbf{B^{ij}}\ \mathbf{N_t^T}\ \mathbf{V^T} \tag{7.21}$$

$$where, \mathbf{U} = [1\ u\ u^2\ u^3], \quad \mathbf{V} = [1\ v\ v^2\ v^3],$$

$$\mathbf{B}^{ij} = \begin{bmatrix} V_{i,j} & V_{i,j+1} & V_{i,j+2} & V_{i,j+3} \\ V_{i+1,j} & V_{i+1,j+1} & V_{i+1,j+2} & V_{i+1,j+3} \\ V_{i+2,j} & V_{i+2,j+1} & V_{i+2,j+2} & V_{i+2,j+3} \\ V_{i+3,j} & V_{i+3,j+1} & V_{i+3,j+3} & V_{i+3,j+3} \end{bmatrix},$$

$$\mathbf{N_s} : as\ defined\ in\ (4.30)\ with\ knot\ spans\ \Delta_i,$$

$$\mathbf{N_t} : as\ defined\ in\ (4.30)\ with\ knot\ spans\ \nabla_i.$$

The overall procedure of constructing a *composite NUB surface* from the input point data of Fig. 7.2 consists of four separate steps:

a) determination of *knot spans* $\{\Delta_i\}$ *and* $\{\nabla_j\}$,
b) determination of *intermediate control points* $\{C_{ij}\}$,
c) determination of *boundary vectors* $\{\mathbf{d}_i, \mathbf{e}_i\}$, and
d) determination of B-spline control vertices $\{V_{ij}\}$.

Once the B-spline control vertices $\{V_{i,j}\}$ are obtained, each patch in the composite NUB surface is evaluated by using (7.21).

1) Estimation of Knot Spans:

For the input data points given in Fig. 7.2, we need to determine knot spans in u-direction $\{\Delta_i : i \in [-2, m+1]\}$ and v-direction knot spans $\{\nabla_j : j \in [-2, n+1]\}$. The correspondence between knot spans and input data are depicted in Fig. 7.11 and Fig. 4.7. As discussed in §4.3.3, a reasonable choice for the knot spans is to make them proportional to the chord-lengths. Thus, the *u-direction* **knot spans** within the actual parameter range (ie, $s \in [s_0, s_m]$) may be given by

$$\Delta_i = \sum_{j=0}^{n} |P_{i+1,j} - P_{i,j}| \quad for \quad i = 0, 1, ..., m-1 \tag{7.22-a}$$

and the *v-direction* **knot spans** by

$$\nabla_j = \sum_{i=0}^{m} |P_{i,j+1} - P_{i,j}| \quad for \quad j = 0, 1, ..., n-1. \tag{7.22-b}$$

The *extended knot spans* may be set to zero (ie, *knot multiplicity of* 3):

$$\Delta_{-2} = \Delta_{-1} = 0 = \Delta_m = \Delta_{m+1}\ ;$$
$$\nabla_{-2} = \nabla_{-1} = 0 = \nabla_n = \nabla_{n+1}. \tag{7.22-c}$$

2) Determination of Intermediate Control Vertices:

At this step, a NUB curve is fitted from each *column* "j" of the input data. That is, the problem here is *(for each column j)*

Given: $\{P_{i(j)} : i = 0, 1..., m\}$: *a sequence mesh points, and*

$\hat{\mathbf{s}}_{0(j)}, \hat{\mathbf{s}}_{m(j)}$: *end boundary tangents.*

Find: $\{C_{i(j)} : i = 0, 1..., m+2\}$: *intermediate control vertices.*

The problem is exactly the one of fitting a *composite NUB curve* from a sequence of data points, which has already been discussed in §4.3. Thus, from (4.33), the **intermediate control vertices** can be obtained from the following linear equation system ($for\ j = 0, ..., n$):

$$\begin{bmatrix} -3 & 3 & 0 & 0 & & & \\ f_0 & (1-f_0-g_0) & g_0 & 0 & & & \\ & & & \cdot & & & \\ & & & & \cdot & & \\ & & & & f_m & (1-f_m-g_m) & g_m \\ & & & & 0 & -3 & 3 \end{bmatrix} \begin{bmatrix} C_{0,j} \\ C_{1,j} \\ \cdot \\ \cdot \\ C_{m+1,j} \\ C_{m+2,j} \end{bmatrix} = \begin{bmatrix} \hat{\mathbf{s}}_{0,j} \\ P_{0,j} \\ \cdot \\ \cdot \\ P_{m,j} \\ \hat{\mathbf{s}}_{m,j} \end{bmatrix} \tag{7.23}$$

$$where,\ f_i = (\Delta_i)^2/(\Delta_{i-1}^2 \Delta_{i-2}^3) \quad and$$

$$g_i = (\Delta_{i-1})^2/(\Delta_{i-1}^2 \Delta_{i-1}^3)\ for\ i = 0, ..., m,$$

$$\Delta_i : \ knot\ spans\ as\ defined\ in (7.22-a).$$

3) Determination of Boundary Vectors:

The next step is to determine the *boundary vectors* $\mathbf{d}_i, \mathbf{e}_i$ from the *boundary tangents* $\{\hat{\mathbf{t}}_{i0}, \hat{\mathbf{t}}_{in} : i = 0, ..., m\}$ and *corner twist* vectors. As depicted in Fig. 7.12, the **boundary vectors** are defined as ($for\ i = 0, ..., m+2$)

$$\mathbf{d}_i = 3(V_{i1} - V_{i0}) \tag{7.24 - a}$$

$$\mathbf{e}_i = 3(V_{i,n+2} - V_{i,n+1}). \tag{7.24 - b}$$

The v-direction *cross boundary tangents* are obtained by evaluating the v-direction derivatives of (7.21). Namely,

$$\mathbf{t}_{i,0} = \frac{\partial}{\partial v}\mathbf{r}^{i0}(0,0) \quad and \quad \mathbf{t}_{i,n} = \frac{\partial}{\partial v}\mathbf{r}^{i,n-1}(0,1). \tag{7.24 - c}$$

By evaluating (7.24-c) and then substituting the relations given in (7.24-a) and (7.24-b) into the result, the following linear equations are obtained:

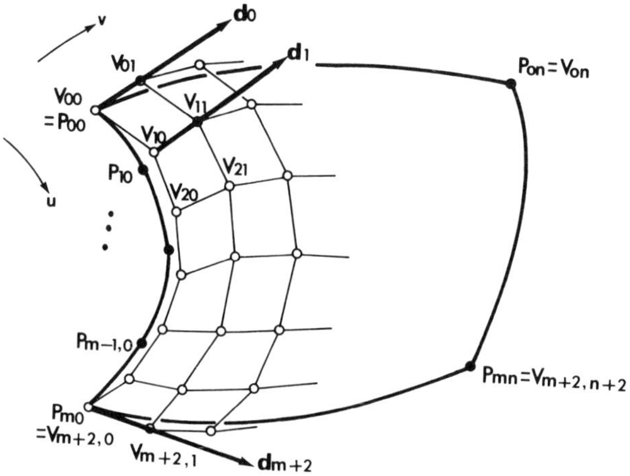

Figure 7.12 Control Vertices and Boundary Vectors of a NUB Surface

$$f_i\mathbf{d}_i + (1 - f_i - g_i)\mathbf{d}_{i+1} + g_i\mathbf{d}_{i+2} = \hat{\mathbf{t}}_{i0}\ ; \quad for \quad i = 0, ..., m \qquad (7.25-a)$$

$$f_i\mathbf{e}_i + (1 - f_i - g_i)\mathbf{e}_{i+1} + g_i\mathbf{e}_{i+2} = \hat{\mathbf{t}}_{in}\ ; \quad for \quad i = 0, ..., m \qquad (7.25-b)$$

where f_i and g_i are as defined in (7.23).

Further, on evaluating the *cross derivatives* of the NUB patch equation (7.21) at the corners, we have

$$3(\mathbf{d}_1 - \mathbf{d}_0) = \hat{\mathbf{x}}_{00}\ ; \quad 3(\mathbf{d}_{m+2} - \mathbf{d}_{m+1}) = \hat{\mathbf{x}}_{m0} \qquad (7.26-a)$$

$$3(\mathbf{e}_1 - \mathbf{e}_0) = \hat{\mathbf{x}}_{0n}\ ; \quad 3(\mathbf{e}_{m+2} - \mathbf{e}_{m+1}) = \hat{\mathbf{x}}_{mn}. \qquad (7.26-b)$$

The "left" boundary vectors $\{\mathbf{d}_i\}$ are then determined from (7.25-a) & (7.26-a), and the "right" boundary vectors $\{\mathbf{e}_i\}$ from (7.25-b) & (7.26-b). Both linear equation systems have the same $\mathbf{M}\ \mathbf{x} = \mathbf{b}$ form as (7.23).

4) Determination of NUB Control Vertices:

The final step is to determine the B-spline control vertices $\{V_{ij}\}$ by fitting a composite NUB curve from each "row" i of *intermediate control vertices.* The problem here is to determine a set of NUB control vertices from a "row" of intermediate control vertices and boundary vectors. Again, the problem is exactly the one of fitting a *composite NUB curve* from a sequence of point data which has already been discussed in §4.3 (*see also Eqn.* 7.23).

Thus, the **NUB control vertices** are obtained by solving the following linear equation system for each "row" i *(for i=0,...,m+2)*:

$$\begin{bmatrix} -3 & 3 & 0 & 0 & & & \\ f_0 & (1-f_0-g_0) & g_0 & 0 & & & \\ & & & \cdot & & & \\ & & & & \cdot & & \\ & & & & f_n & (1-f_n-g_n) & g_n \\ & & & & 0 & -3 & 3 \end{bmatrix} \begin{bmatrix} V_{i,0} \\ V_{i,1} \\ \cdot \\ \cdot \\ V_{i,n+1} \\ V_{i,n+2} \end{bmatrix} = \begin{bmatrix} \mathbf{d}_i \\ C_{i,0} \\ \cdot \\ \cdot \\ C_{i,n} \\ \mathbf{e}_i \end{bmatrix} \tag{7.27}$$

where, f_j *and* g_j *are as defined in* (7.23) *with* ∇_j *(instead of* Δ_i*),*

$\{C_{ij}\}$*: intermediate control vertices from* (7.23),

$\mathbf{d}_i$*: "left" boundary vectors from* $(7.25-a)$ *&* $(7.26-a)$,

$\mathbf{e}_i$*: "right" boundary vectors from* $(7.25-b)$ *&* $(7.26-b)$.

The procedure for fitting a composite NUB surface may be summarized as follows:

a) Determine the *knot spans* from (7.22).
b) Find *intermediate control vertices* $\{C_{ij}\}$ by solving (7.23).
c) Determine *boundary vectors* $\mathbf{d}_i, \mathbf{e}_i$ from (7.25) and (7.26).
d) Determine B-spline control points $\{V_{ij}\}$ by solving (7.27).

The resulting composite surface consists of an array of bicubic NUB patches each having the patch equation (7.21). It is a C^2 surface and is completely defined in terms of the following *geometric handles*:

$\mathbf{S} = \{V_{ij} : i = 0, 1, ..., m+2;\ j = 0, 1, ..., n+2\}$: control vertices,
$\mathbf{D_s} = \{\Delta_{-2}, \Delta_{-1}, ..., \Delta_{m+1}\}$: u-direction (or s-direction) knot span vector,
$\mathbf{D_t} = \{\nabla_{-2}, \nabla_{-1}, ..., \nabla_{n+1}\}$: v-direction (or t-direction) knot span vector.

The composite NUB method introduced in this section produces *aesthetically smooth* surface when the physical spacing of data points is *semi-even*. This is as far as we can get with the *tensor product* methods. When the data point spacing is *uneven*, we need a completely different construction scheme which is the subject of the next section.

7.6 COMPOSITE BEZIER SURFACE FITTING

If the physical spacing of data points is *completely uneven* as the one shown in Fig. 7.13b, all the *tensor product* methods (ie, the Ferguson methods and B-spline methods) suffer from local flatness and bulges because the differences in individual chord-lengths can not be taken into account. This section presents a method, due to Choi *et al* (1990a), of constructing a G^1 composite surface which is free of local flatness and bulges even with a very uneven spacing of data points.

The *composite Bezier surface fitting method* to be introduced in this section consists of the three separate steps below:

- *construction of Bezier mesh curves,*
- *determination of "off-boundary" control points by using a G^1 condition, and*
- *determination of "internal" control points.*

It is a localized curve net interpolation scheme based on a *geometric continuity condition* between rectangular Bezier patches.

Localized surface interpolation schemes have been studied quite extensively. Chiyokura (1986) and Yoo (1987) proposed methods of constructing G^1 composite surfaces consisting of rational cubic patches. A G^1 surface construction method based on subdivision techniques is presented in Shirman and Sequin (1987). A general method of determining "twist points" is proposed in Sarraga (1987). More details on the subject of curve net interpolation may be found in Chapter 10.

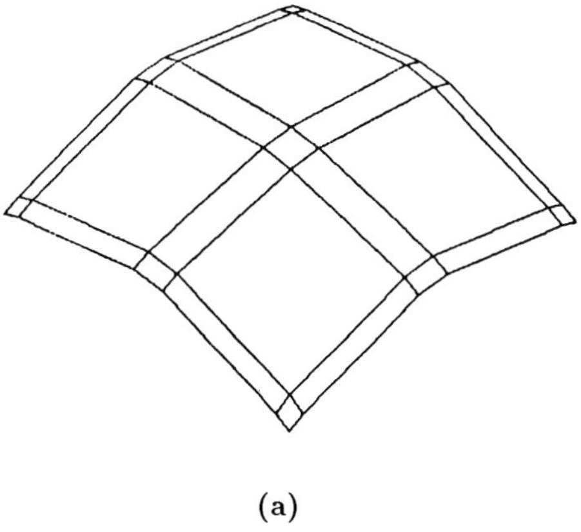

(a)

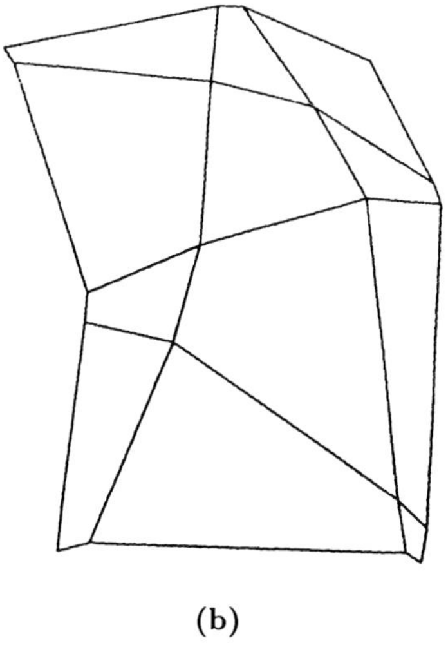

(b)

Figure 7.13 Semi-evenly and Unevenly Spaced Data Points

Before proceeding with the surface construction steps, basics of Bezier geometry from the previous chapter are reviewed. A Bezier curve is defined as

$$\mathbf{r}(u) = \sum_{i=0}^{n} B_i^n(u) V_i \ ; \quad 0 \le u \le 1 \tag{7.28}$$

$$where, \ B_i^n(u) = \frac{n!}{(n-i)! \ i!} u^i (1-u)^{n-1}.$$

A Bezier curve of degree n can be *degree-elevated as*

$$V_i^+ = \frac{i}{n+1} V_{i-1} + (1 - \frac{i}{n+1}) V_i \ ; \quad i = 0, 1, ..., n+1 \tag{7.29}$$

where, V_i : *control vertices of a given Bezier curve of degree n,*

V_i^+ : *control vertices of the degree elevated Bezier curve.*

A Bezier patch of degrees m, n is expressed as

$$\mathbf{r}^{m,n}(u,v) = \sum_{i=0}^{m} \sum_{j=0}^{n} B_i^m(u) B_j^n(v) V_{ij}. \tag{7.30}$$

The derivative of the *Bernstein polynomial* is given by

$$\frac{d}{du} B_i^n(u) = n\big(B_{i-1}^{n-1}(u) - B_i^{n-1}(u)\big). \tag{7.31}$$

The following identity relation is very useful in deriving a geometric continuity condition between Bezier patches:

$$B_i^m(u) \sum_{j=0}^{n} V_j B_j^n(u) \equiv \frac{m! \ n!}{(m+n)!} \sum_{j=0}^{m+n} \binom{j}{i} \binom{m+n-j}{m-i} V_{j-i} B_j^{m+n}(u) \tag{7.32}$$

where the terms having negative indices or undefined combinations are all set to zero.

Shown in Fig. 7.14 are two surface patches $\mathbf{r}^a(u,v)$ and $\mathbf{r}^b(u,v)$ sharing a common boundary curve, $\mathbf{r}^a(0,v) = \mathbf{r}^b(0,v)$ for $v \in [0,1]$. The designation of the parameter directions is made such that the "u=0 curve" of both patches becomes the common boundary curve. Let us define the derivatives of the surface patches on the common boundary curve as

$$\mathbf{a}(v) = \mathbf{r}_u^a(0,v) \tag{7.33 – a}$$

$$\mathbf{b}(v) = \mathbf{r}_u^b(0,v) \tag{7.33 – b}$$

$$\mathbf{c}(v) = \mathbf{r}_v^a(0,v) = \mathbf{r}_v^b(0,v) \tag{7.33 – c}$$

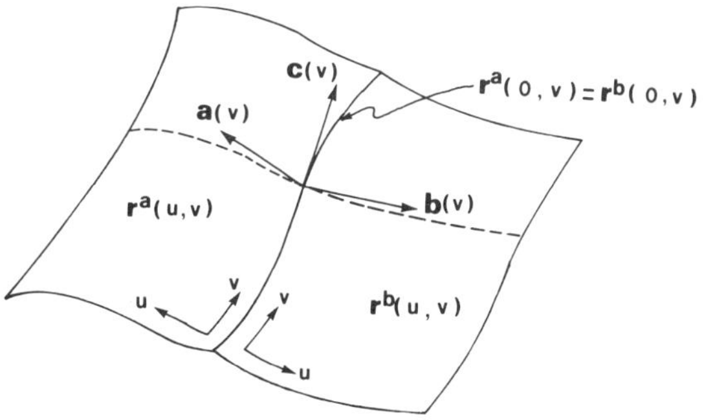

Figure 7.14 Construction of G^1-condition between Two Patches

The derivative $\mathbf{c}(v)$ is the tangent vector of the common boundary curve, and $\mathbf{a}(v)$ and $\mathbf{b}(v)$ are *cross-boundary* tangent vectors of $\mathbf{r}^a(u,v)$ and $\mathbf{r}^b(u,v)$, respectively.

The two patches are said to be G^1 *(geometric C^1)-continuous* if the tangent planes of the two patches along the common boundary curve coincide. If the cross-boundary tangent vectors of the two patches along the common boundary curve are collinear, they are "collinear G^1-continuous". This "tangent plane continuity" is also called a "gradient continuity" or "visual continuity". The tangent plane continuity requires that the three tangent vectors $\mathbf{a}(v), \mathbf{b}(v), \mathbf{c}(v)$ lie on the same plane, which means that there exist three scalar functions $p(v), q(v), r(v)$ which are not zero at the same time such that the following holds:

$$p(v)\mathbf{a}(v) + q(v)\mathbf{b}(v) + r(v)\mathbf{c}(v) = \mathbf{0} \tag{7.34}$$

This is a general G^1-condition between two rectangular patches.

7.6.1 Construction of Bezier Mesh Curves

The first step of constructing a G^1 composite Bezier surface is to construct a smooth curve-net from the input data of Fig. 7.2. The main idea here is to fit a smooth composite curve from each "row" and "column" of input data. Since the physical spacing of data points is uneven, we have a curve fitting problem for unevenly spaced point data. Thus, any one of the curve fitting methods introduced in §4.3 may be employed. If the *chord-length spline fitting* method is used, each of the *Ferguson curve* in the chord-length spline is then converted to a *cubic Bezier curve* (see Eqn. (3.10)).

The chord-length spline fitting method of §4.3.2 is reproduced here. For example, a chord-length spline curve consisting of Ferguson curves can be fitted from row "i" of input data

$$\hat{\mathbf{t}}_{i0}, P_{i0}, P_{i1}, ..., P_{in}, \hat{\mathbf{t}}_{in}$$

by solving the following linear equation system (See Eqn. (4.27)):

$$\begin{bmatrix} 1 & 0 & 0 & & & \\ \omega_1^2 & 2\omega_1 + 2\omega_1^2 & 1 & & & \\ & & \cdot & & & \\ & & & \cdot & & \\ & & & \omega_{n-2}\omega_{n-1}^2 & 2\omega_{n-1} + 2\omega_{n-1}^2 & 1 \\ & & & 0 & 0 & 1 \end{bmatrix} \begin{bmatrix} \mathbf{t}_{i,0} \\ \mathbf{t}_{i,1} \\ \cdot \\ \cdot \\ \mathbf{t}_{i,n-1} \\ \mathbf{t}_{i,n} \end{bmatrix} = \begin{bmatrix} 2\hat{\mathbf{t}}_{i0} \\ Q_1 \\ \cdot \\ \cdot \\ Q_{n-1} \\ 2\hat{\mathbf{t}}_{in} \end{bmatrix} \tag{7.35}$$

$$where, \quad Q_j = 3\{P_{i,j+1} + (\omega_j^2 + 1)P_{i,j} - \omega_j^2 P_{i,j-1}\},$$

$$\omega_j = |P_{i,j+1} - P_{i,j}|/|P_{i,j} - P_{i,j-1}| \quad for \quad j = 1, 2, ..., n-1,$$

$$\omega_0 = 1.$$

Once all the $\mathbf{t}_{(i)j}$'s in (7.35) have been determined, the jth curve segment spanning $[P_{i,j-1}, P_{i,j}]$ is defined in terms of the following *end condition vector*:

$$\mathbf{S}^j = [P_{i,j-1} \quad P_{i,j} \quad \omega_{j-1}\mathbf{t}_{i,j-1} \quad \mathbf{t}_{i,j}]^T.$$

The Bezier control vertices V_0, V_1, V_2, V_3 for this cubic segment are then obtained as (from the result of Example 4.1 in §4.2.3)

$$V_0 = P_{i,j-1}$$

$$V_1 = P_{i,j-1} + \omega_{j-1}\mathbf{t}_{i,j-1}/3$$

$$V_2 = P_{i,j} - \mathbf{t}_{i,j}/3$$

$$V_3 = P_{i,j}.$$

Finally, the cubic control vertices are converted to *sextic* control vertices by applying the degree elevation formula (7.29) three times. The reason for this degree elevation will be explained later.

By repeating the above steps for each curve segment, row-wise and column-wise, a smooth mesh of *sextic (degree 6) Bezier curves* is obtained. Thus, a rectangular region in the grid mesh is surrounded by four sextic Bezier curves. If each rectangular region is modeled as a *bisextic Bezier patch* while maintaining required geometric continuity across patch boundaries, the curve mesh would become a *composite Bezier surface*. In other words, it is necessary to provide internal Bezier control vertices in order to make each region a bisextic Bezier patch.

The control vertices defining the mesh curves are called **boundary control vertices**, and those directly connected to the *boundary control vertices* are called **off-boundary control points**. There are 49 control vertices (or control points) in a bisextic Bezier patch. Among them, 24 are *boundary control points* and 16 are *off-boundary control points*. The remaining ones at the center of the patch are called **internal control points**.

7.6.2 Determination of Off-boundary Control Points

In order to determine the *off-boundary control points* so that each pair of Bezier patches are G^1-continuous, we need to have a suitable G^1 condition. Let $A(u,v)$ and $B(u,v)$ denote two adjacent rectangular Bezier patches as shown in Fig. 7.15. The parametric directions of the two Bezier patches are defined such that the common boundary curve becomes $A(0,v) \equiv B(0,v)$. For notational simplicity, we introduce "control vectors" as depicted in Fig. 7.15:

$$\mathbf{a}_i = R_i - S_i \tag{7.36 – a}$$

$$\mathbf{b}_i = T_i - S_i \tag{7.36 – b}$$

$$\mathbf{c}_i = S_{i+1} - S_i \tag{7.36 – c}$$

where, $S_i = V^a_{0i} \equiv V^b_{0i}$: *common boundary control points,*

$R_i = V^a_{1i}$: *off-boundary control points of A(u,v),*

$T_i = V^b_{1i}$: *off-boundary control points of B(u,v).*

If the degrees of $A(u,v)$ and $B(u,v)$ are m and n, their surface equations are given by

$$A(u,v) = \sum_{i=0}^{m}\sum_{j=0}^{n} V^a_{ij} B^m_i(u) B^n_j(v)$$

$$B(u,v) = \sum_{i=0}^{m}\sum_{j=0}^{n} V^b_{ij} B^m_i(u) B^n_j(v).$$

From the results of (7.31) and (7.36), the u-directional derivative of $A(u,v)$ along the common boundary curve is obtained as

$$\begin{aligned} \mathbf{a}(v) &= A_u(0,v) \\ &= m \sum_{j=0}^{n} \mathbf{a}_j B^n_j(v). \end{aligned} \tag{7.37 – a}$$

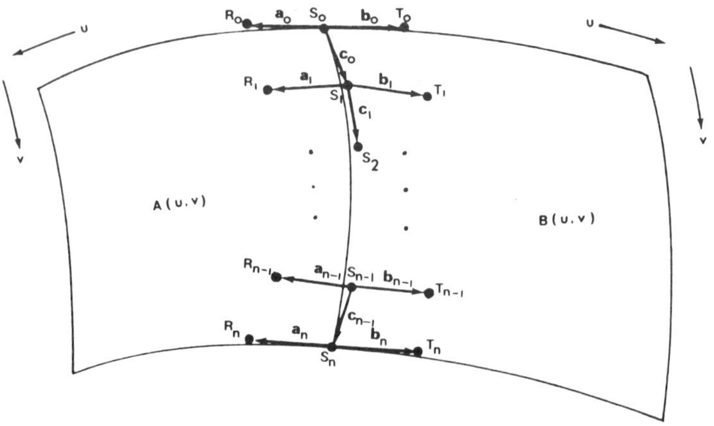

Figure 7.15 G^1 Construction between Bezier Patches

Similarly, the u-directional derivative of $B(u, v)$ at $u = 0$ is expressed as

$$\mathbf{b}(v) = B_u(0, v) = m \sum_{j=0}^{n} \mathbf{b}_j B_j^n(v). \qquad (7.37 - b)$$

And the tangent vector of the common boundary curve is also expressed similarly as

$$\mathbf{c}(v) = A_v(0, v) = n \sum_{j=0}^{n-1} \mathbf{c}_j B_j^{n-1}(v). \qquad (7.37 - c)$$

Since the degree of $\mathbf{c}(v)$ in (7.37) is one lower than that of $\mathbf{a}(v)$ and $\mathbf{b}(v)$, the degree of $r(v)$ in (7.34) may be set to one higher than that of $p(v)$ and $q(v)$, or zero (ie, $r(v) \equiv 0$). For computational simplicity, we employ the following Bezier form for the scalar functions in (7.34):

$$p(v) = \sum_{j=0}^{k} p_j B_j^k(v) \qquad (7.38 - a)$$

$$q(v) = \sum_{j=0}^{k} q_j B_j^k(v) \qquad (7.38 - b)$$

$$r(v) = \sum_{j=0}^{k+1} r_j B_j^{k+1}(v) \quad (or\ r(v) \equiv 0). \qquad (7.38 - c)$$

On substituting (7.37) and (7.38) into the G^1 equation (7.34), and after simplifying the result by using the identity relation (7.32), the following G^1 condition is obtained.

$$\sum_{i=0}^{k} d_{ij}(p_i\mathbf{a}_{j-i} + q_i\mathbf{b}_{j-i}) + \sum_{i=0}^{k+1} e_{ij}(r_i\mathbf{c}_{j-i}) = \mathbf{0} \quad for \quad j = 0, 1, ..., n+k \tag{7.39}$$

$$where, \quad d_{ij} = \begin{cases} \binom{j}{i}\binom{n+k-j}{k+1-i} & \text{if } max(0, j-n) \leq i \leq min(j,k), \\ 0, & \text{otherwise.} \end{cases}$$

$$e_{ij} = \begin{cases} \frac{k+1}{n}\binom{j}{i}\binom{n+k-j}{k+1-i} & \text{if } max(0, j-n+1) \leq i \leq min(j, k+1), \\ 0, & \text{otherwise.} \end{cases}$$

p_i, q_i, r_i : *coefficients of the scalar functions in* (7.38),

$\mathbf{a}_i, \mathbf{b}_i, \mathbf{c}_i$: *Bezier vectors as defined in* (7.36) *and Fig.* 7.15.

In the G^1 equation (7.39), the u-direction degree m is set to be equal to the v-direction degree n. That is, $m = n$. There are many possibilities as to the choices of n, k, p_i, q_i, and r_i in (7.39). For the reasons explained in Choi *et al* (1990a), we set $k = 3$, $n = 6$, and $r_i = 0$. With these values of k, n, and r_i, the G^1 equation (7.39) may be evaluated as follows:

$$p_0\mathbf{a}_0 + q_0\mathbf{b}_0 = 0 \tag{7.40 - a}$$
$$6(p_0\mathbf{a}_1 + q_0\mathbf{b}_1) + 3(p_1\mathbf{a}_0 + q_1\mathbf{b}_0) = 0 \tag{7.40 - b}$$
$$15(p_0\mathbf{a}_2 + q_0\mathbf{b}_2) + 18(p_1\mathbf{a}_1 + q_1\mathbf{b}_1) + 3(p_2\mathbf{a}_0 + q_2\mathbf{b}_0) = 0 \tag{7.40 - c}$$
$$20(p_0\mathbf{a}_3 + q_0\mathbf{b}_3) + 45(p_1\mathbf{a}_2 + q_1\mathbf{b}_2) + 18(p_2\mathbf{a}_1 + q_2\mathbf{b}_1) + (p_3\mathbf{a}_0 + q_3\mathbf{b}_0) = 0 \tag{7.40 - d}$$
$$15(p_0\mathbf{a}_4 + q_0\mathbf{b}_4) + 60(p_1\mathbf{a}_3 + q_1\mathbf{b}_3) + 45(p_2\mathbf{a}_2 + q_2\mathbf{b}_2) + 6(p_3\mathbf{a}_1 + q_3\mathbf{b}_1) = 0 \tag{7.40 - e}$$
$$6(p_0\mathbf{a}_5 + q_0\mathbf{b}_5) + 45(p_1\mathbf{a}_4 + q_1\mathbf{b}_4) + 60(p_2\mathbf{a}_3 + q_2\mathbf{b}_3) + 15(p_3\mathbf{a}_2 + q_3\mathbf{b}_2) = 0 \tag{7.40 - f}$$
$$(p_0\mathbf{a}_6 + q_0\mathbf{b}_6) + 18(p_1\mathbf{a}_5 + q_1\mathbf{b}_5) + 45(p_2\mathbf{a}_4 + q_2\mathbf{b}_4) + 20(p_3\mathbf{a}_3 + q_3\mathbf{b}_3) = 0 \tag{7.40 - g}$$
$$3(p_1\mathbf{a}_6 + q_1\mathbf{b}_6) + 18(p_2\mathbf{a}_5 + q_2\mathbf{b}_5) + 15(p_3\mathbf{a}_4 + q_3\mathbf{b}_4) = 0 \tag{7.40 - h}$$
$$3(p_2\mathbf{a}_6 + q_2\mathbf{b}_6) + 6(p_3\mathbf{a}_5 + q_3\mathbf{b}_5) = 0 \tag{7.40 - i}$$
$$p_3\mathbf{a}_6 + q_3\mathbf{b}_6 = 0 \tag{7.40 - j}$$

Recall that the *boundary control vectors* $\mathbf{a}_0, \mathbf{a}_6, \mathbf{b}_0, \mathbf{b}_6$ have already been determined. Note also that (7.40) in fact is a *collinear* G^1 condition because $r(v) \equiv 0$. It can be shown that a satisfactory choice for the coefficients of (7.40) is given by

$$p_0 = p_1 = b_0; \quad p_2 = p_3 = b_6; \quad q_0 = q_1 = a_0; \quad q_2 = q_3 = a_6 \tag{7.41}$$

$$where, \quad a_i = |\mathbf{a}_i|, b_i = |\mathbf{b}_i|.$$

There are five *off-boundary control vectors* that have to be fixed in each of the Bezier patches;

$$\mathbf{a}_1 \sim \mathbf{a}_5 \; in \; A(u,v) \quad and \quad \mathbf{b}_1 \sim \mathbf{b}_5 \; in \; B(u,v).$$

From the results in (7.41) and (7.40-a), the second G^1 equation (7.40-b) reduces to

$$(b_0\mathbf{a}_1 + a_0\mathbf{b}_1) = 0. \tag{7.42}$$

A solution to (7.42) is obtained, again from (7.40-a), by setting

$$\mathbf{a}_1 = \mathbf{a}_0 \quad and \quad \mathbf{b}_1 = \mathbf{b}_0. \tag{7.43}$$

The control points $\mathbf{a}_1$ and $\mathbf{b}_1$ (actually R_1 and T_1 in Fig. 7.15) are termed as **twist points** in Sarraga (1987), and they could have been determined by solving a set of (under-determined) linear equations of the type (7.42) formed by the four patches sharing the "corner point" S_0 of Fig. 7.15. Actually, Sarraga's linear equations are satisfied by the solution (7.43) which probably would be the best choice. An implication of (7.43) is that it gives a complete localness to the surface construction scheme. Due to the symmetry, the other *twist points* $\mathbf{a}_5$ and $\mathbf{b}_5$ are obtained from $(7.40-i)$ as

$$\mathbf{a}_5 = \mathbf{a}_6 \quad and \quad \mathbf{b}_5 = \mathbf{b}_6. \tag{7.44}$$

The remaining off-boundary control points are inter-related by the middle G^1 equation (7.40-c) through (7.40-h). By substituting the results of (7.41) into these six G^1 equations, and after rearranging the terms, we obtain the following set of linear equations:

$$\begin{aligned}
&5(3b_0\mathbf{a}_2 + 3a_0\mathbf{b}_2) = -3(b_6\mathbf{a}_0 + a_6\mathbf{b}_0)\\
&5(9b_0\mathbf{a}_2 + 9a_0\mathbf{b}_2 + 4b_0\mathbf{a}_3 + 4a_0\mathbf{b}_3) = -19(b_6\mathbf{a}_0 + a_6\mathbf{b}_0)\\
&5(9b_6\mathbf{a}_2 + 9a_6\mathbf{b}_2 + 12b_0\mathbf{a}_3 + 12a_0\mathbf{b}_3 + 3b_0\mathbf{a}_4 + 3a_0\mathbf{b}_4) = -6(b_6\mathbf{a}_0 + a_6\mathbf{b}_0)\\
&5(3b_6\mathbf{a}_2 + 3a_6\mathbf{b}_2 + 12b_6\mathbf{a}_3 + 12a_6\mathbf{b}_3 + 9b_0\mathbf{a}_4 + 9a_0\mathbf{b}_4) = -6(b_0\mathbf{a}_6 + a_0\mathbf{b}_6)\\
&5(4b_6\mathbf{a}_3 + 4a_6\mathbf{b}_3 + 9b_6\mathbf{a}_4 + 9a_6\mathbf{b}_4) = -19(b_0\mathbf{a}_6 + a_0\mathbf{b}_6)\\
&5(3b_6\mathbf{a}_4 + 3a_6\mathbf{b}_4) = -3(b_0\mathbf{a}_6 + a_0\mathbf{b}_6)
\end{aligned}$$

After a series of row operations, the above linear equation system can be expressed in a matrix form as

$$\mathbf{M}\;\mathbf{X} = \mathbf{D} \tag{7.45}$$

where

$$\mathbf{M} = \begin{bmatrix} b_0 & a_0 & 0 & 0 & 0 & 0 \\ 0 & 0 & 0 & 0 & b_6 & a_6 \\ 0 & 0 & 4b_0 & 4a_0 & 0 & 0 \\ 0 & 0 & 4b_6 & 4a_6 & 0 & 0 \\ 3b_6 & 3a_6 & 0 & 0 & b_0 & a_0 \\ b_6 & a_6 & 0 & 0 & 3b_0 & 3a_0 \end{bmatrix} ; \mathbf{X} = \begin{bmatrix} \mathbf{a}_2 \\ \mathbf{b}_2 \\ \mathbf{a}_3 \\ \mathbf{b}_3 \\ \mathbf{a}_4 \\ \mathbf{b}_4 \end{bmatrix} ; \mathbf{D} = \frac{1}{5} \begin{bmatrix} -\mathbf{d}_1 \\ -\mathbf{d}_2 \\ -10\mathbf{d}_1 \\ -10\mathbf{d}_2 \\ 8\mathbf{d}_1 \\ 8\mathbf{d}_2 \end{bmatrix},$$

$$\mathbf{d}_1 = (b_6\mathbf{a}_0 + a_6\mathbf{b}_0),$$

$$\mathbf{d}_2 = (b_0\mathbf{a}_6 + a_0\mathbf{b}_6).$$

The determinant of the coefficient matrix M in (7.45) is computed as

$$det(\mathbf{M}) = K\{b_0 b_6(\alpha - \beta)\}^3, \tag{7.46}$$

where K: constant, $\alpha = a_0/b_0$ and $\beta = a_6/b_6$. Note in (7.46) that the ratios α, β of the *magnitudes of the tangent vectors at the two corner points* play a key role in solving the linear system because the rank of the coefficient matrix M is full only when $\alpha \neq \beta$.

When $\alpha \neq \beta$, the linear equation system (7.45) yields a unique solution. Since the rank of M is *full* and M is a sparse matrix, (7.45) may easily be solved for the unknown **off-boundary control points**. The solution in this case is given by

$$\mathbf{a}_2 = \{(3 + a_6/a_0)\mathbf{a}_0 + (a_0/a_6)\mathbf{a}_6\}/5 \tag{7.47 – a}$$

$$\mathbf{a}_3 = \{(a_6/a_0)\mathbf{a}_0 + (a_0/a_6)\mathbf{a}_6\}/2 \tag{7.47 – b}$$

$$\mathbf{a}_4 = \{(a_6/a_0)\mathbf{a}_0 + (3 + a_0/a_6)\mathbf{a}_6\}/5 \tag{7.47 – c}$$

$$\mathbf{b}_2 = \{(3 + b_6/b_0)\mathbf{b}_0 + (b_0/b_6)\mathbf{b}_6\}/5 \tag{7.48 – a}$$

$$\mathbf{b}_3 = \{(b_6/b_0)\mathbf{b}_0 + (b_0/b_6)\mathbf{b}_6\}/2 \tag{7.48 – b}$$

$$\mathbf{b}_4 = \{(b_6/b_0)\mathbf{b}_0 + (3 + b_0/b_6)\mathbf{b}_6\}/5 \tag{7.48 – c}$$

where $a_i = |\mathbf{a}_i|$ and $b_i = |\mathbf{b}_i|$. A nice feature of the solution procedure is that (7.47) and (7.48) are also valid for $\alpha = \beta$. Further, the **off-boundary** control points of a patch are determined solely from its own *boundary control points*, which makes the surface construction scheme completely local (and symmetric).

7.6.3 Determination of Internal Control Points

The composite G^1 surface interpolating over an array of $(m+1)\times(n+1)$ data points is composed of $m \times n$ bisextic Bezier patches. The *boundary control vectors* have already been determined during the mesh curve construction phase in §7.6.1, and the **off-boundary control vectors** are determined from (7.43) and (7.47) of §7.6.2. Figure 7.16a is a typical bisextic Bezier patch defined by 49 control points. The **internal control points** V_{ij} with $i, j \in [2,4]$ are yet to be determined. This section presents a simple method of estimating the internal control points.

Shown in Fig. 7.16b is a sextic Bezier curve whose internal control points V_2, V_3, V_4 are unknown. Since the given information about the curve $\mathbf{r}(u)$ are its "end conditions", it may be regarded as a cubic Bezier curve having four control points $\mathbf{e}_0, \mathbf{e}_1, \mathbf{e}_2, \mathbf{e}_3$. From the end conditions, we immediately have the following relations:

$$\mathbf{e}_0 = V_0; \quad \mathbf{e}_3 = V_6; \quad \mathbf{e}_1 = 2V_1 - V_0; \quad \mathbf{e}_2 = 2V_5 - V_6 \tag{7.49}$$

By elevating the degree of the cubic curve three times (using Eqn. (7.29)), a new set of control points $\mathbf{e}_i^*$ for $i = 0, 1, ..., 6$ are obtained. Then the unknown control points in Fig. 7.16b are given by $V_i = \mathbf{e}_i^*$ for $i = 2, 3, 4$. The internal control points of Fig. 7.16a are determined by

a) applying the above procedure for each row of the control vertices,
b) repeating it for each column, and
c) taking the average of the two.

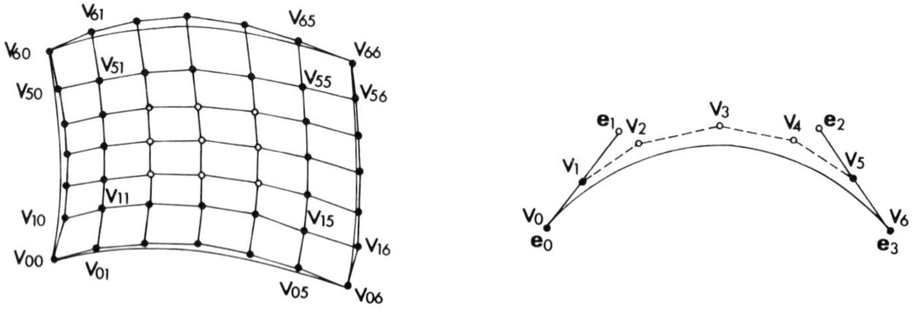

(a) (b)

Figure 7.16 Geometric Estimation of Internal Control Points

Figure 7.17a and 7.17b, respectively, are a NUB surface and a G^1 composite Bezier surface for the same input data given in Fig. 7.13b.

The proposed G^1 surface interpolation scheme is quite efficient in terms of both computation time and storage requirements. In addition, the surface construction scheme has some distinctive features, for example: a) it is a completely local scheme meaning that all the control points inside a patch are determined from its boundary control points only, b) isoparametric curves of the composite surface are smooth throughout the entire surface, and c) the surface consists of (nonrational) Bezier patches.

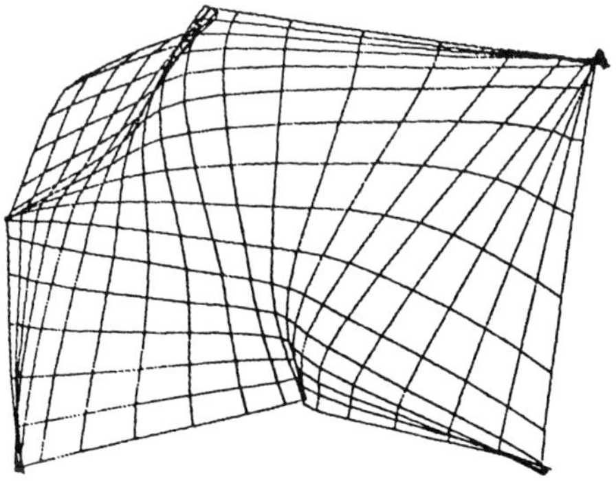

(a)

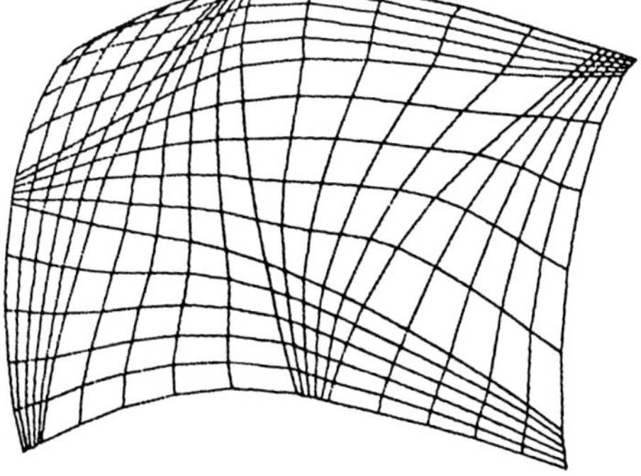

(b)

Figure 7.17 NUB Surface(a) and G^1 Bezier Surface(b)

7.7 INPUT DATA PREPARATION FOR COMPOSITE SURFACE FITTING

Throughout the discussion, it was assumed that input data were given in the form of Fig. 7.2, namely, an $(m+1) \times (n+1)$ array of *data points* $\{P_{ij}\}$, *cross boundary tangents* at the data points on the surface boundaries, and *twist vectors* at the corners. Further it was implicitly assumed that the input data were "true" coordinate values on the surface. In practice, however, the input data are usually given as a collection of *data paths* (possibly obtained from a coordinate measuring machine). Individual data paths may consist of unequal number of data points which are subject to "errors".

Thus, in order to construct a *smooth* (composite) surface from a collection of (measured) data paths, four phases of input data preparations are required. First, each data path has to be *faired* (or *smoothed* to form a smooth composite curve). This phase is termed **curve fairing**. The next step is called **remeshing** in which a rectangular mesh of data points is constructed from the fitted curves. Thirdly, once an array of data points is obtained, **surface fairing** is carried out if necessary. Finally, the *boundary tangents* and *corner twist vectors* are estimated (if they are not given).

7.7.1 Curve and Surface Fairing

Fairing (or smoothing) is a term used to mean the adjustment of the position of input data points in order to improve the smoothness of interpolating curves and surfaces. Fairing may be carried out either interactively or automatically. As it is not easy to give an exact definition of surface smoothness, interactive fairing methods have widely been used. In interactive smoothing, the surface is rendered (ie, displayed) on graphics screen and the user adjusts selected data points until he is satisfied with the smoothness of the resulting surface displayed. Surface rendering methods proposed in the literature include: display of *Gaussian curvature* on color graphics screen (Dill, 1981; Dill and Rogers, 1982), display of divided differences of feature curves (Renz, 1982), display of reflection lines (Klass, 1980; Klass, 1988), use of orthotomic curves (Hoschek, 1985) or polarity (Hoschek, 1984) to detect inflection points, and display of isophotes (Poechl, 1984).

For an automatic fairing, some form of smoothness criteria has to be defined first. Widely accepted fairing criteria are

a) surface continuity (C^1 or C^2),
b) no unwanted inflections, and
c) gradual change of curvature

as discussed in Ding and Davies (1987). Automatic fairing methods reported in the literature may be classified into energy method and local spring back method. In **energy methods**

(Kjellander, 1983a; Kjellander, 1983b), the curve is modeled as a physical beam and then fairing equations are derived from an objective function minimizing the internal bending energy of the beam.

The local **spring back method** (Kjellander, 1983a; Farin *et al*, 1987) simulates the behavior of the physical spline supported by a sequence of "weights" (ie, ducks). A suspected point is faired by removing the duck (or setting the weight of the duck to zero) on that point. This fairing action is equivalent to minimizing the difference in third derivatives on both sides of the suspected data point. For the standard cubic spline curve (see §4.2.2), local spring back method becomes a version of the energy method. A third method of fairing, called the **circle rate method**, is also reported in the literature (Ding and Davies, 1987). In this method, a circle is fitted from three consecutive data points, and the circle rate (ie, changes in the radii of successive circles) is taken as a measure of smoothness.

Let us consider the two curve segments $\mathbf{r}^a(u)$ and $\mathbf{r}^b(u)$ shown in Fig. 7.18. The middle point P_i is free to move so that the overall curve becomes smoother. In the local spring back method, the difference in third derivatives on both sides of the "suspected" point is to be minimized. That is, the *fairing criterion* in this case is given by

$$d^3\mathbf{r}^a(u)/du^3|_{u=1} - d^3\mathbf{r}^b(u)/du^3|_{u=0} = \mathbf{0}. \tag{7.50}$$

In the energy method, the total bending energy of the two curve segments is minimized. The total bending energy is approximated as

$$E = \int_0^1 [\ddot{\mathbf{r}}^a(u) \cdot \ddot{\mathbf{r}}^a(u)]du + \int_0^1 [\ddot{\mathbf{r}}^b(u) \cdot \ddot{\mathbf{r}}^b(u)]du. \tag{7.51}$$

Thus, the value of the *optimum point* P_i^* minimizing the energy is obtained by solving the following equation for P_i:

$$\partial E/\partial P_i = \mathbf{0}. \tag{7.52}$$

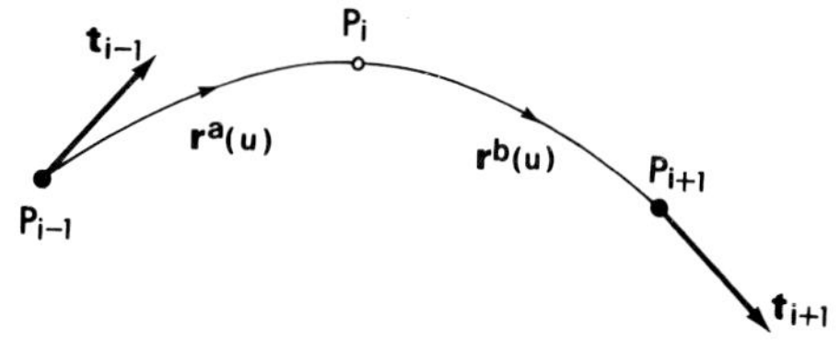

Figure 7.18 Curve Fairing Formulation

When the curve segments, $\mathbf{r}^a(u)$ and $\mathbf{r}^b(u)$, are Ferguson curves given by (4.7), both fairing criteria (7.50) and (7.52) give the same solution which is given by

$$P_i^* = [(P_{i-1} + \mathbf{t}_{i-1}/2) + (P_{i+1} - \mathbf{t}_{i+1}/2)]/2. \tag{7.53}$$

If a "damping factor" α is multiplied to the above equation, it becomes the fairing equation (4.54) introduced earlier in Chapter 4.

Let P_i° denote the original (measured) data point. In practice, the amount of *correction* is not allowed to exceed a specified amount ε_i so that the actual position of the data point P_i is computed from

$$P_i = P_i^\circ + \lambda(P_i^* - P_i^\circ)/|P_i^* - P_i^\circ| \tag{7.54}$$

where, $\lambda = min\{\varepsilon_i, |P_i^* - P_i^\circ|\}$,
$P_i^* = \alpha[(P_{i-1} + \mathbf{t}_{i-1}/2) + (P_{i+1} - \mathbf{t}_{i+1}/2)]/2$,
P_i°*: original input data point,*
ε_i*: maximum correction allowance for* P_i*,*
α*: damping factor.*

Fairing is carried out by applying (7.54) repeatedly to each point on a path.

Shown in Fig. 7.19 is a data point P_{ij} to be corrected. From a *fairing criterion* similar to (7.50) or (7.52), the optimum point is obtained as

$$\begin{aligned} P_{ij}^* =&(P_{i-1,j} + P_{i+1,j} + P_{i,j-1} + P_{i,j+1})/4 \\ &+ (\mathbf{s}_{i-1,j} - \mathbf{s}_{i+1,j} + \mathbf{t}_{i,j-1} - \mathbf{t}_{i,j+1})/8. \end{aligned} \tag{7.55}$$

From (7.55), a surface fairing equation of the form (7.54) may be derived. And then actual *surface fairing* is carried out by applying it to each data point.

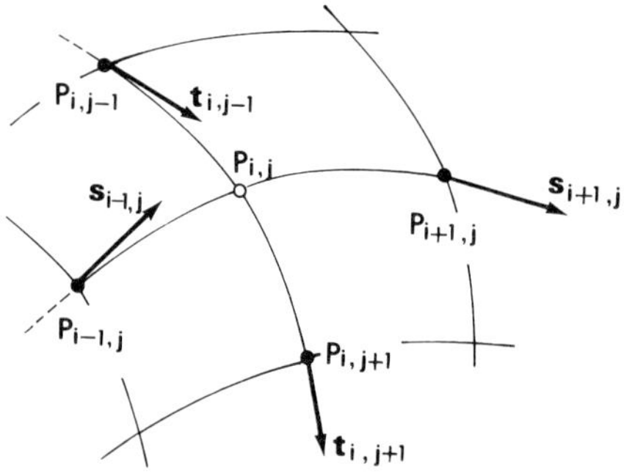

Figure 7.19 Surface Fairing Formulation

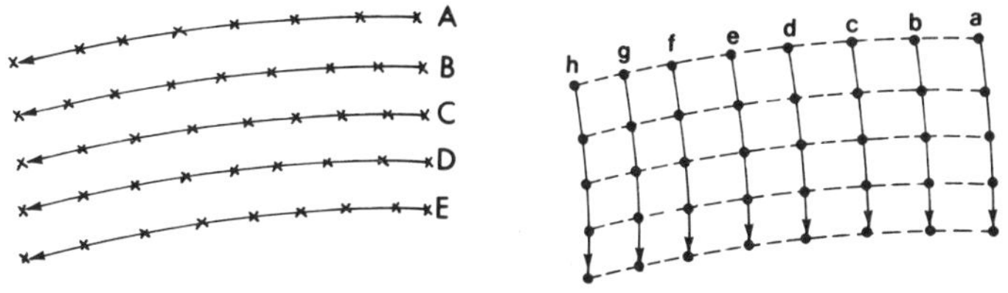

(a) (b)

Figure 7.20 Data Paths(a) and Remeshing(b)

7.7.2 Remeshing and Boundary Condition Estimation

If input data is given in the form of *data paths* as depicted in Fig. 7.20a, the input data has to be remeshed. First, each data path is fitted as a composite curve (see Chapter 4), accompanied by a subsequent curve fairing if necessary. And then *mesh intersection points* are sampled by defining a series of **cross-mesh curves** as depicted in Fig. 7.20b.

The data array obtained from the remeshing may undergo surface fairing operations if necessary. Finally, end tangents of each mesh curves are estimated (see §4.2.2). The corner twist vectors are usually set to zero.

7.8 DISCUSSIONS

Introduced in this chapter are five methods of constructing a "smooth" surface interpolating over an array of data points. They are

a) the FMILL surface fitting method,

b) Ferguson surface fitting method,

c) uniform B-spline surface fitting method,

d) NUB (non-uniform B-spline) surface fitting method, and

e) G^1 *composite Bezier surface fitting method.*

When the physical spacing of data points is even as shown in Fig. 7.5, all the five methods would produce similar surfaces. If the spacing is "semi-even" (see Fig. 7.13a), the NUB fitting should be the choice. The "quality" of the resulting surface may be measured in terms "patch boundary smoothness" and "aesthetic smoothness". The FMILL method results in a C^1-surface, while the tensor product methods (ie, Ferguson and B-spline methods) give C^2-surfaces.

The NUB fitting method gives the most general result under the tensor product approach. All the tensor product methods generate a composite surface composed of bicubic patches and share the same underlying algebraic structure (meaning that the surface fitting problem becomes the problem of solving the linear equation system introduced in Appendix A).

The fifth method, G^1 composite Bezier fitting, should be used as a last resort because the degree of the resulting surface is quite high (ie, six). The first four methods produce bicubic surface patches all of which can easily be converted to bicubic Bezier patches. B-spline surfaces are easily converted to Bezier surfaces by inserting knots (Boehm, 1980) as discussed in Chapter 6.

CHAPTER 8

SURFACE CONSTRUCTION FROM SCATTERED 3D DATA

8.1 INTRODUCTION

This chapter presents a procedure for constructing a smooth composite surface interpolating to a set of *scattered data points*. The interpolating surface is composed of *triangular Bezier patches*. The interpolation scheme defined over triangles consists of three steps:

- *Triangulation: construction of a triangular grid (ie, polyhedron consisting of triangular faces) by connecting data points.*
- *Curve-net interpolation: construction of a smooth curve-net whose domain is the collection of "edges" of the triangular grid.*
- *Surface interpolation: construction of a smooth composite surface by "filling" in the curve-net with triangular Bezier patches.*

1) Triangulation:

The input data for the triangular interpolation scheme is a collection of 3D data points $\{P_i\}$ which may have come from 3D coordinate measuring machines. By *triangulating* the input data, connectivity (ie, topology) is established among the data points. The resulting triangular grid becomes a polyhedron having triangular "faces". The polyhedron itself may be regarded as a "first order" model (ie, C^0-surface) of the "final" surface to be constructed.

2) Curve-net Construction:

In the second step, only the "edges" of the *triangular polyhedron* are refined leaving the "faces" undefined. Each edge is replaced by a cubic curve segment so that the overall curve-net becomes smooth. For the curve-net interpolation, the cubic Bezier curve model introduced in §3.3.3 is used.

3) G^1-surface Construction:

The undefined faces in the cubic curve-net is then filled with triangular surface patches while ensuring *geometric (or visual)* continuity between adjoining patches. For this we need a *G^1-condition* similar to the one introduced in §7.6. Each triangular patch is modeled as a cubic triangular Bezier patch introduced in §5.2.5. To ensure the required continuity across patch boundaries, each cubic patch is *subdivided* into three subpatches, their degrees are *elevated* to quartic, and then the G^1-condition is applied to each common boundary curve.

8.2 TRIANGULATION OF INPUT DATA

Presented in this section is a method due to Choi *et al* (1988b) for triangulating scattered data in 3D space. The procedure for constructing an (optimal) triangular grid from input data $\{P_i\}$ consists of the following five steps:

- *Preprocessing (sorting) of input data.*
- *Initialization: initial triangulation.*
- *Main triangulation.*
- *Trimming of surface boundary.*
- *Optimization: grid improvement.*

8.2.1 Preprocessing of Input Data

Let S be a smooth (imaginary) surface on which the input data points $\{P_i\}$ are located. It is assumed that

(a) there exists a 3D point C from which all the points on S can be seen, and
(b) a *convex cone* can be defined, with its *apex* at C and unit *axis vector* $\mathbf{u}$, that contains $\{P_i\}$ in it.

The class of surfaces satisfying the above assumptions includes plane and *closed convex surfaces* such as sphere and ellipsoid. For a closed convex surface, the *apex point* C is taken a point on the surface. If a surface S satisfies assumption (a) only, the $\{P_i\}$ are projected on a sphere centered at C and then the projected points are triangulated. The preprocessing of input data is carried out as follows:

Procedure_Preprocess($\{P_i\} \Rightarrow \{\mathbf{v}_i\}$) :

1) Read (or compute) the apex point C and unit axis vector $\mathbf{u}$.
2) $\mathbf{v}_i = P_i - C$ for all i.
3) Sort $\{\mathbf{v}_i\}$ in ascending order of the angle θ_i between $\mathbf{u}$ and $\mathbf{v}_i$.

Providing the apex point C and axis vector $\mathbf{u}$ may not be an easy task, and it may be determined interactively. Since $0 \leq \theta_i \leq 90°$, the sorting may be carried out in descending order of $\mathbf{u} \cdot \mathbf{v}_i/|\mathbf{v}_i|$ where "·" indicates a *dot (scalar) product.* The preprocessed 3D points $\{\mathbf{v}_i\}$ are stored in a static array called a "vertex list".

Before describing the triangulation procedure, relevant *terminologies and conventions* are introduced: A 3D point $\mathbf{v}_i$ in the *vertex list* is called a vertex which is treated as a vector as well as a 3D point. Let A, B, C be three *vertices* forming a *triangle* as depicted in Fig. 8.1. An edge AB is defined as a directed line segment from A to B. The edges of a triangle have anti-clockwise directions viewing toward the origin O.

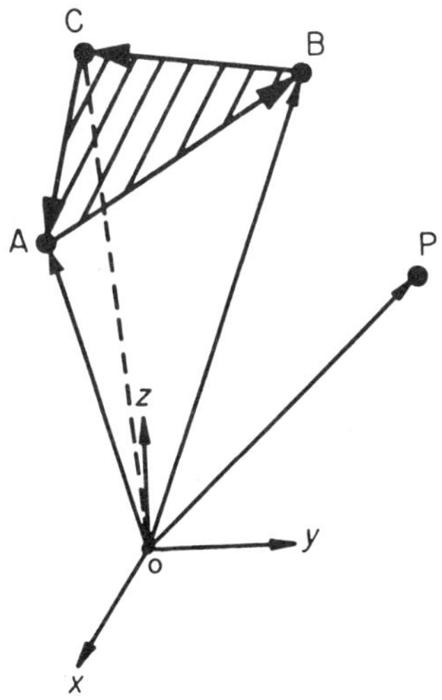

Figure 8.1 Visibility of an Edge in a Triangle

The three vertices A, B, C in Fig. 8.1 form a triangle in 3D space if they are not collinear. This condition is satisfied when the *determinant* of the three vectors is not zero, namely, if $det\ (A, B, C) \neq 0$. One may easily verify that the determinant is equivalent to a **triple scalar product** (See Chap. 2):

$$det(A, B, C) = (A \times B) \cdot C. \tag{8.1}$$

A geometric interpretation of (8.1) leads to the following definitions for the relative direction of a 3D point P with respect to an edge AB of a triangle.

$$\begin{aligned} &(P.\underline{\text{right}}.AB) \text{ is true if } det(A, B, P) < 0 \\ &(P.\underline{\text{on}}.AB) \text{ is true if } det(A, B, P) = 0 \\ &(P.\underline{\text{left}}.AB) \text{ is true if } det(A, B, P) > 0. \end{aligned} \tag{8.2}$$

The edge AB is <u>visible</u> from the point P iff (P.<u>right</u>.AB) is true, which is the case in Fig. 8.1. When an edge is visible from a 3D point, the end vertices of the edge are also *visible* from the point.

A loop formed by a sequence of *edges* is called a <u>polygon</u> if the loop has no self-intersections viewing from the origin O. Let the loop have an anti-clockwise direction (viewing toward the origin O) and let P, Q, R be three consecutive vertices in that order. Then, Q is a <u>convex vertex</u> if $det(Q, R, P) \geq 0$, and it becomes a <u>strictly convex</u> vertex if a strict inequality holds. A (3D) polygon is called a <u>strictly convex polygon</u> if all the vertices are strictly convex.

8.2.2 Initial Triangulation

A *triangle* in a triangular grid is represented as a tuple consisting of three vertices and three indices (ie, pointers) to neighboring triangles, and is to be stored in an integer array TLIST. That is, TLIST (triangle-list) consists of an array of "rows", each *row* containing six integers for a triangle (Lawson, 1977): The first three integers are the triangle's vertex indices (ie, indices of *vertex-list*) in anti-clockwise order; the last three integers are the indices of adjacent triangles in anti-clockwise order. The first *vertex* corresponds to the first *adjacent triangle*, which is located on the opposite side of the vertex, and so on.

Shown in Fig. 8.2b is a triangular grid having four triangles formed by five vertices, with the numeric values of the TLIST as given in Fig. 8.2a. When a 3D triangle is drawn on paper, the triangle is assumed to be viewed toward the origin O (See Fig. 8.1).

(a)

triangle index	Vertex index			Adjacent triangle index		
1	3	1	2	4	2	3
2	4	3	2	1	0	0
3	5	1	3	1	0	4
4	5	2	1	1	3	0
5	-	-	-	-	-	-

(b)

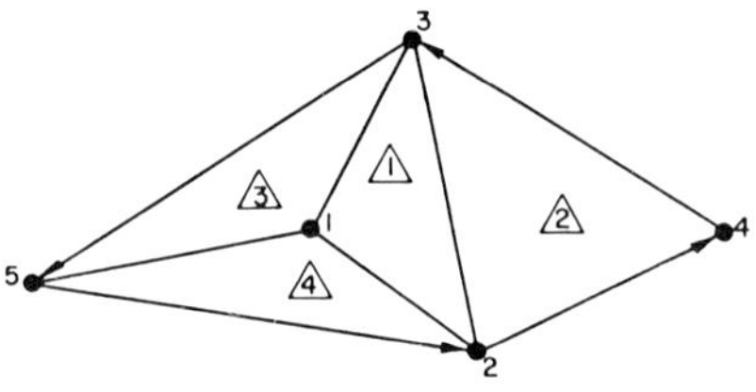

(c)

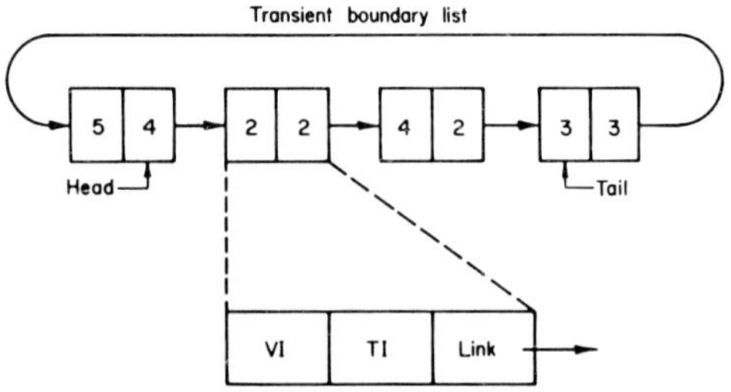

Figure 8.2 A Triangular Grid and its Transient Boundary List

Shown in Fig. 8.2c is a linked list for the *boundary loop* of the triangular grid. The linked list is called *transient boundary list* (TBL) which is pointed by two pointers "Head" and "Tail". Each *node* in the TBL stores an edge's starting vertex index (VI) and its left triangle index (TI), as indicated in Fig. 8.2c.

Now we are ready to describe a procedure for constructing an initial triangular grid from the first few vertices in the *vertex-list*. Shown in Fig. 8.3a is a case where the first four vertices in the *vertex-list* are coplanar and the fifth vertex is located *left* side of the edge V_1V_2. The initial triangulation procedure may be formally described as:

Procedure_initial-triangulation:

1) Get the first two vertices V_1, V_2 and put in TBL (Fig. 8.3b);

2) Get a next vertex V_i;

3) While (V_i .On. V_1V_2) do {

if V_i is closer to "Head" vertex then store in front (Fig.8.3c);

if V_i is closer to "Tail" vertex then store at end (Fig. 8.3d);

get a next vertex V_i};

4) If (V_i .Right. V_1V_2) then reverse the direction of the TBL;

5) Insert V_i in front and store the triangle numbers (Fig. 8.3e);

6) Construct a TLIST for the initial triangular grid.

The information stored in the TBL (transient boundary list) and TLIST (triangle list) is to be used in the main triangulation stage.

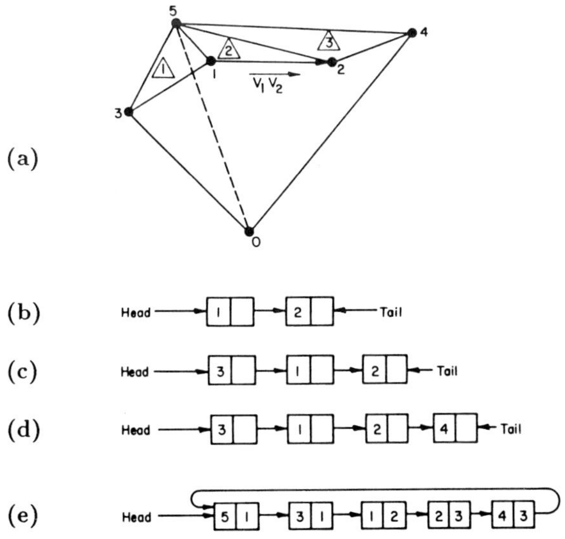

Figure 8.3 Initial Triangulation

8.2.3 Main Triangulation

An example of *main triangulation* is shown in Fig. 8.4. Figure 8.4a shows the initial triangular grid of Fig. 8.3 together with the contents of TBL (transient boundary list) and TLIST (triangle-list). Also indicated in Fig. 8.4a is that vertices 3,1,2,4 are *visible* from vertex 6. Note that vertex 3 is the *left-most visible node* (LVN) and vertex 4 is the *right-most visible node* (RVN). The main triangulation process may be described as:

Procedure_main-triangulation:

1) *Get a next vertex and put it in the new node NN (See Fig. 8.4a).*
2) *Find the left-most and right-most visible nodes LVN, RVN (See Fig. 8.4a).*
3) *Construct a triangle from LVN, NN, and the node that follows LVN, and then insert NN into TBL right after the LVN node (See Fig. 8.4b).*
4) *Keep constructing triangles until NN is directly connected to RVN so that the nodes between LVN and RVN are removed from TBL (See Fig. 8.4c).*
5) *If the vertex-list is not empty go to 1), else stop.*

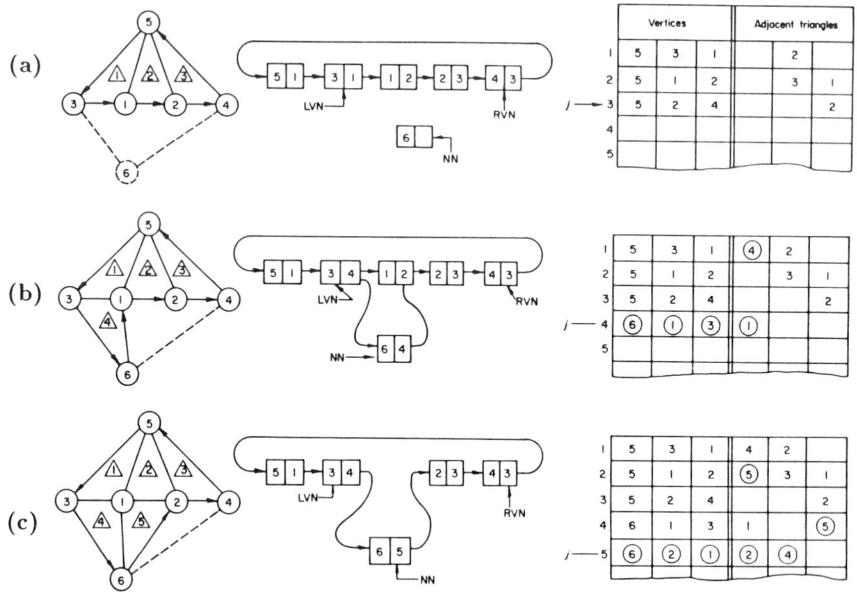

Figure 8.4 Main Triangulation Example

8.2.4 Surface Boundary Trimming

We assume that the surface to be constructed is an *open surface without any holes* in it. Let $\{B_i\}$ denote the vertices on the *boundary* of the surface. If the *polygon* obtained by sequentially connecting the *boundary vertices* $\{B_i\}$ is not *convex*, the *main triangulation* will produce some triangles outside the actual surface boundary. For *boundary trimming*, it is necessary to maintain edge information in a separate place.

All the *edges* in the triangular grid are stored in a static array ELIST (*edge-list*). For this, we may need to know in advance the size of the ELIST. For an open triangular grid, number e of edges is expressed, in terms of numbers of *interior and boundary* vertices, as follows (Choi *et al*, 1988b):

$$e = 3i + 3b - 3 \tag{8.3}$$

where i and b are numbers of interior and boundary vertices, respectively. One row in an ELIST consists of *starting vertex index, ending vertex index, left triangle index,* and *right triangle index.* The following procedure may be used in constructing an ELIST from a given TLIST:

Procedure_construct-ELIST:

```
e := 1 ; //initialize//
For all row i of TLIST do
    for j := 1 to 3 do begin
      k := TLIST(i, j + 3) ; //retrieve triangle indices//
      if (k > i) or (k = 0) then begin
        j1 := (j mod 3) + 1 ;
        j2 := (j + 1 mod 3) + 1 ;
        ELIST(e, 1) := TLIST(i, j1); //starting vertex//
        ELIST(e, 2) := TLIST(i, j2); //ending vertex//
        ELIST(e, 3) := i ; //left triangle//
        ELIST(e, 4) := k; //right triangle//
      end; //if//
      e := e + 1;
    end //for//
```

An example of boundary trimming is shown in Fig. 8.5 where the three triangles 1,2,3 need to be removed. It is assumed that the given sequence of *boundary vertices* $\{B_i\}$ forms a closed loop, and that each pair of consecutive boundary vertices has already been connected to form an edge. In Fig. 8.5, the boundary vertices B_1 through B_7 are connected to form a "path". The basic idea here is to construct an anti-clockwise *transient boundary list* (TBL) from the boundary vertices, and then recursively remove the triangles located on the right-hand side of

(a)

(b)

(c)

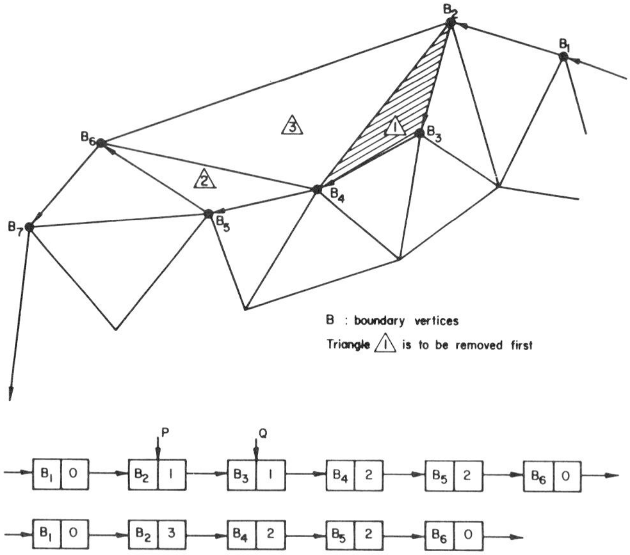

Figure 8.5 Surface Boundary Trimming

the TBL. In the figure, triangle 1 is the *right triangle* of the edge B_2B_3. Once this edge is removed, B_3 is also deleted so that B_2B_4 becomes a new edge whose right triangle is now triangle 3. The boundary trimming scheme may be summarized as follows:

Procedure_boundary-trimming:

1) Construct-ELIST.

2) Construct an anti-clockwise TBL, for the boundary loop $\{B_i\}$ *using the information in the ELIST, such that the right triangle of each edge is stored in the triangle index (TI) field (See Fig. 8.5b).*

3) //Keep removing right triangles until no more left//

Repeat //P is predecessor node of Q//

Find a node Q such that $(P\uparrow.TI = Q\uparrow.TI = j > 0)$;

If Found then begin

Set $P\uparrow.TI$ *to the adjacent triangle of "j" facing the vertex* $Q\uparrow.VI$;

Delete node Q from TBL (See Fig. 8.5c);

Remove triangle j from the TLIST;

end //if//

Until (Not Found)

8.2.5 Triangulation Criteria and Grid Improvement

The triangular grid constructed so far is only a feasible one which needs to be improved. Grid improvement is carried out by *swapping* edges. As shown in Fig. 8.6, an interior edge e defines a quadrilateral V_1, V_2, V_3, V_4 which is surrounded by four triangles t_3, t_4, t_5, t_6. The following macro-operation will *swap* the diagonal of the quadrilateral, namely, the edge e in Fig. 8.6 is deleted and a new edge joining V_2 and V_4 is created.

Procedure_Swap(e):

1) *//Retrieve the vertices and triangles from ELIST//*
   $V_1 := ELIST(e,1)$; $V_3 := ELIST(e,2)$;
   $t_1 := ELIST(e,3)$; $t_2 := ELIST(e,4)$;
2) *Retrieve* V_4, t_5, t_6 *from* $TLIST(t_1,*)$.
3) *Retrieve* V_2, t_3, t_4 *from* $TLIST(t_2,*)$.
4) *Locate the edge indices* e_1, e_2, e_3, e_4 *of ELIST using* V_1, V_2, V_3, V_4.
5) *//Swap the triangles* t_1, t_2*//*
   $ELIST(e,1) := V_2$; $ELIST(e,2) := V_4$;
   Update $TLIST(t_1,*)$ *and* $TLIST(t_2,*)$;
6) *Update* $TLIST(t_3,*)$, $TLIST(t_4,*)$, $TLIST(t_5,*)$, *and* $TLIST(t_6,*)$.
7) *Update* $ELIST(e_1,*)$, $ELIST(e_2,*)$, $ELIST(e_3,*)$, *and* $ELIST(e_4,*)$.

Now the question is "when a swap is preferred?" To answer the question, a criterion for "optimum triangulation" should be established first. An optimal triangulation is known to be equivalent to the dual of a Thiessen region diagram (Cline and Renka, 1984) when triangulation is performed on a planar domain.

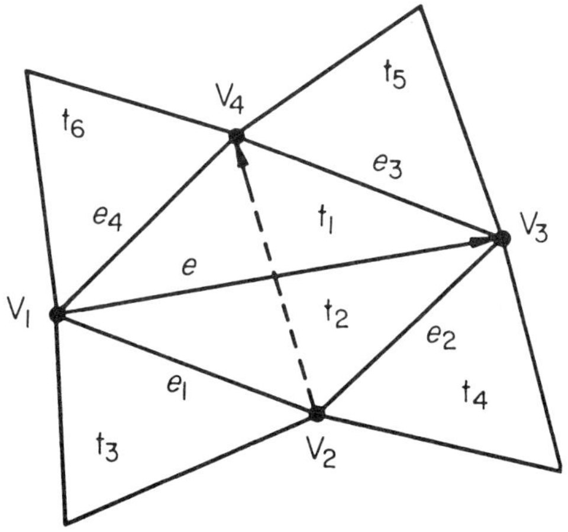

Figure 8.6 Quadrilateral and Swap Operation

Figure 8.7 Max-Min Angle Criterion

In this case, the *max-min angle criterion* gives an optimal triangulation (Lawson, 1977). Shown in Fig. 8.7 is a quadrilateral composed of two triangles. In Fig. 8.7a, the angle ∠bdc is the *minimum interior angle* of the two triangles, while ∠bac is the *minimum interior angle* in Fig. 8.7b. Since ∠bac is larger than ∠bdc, the triangulation in Fig. 8.7b is preferred. The max-min angle criterion may be stated as follows:

*Definition(***Max-min angle criterion***):*
For a quadrilateral formed by merging two triangles, the triangulation that maximizes the minimum interior angle of the two resulting triangles is chosen.

The max-min angle criterion does not have a clear geometric meaning for a general 3D triangulation. Thus, an intuitive criterion called a *smoothness criterion* is introduced in Choi *et al* (1988b). Consider a *strictly convex polygon* abcd shown in Fig. 8.8. The polygon (ie, quadrilateral) is bounded by four *edges* e_1, e_2, e_3, e_4. Then the smoothness criterion is stated as:

*Definition(***Smoothness criterion***):*
The smoothness criterion is the choice of a diagonal that maximizes the minimum of the four scalar products of unit normal vectors of the triangles, one scalar product for each edge e_1 *of the strictly convex quadrilateral.*

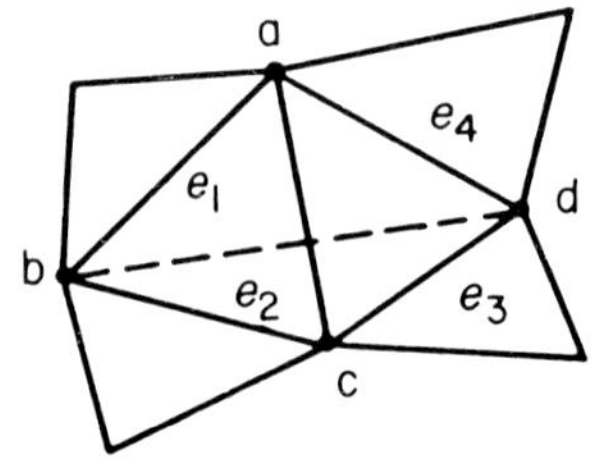

Figure 8.8 Smoothness Criterion

Now we are ready to present a procedure for improving the triangular grid. The grid improvement procedure may be formalized as follows:

Procedure_grid-improvement:

1) Construct-ELIST.

2) For all interior edges "e" do

if the smoothness criterion prefers a "swap" then Swap(e)

else if the minimum of the four scalar products are (almost) identical

and the max-min angle criterion prefers a swap then Swap(e).

The grid-improvement procedure may have to be applied repeatedly until no significant improvement is observed. Figure 8.9 illustrates the progress of the grid improvement process. Another example is shown in Fig. 8.10.

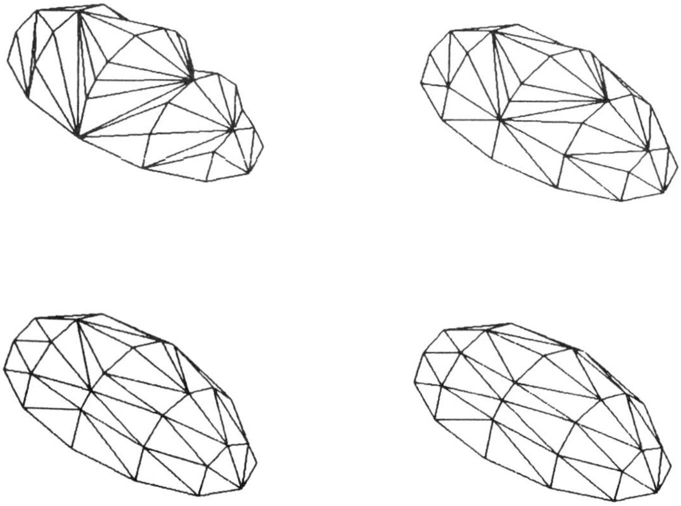

Figure 8.9 Progress of Grid Improvement Process

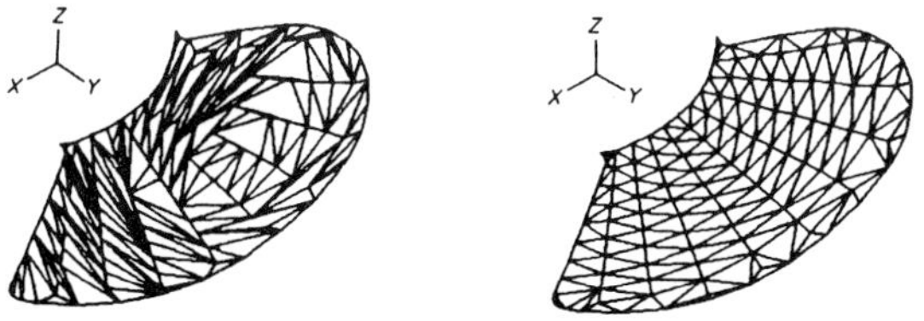

Figure 8.10 Triangulation of 3D Points on a Fan Blade

8.3 CONSTRUCTION OF CUBIC CURVE-NET

This section considers the problem of constructing a smooth curve-net which interpolates to the **triangular grid** stored in the static array TLIST. The curve-net interpolation procedure consists of two steps: 1) Estimation of *surface normal* at each *vertex* of the triangular grid, and 2) construction of a cubic Bezier curve for each *edge* of the triangular grid.

8.3.1 Surface Normal Estimation

For the estimation of surface normals from scattered data, the so called **triangular Shepard method** (Little, 1983) is widely used. Due to the reasons to be explained later, we use a modified version of the Shepard method. Let $\mathbf{n}_i$ be the unit normal vector of the triangle i one of whose vertex is the "current" point P as depicted in Fig. 8.11a. Let d_i denote the distance from the vertex P to the middle point of the "opposite" edge of triangle i (see the figure). Then, the *unit surface normal* $\mathbf{n}_p$ at the data point P can be estimated as a *weighted average of face normals* $\mathbf{n}_i$ *of surrounding triangles*:

$$\mathbf{n}_p = \left\{\sum_i (\mathbf{n}_i/d_i^2)\right\} \Big/ \left|\sum_i (\mathbf{n}_i/d_i^2)\right|. \tag{8.4}$$

In the Shepard method, the product of lengths of the two edges sharing P is used in place of d_i in (8.4). According to Stead (1984), however, the influence of the interior angle θ_i of triangle i at the vertex P should also be accounted for in addition to that of the edge lengths. If the interior angle θ_i is increased while the lengths of the edges sharing P are fixed, the influence of $\mathbf{n}_i$ needs to be increased. Other methods are also proposed in the literature (Lawson, 1984). The *normal* estimation method (8.4) is not directly applicable to *boundary vertices, however.* The vertices O, P, S in Fig. 8.11b are located on the boundary of the surface (ie, triangular

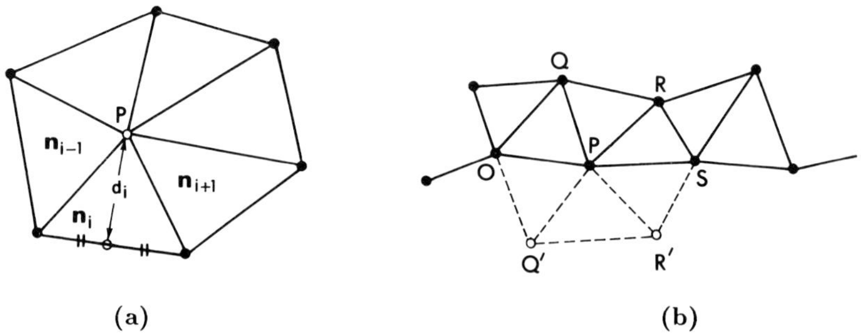

Figure 8.11 Estimation of Surface Normal Vectors

grid). In order to use (8.4) properly, we need to add up triangles beyond the boundary of the triangular grid. Let us define

$angle(PQ)$: angle between face normals of the two triangles sharing the edge PQ,
$length(PQ)$: length of the edge PQ.

The *pseudo* triangle POQ' in Fig. 8.11b is constructed such that

$$angle(OP) = \{angle(OQ) + angle(PQ)\}/2 \qquad (8.5-a)$$

$$length(PQ) = length(PQ'); \; length(OQ) = length(OQ'). \qquad (8.5-b)$$

The triangle $PR'S$ is similarly constructed. Then, the third triangle $PQ'R'$ is automatically obtained. Now we apply (8.4) at the vertex P.

8.3.2 Construction of Cubic Bezier Mesh Curves

The next step is to construct a cubic Bezier curve at each *edge* of the triangular grid. A cubic Bezier curve is defined in terms of four control points V_0, V_1, V_2, V_3. Shown in Fig. 8.12 is an edge PQ together with the *tangent plane* π_p at P ($\mathbf{n}_p$ is a *unit surface normal*). Let R denote the projected point of Q on π_p. Then, it is given by

$$R = Q + ((P - Q) \cdot \mathbf{n}_p)\mathbf{n}_p. \qquad (8.6)$$

The *end control vertices* of the cubic Bezier curve may be designated as

$$V_0 = P \quad ; \quad V_3 = Q$$

and the *interior control vertex* V_1 may be determined as (Piper, 1987)

$$V_1 = V_0 + \{1/3 + (h/9)/|R - P|\}(R - P), \qquad (8.7)$$

where $h = |R - Q|$. The remaining control point V_2 can be determined the same way.

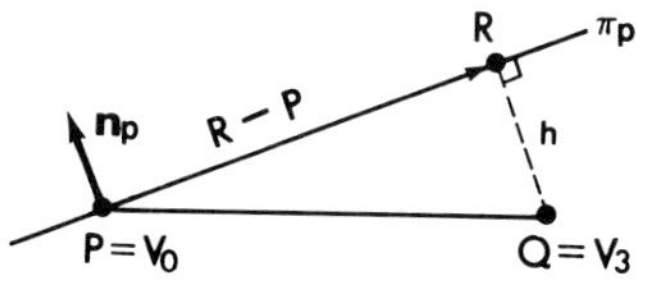

Figure 8.12 Determination of Bezier Control Points

8.4 PROPERTIES OF TRIANGULAR BEZIER PATCHES

This section presents some useful properties of triangular Bezier patches which are essential in constructing composite triangular Bezier surfaces. The topics to be discussed in this section are definition of triangular Bezier patch, directional derivatives, degree elevation, and subdivision.

8.4.1 Definition of Triangular Bezier Patches

As discussed in Chapter 6, a Bezier curve $\mathbf{r}(u)$ of degree n has the following general form:

$$\mathbf{r}(u) = \textstyle\sum_{i=0}^{n} B_i^n(u) V_i \tag{8.8}$$

where $B_i^n(u)$ are *univariate Bernstein polynomials*. A rectangular Bezier patch is defined in terms of $\{V_{ij}\}$ as

$$\mathbf{r}(u,v) = \sum_{i=0}^{m} \sum_{j=0}^{n} B_i^m(u) B_j^n(v) V_{ij}. \tag{8.9}$$

In a similar way, a degree-n **triangular Bezier patch** (TBP) is obtained by generalizing the cubic triangular patch of §5.2.6. That is, a TBP is defined, in terms of control vertices $\{\mathbf{p}_{ijk}\}$, as follows:

$$\mathbf{r}(u,v) = \sum_{i+j+k=n} B_{ijk}^n(u,v)\mathbf{p}_{ijk} \tag{8.10}$$

where $B_{ijk}^n(u)$ are **bivariate Bernstein polynomials** given by

$$B_{ijk}^n(u,v) = \frac{n!}{i!\;j!\;k!} u^i v^j (1-u-v)^k \tag{8.11}$$

A cubic TBP with its control vertices is shown in Fig. 8.13.

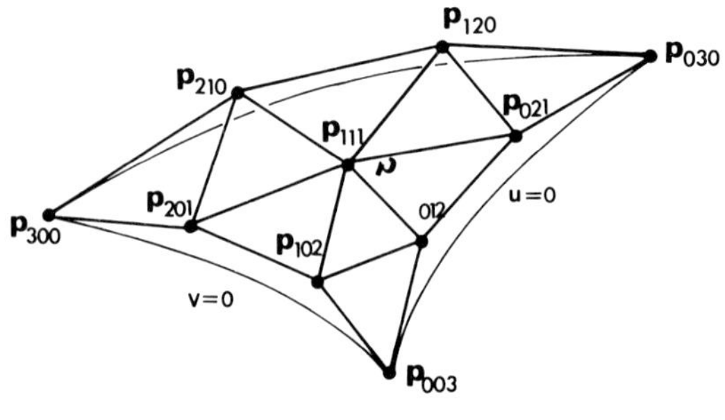

Figure 8.13 Cubic Bezier Patch and its Control Vertices

8.4.2 Directional Derivatives of Triangular Bezier Patch

It is essential to be able to find the tangent vectors at any point on a triangular Bezier patch. Let $\mathbf{d} = (d_1, d_2)$ denote a direction vector at a point $\mathbf{u} = (u, v)$ on the **domain triangle** as depicted in Fig. 8.14a. Then, the tangent vector of the surface $\mathbf{r}(u, v)$ corresponding to the domain vector $\mathbf{d}$ is given by (Piper, 1987)

$$\begin{aligned} \mathbf{D_d r}(u,v) &= \frac{d}{dt}\Bigg(\sum_{i+j+k=n} B^n_{ijk}(u + d_1 t, v + d_2 t)\mathbf{p}_{ijk} \Bigg)_{t=0} \\ &= n \sum_{i+j+k=n-1} B^{n-1}_{ijk}(u,v)\{d_1 \mathbf{p}_{i+1,j,k} + d_2 \mathbf{p}_{i,j+1,k} - (d_1 + d_2)\mathbf{p}_{i,j,k+1}\}. \end{aligned} \tag{8.12}$$

The u-direction partial derivative $\mathbf{r}_u(u, v)$ is obtained from (8.12) with $\mathbf{d} = (1, 0)$. Similarly, $\mathbf{r}_v(u, v)$ is obtained by setting $\mathbf{d} = (0, 1)$. That is,

$$\mathbf{r}_u(u,v) = n \sum_{i+j+k=n-1} B^{n-1}_{ijk}(u,v)\{\mathbf{p}_{i+1,j,k} - \mathbf{p}_{i,j,k+1}\} \tag{8.13 − a}$$

$$\mathbf{r}_v(u,v) = n \sum_{i+j+k=n-1} B^{n-1}_{ijk}(u,v)\{\mathbf{p}_{i,j+1,k} - \mathbf{p}_{i,j,k+1}\} \tag{8.13 − b}$$

The same results would have been obtained by directly taking partial derivatives of the surface equation (8.10). The **isoparametric derivatives** in (8.13) simplify somewhat when they are evaluated at a patch boundary. The isoparametric derivatives at $u = 0$ boundary, for example, are given by

$$\mathbf{r}_u(0,v) = n \sum_{j=0}^{n-1} B^{n-1}_j(v)\{\mathbf{p}_{1,j,n-j-1} - \mathbf{p}_{0,j,n-j}\} \tag{8.14 − a}$$

$$\mathbf{r}_v(0,v) = n \sum_{j=0}^{n-1} B^{n-1}_j(v)\{\mathbf{p}_{0,j+1,n-j-1} - \mathbf{p}_{0,j,n-j}\} \tag{8.14 − b}$$

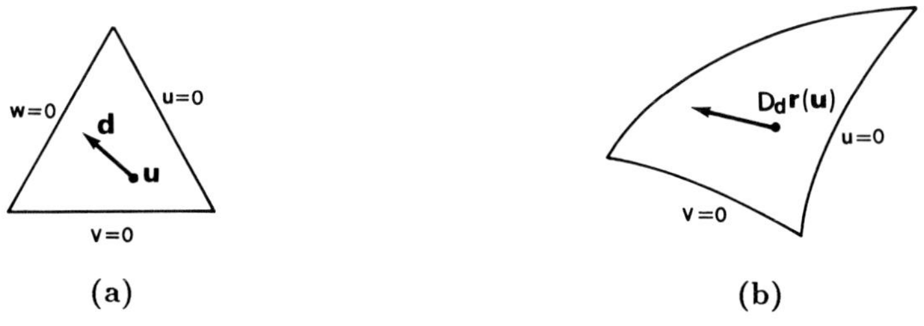

Figure 8.14 Directional Derivatives of Triangular Bezier Patch

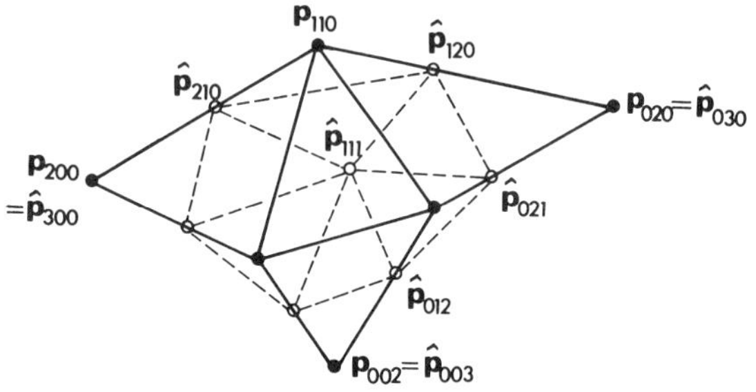

Figure 8.15 Degree Elevation of a Triangular Bezier Patch

8.4.3 Degree Elevation of Triangular Bezier Patch

The degree of a triangular Bezier patch can be elevated the same way the degree of a Bezier curve is elevated (see §6.4). Let $\{\mathbf{p}_{ijk}\}$ denote the control vertices of a triangular Bezier patch of degree $n-1$, then the control vertices $\{\hat{\mathbf{p}}_{ijk}\}$ of a degree-n Bezier patch are obtained from the following recursive formula $(i+j+k=n)$:

$$\hat{\mathbf{p}}_{ijk} = \frac{i}{n}\mathbf{p}_{i-1,j,k} + \frac{j}{n}\mathbf{p}_{i,j-1,k} + \frac{k}{n}\mathbf{p}_{i,j,k-1}. \tag{8.15}$$

Then, the degree-elevated patch represents the same surface as the original patch (Farin, 1986). That is, for all $u, v \in [0,1]$, we have

$$\sum_{i+j+k=n} B^n_{ijk}(u,v)\hat{\mathbf{p}}_{i,j,k} = \sum_{i+j+k=n-1} B^{n-1}_{ijk}(u,v)\mathbf{p}_{i,j,k}. \tag{8.16}$$

A degree-elevation example, from quadratic to cubic, is given in Fig. 8.15.

8.4.4 Subdivision of Triangular Bezier Patch

Consider the following recursive function for the control vertices of triangular Bezier patches.

$$\mathbf{p}^r_{ijk} = u\ \mathbf{p}^{r-1}_{i+1,j,k} + v\ \mathbf{p}^{r-1}_{i,j+1,k} + w\ \mathbf{p}^{r-1}_{i,j,k+1} : i+j+k=n-r \tag{8.17}$$
$$with\ \ \mathbf{p}^0_{ijk} \equiv \mathbf{p}_{ijk},$$

where $w = 1-u-v$. Note that $\mathbf{p}^r_{ijk}$ is in fact a function of $\mathbf{u} = (u,\ v,\ w)$. This recursive relation is known as **de Casteljau algorithm** which is another way of defining a triangular

Bezier patch (Farin, 1988). That is, at a given recursion level r, the triangular Bezier patch $\mathbf{r}(u,v)$ of degree n is expressed as

$$\mathbf{r}(u,v) = \sum_{i+j+k=n-r} B_{ijk}^{n-r}(u,v)\mathbf{p}_{i,j,k}^{r}(\mathbf{u}) \qquad (8.18-a)$$

Finally, at the end of the recursion, we have

$$\mathbf{r}(u,v) = \mathbf{p}_{0,0,0}^{n}(\mathbf{u}) \qquad (8.18-b)$$

where $\mathbf{u} = (u,\ v,\ w)$ with $w = (1-u-v)$.

The de Casteljau algorithm (8.17) can be used in subdividing a triangular Bezier patch into three subpatches. At each iteration of (8.17), an *intermediate control vertex* $\mathbf{p}_{ijk}^{r}(\mathbf{u}_0)$ is determined from three control vertices at level $r-1$ as an *affine map* of the domain triangle as depicted in Fig. 8.16. Let $\mathbf{r}^{\alpha}(u,v)$ for $\alpha = u, v, w$ denote a sub-triangular patch containing the boundary $\alpha = 0$, then the three sub-patches subdivided at $\mathbf{u}_0 = (u_0, v_0, w_0)$ are expressed as

$$\mathbf{r}^{u}(u,v) = \sum_{i+j+k=n} B_{ijk}^{n}(u,v)\ \mathbf{p}_{0jk}^{i}(\mathbf{u}_0) \qquad (8.19-a)$$

$$\mathbf{r}^{v}(u,v) = \sum_{i+j+k=n} B_{ijk}^{n}(u,v)\ \mathbf{p}_{i0k}^{j}(\mathbf{u}_0) \qquad (8.19-b)$$

$$\mathbf{r}^{w}(u,v) = \sum_{i+j+k=n} B_{ijk}^{n}(u,v)\ \mathbf{p}_{ij0}^{k}(\mathbf{u}_0) \qquad (8.19-c)$$

The shaded sub-patch in Fig.8.16 represents $\mathbf{r}^{w}(u,v)$ with $n=2$. Thus, this sub-patch is defined in terms of the following six intermediate control vertices.

$$\mathbf{p}_{200}^{0}(\mathbf{u}_0) \equiv \mathbf{p}_{200},\ \mathbf{p}_{110},\ \mathbf{p}_{020},\ \mathbf{p}_{100}^{1}(\mathbf{u}_0),\ \mathbf{p}_{010}^{1}(\mathbf{u}_0),\quad and\quad \mathbf{p}_{000}^{2}(\mathbf{u}_0).$$

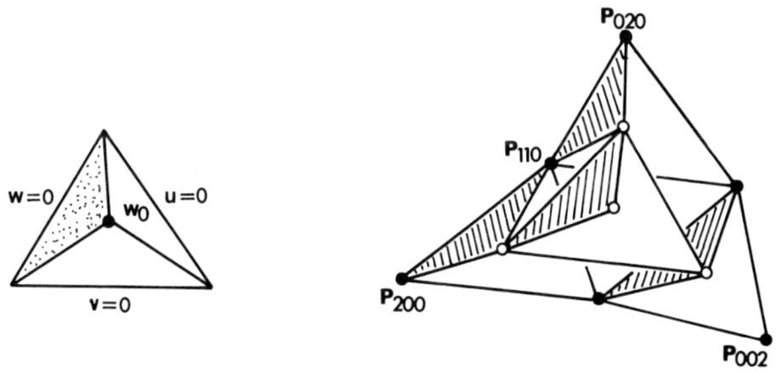

Figure 8.16 Subdivision Example

8.5 CONSTRUCTION OF COMPOSITE TRIANGULAR BEZIER SURFACE

This section describes the final phase, the *surface interpolation phase.* Since each triangular region in the *curve-net* is bounded by three cubic Bezier curves, it can be modeled as a cubic *triangular Bezier patch* (abbreviated as TBP) by providing a suitable *center control vertex* $\mathbf{p}_{111}$ (See Fig. 8.13). However, the resulting composite surface may not be smooth because adjacent triangular patches are only C^0-continuous across their common boundaries. The overall procedure for constructing smooth composite surface interpolating to the curve-net data consists of the following steps:

a) Determination of the center control vertex for each triangle.

b) Subdivision and degree-elevation.

c) G^1 correction of interior control vertices.

We first introduce a *geometric first order continuity condition* (abbreviated as G^1-condition) between TBPs.

8.5.1 G^1-Condition between Triangular Bezier Patches

Consider the two TBPs (triangular Bezier patches), $A(u,v)$ and $B(u,v)$, shown in Fig. 8.17. For both patches, the common boundary curve corresponds to $u = 0$ boundaries. That is, $A(0,v) \equiv B(0,v)$ as depicted in the figure. Now define *isoparametric derivatives* on the common boundary curve as follows

$$\mathrm{a}(v) = \partial A(0,v)/\partial u \qquad (8.20-a)$$

$$\mathrm{b}(v) = \partial B(0,v)/\partial u \qquad (8.20-b)$$

$$\mathrm{c}(v) = \partial A(0,v)/\partial v = \partial B(0,v)/\partial v \qquad (8.20-c)$$

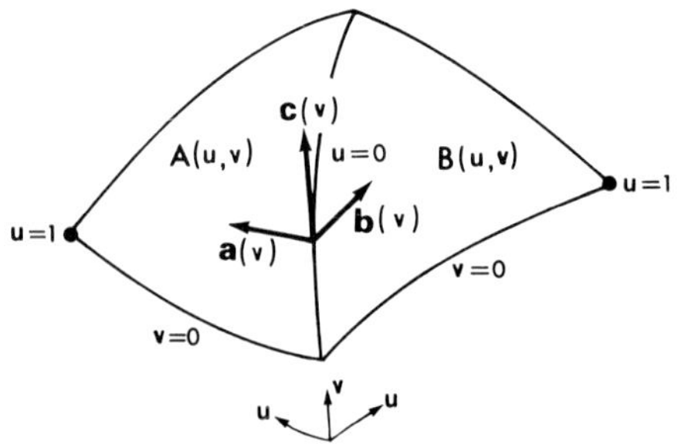

Figure 8.17 Construction of G^1-Condition between TBPs

Then, the two patches $A(u,v)$, $B(u,v)$ are G^1-continuous across their common boundary curve if the three tangent vectors $\mathbf{a}(v), \mathbf{b}(v), \mathbf{c}(v)$ are on the same plane (See §7.6 for details). This **coplanar condition** is satisfied if the following relation holds

$$p(v)\mathbf{a}(v) + q(v)\mathbf{b}(v) + r(v)\mathbf{c}(v) = \mathbf{0} \tag{8.21}$$

for some scalar functions $p(v), q(v), r(v)$ which are not identically zero at the same time. The G^1*-condition* (8.21) is exactly the same as the one given by Eqn. (7.34) in §7.6. Let us define

$\{\mathbf{p}_{ijk} : i+j+k=n\}$: *control points of the TBP A(u,v), and*

$\{\mathbf{q}_{ijk} : i+j+k=n\}$: *control points of the TBP B(u,v).*

Then, the two TBPs are expressed as (See Eqn. (8.10))

$$A(u,v) = \sum_{i+j+k=n} B^n_{ijk}(u,v)\ \mathbf{p}_{ijk}, \text{ and}$$

$$B(u,v) = \sum_{i+j+k=n} B^n_{ijk}(u,v)\ \mathbf{q}_{ijk},$$

where $B^n_{ijk}(u,v)$ are *bivariate Bernstein polynomials* given by (8.11). Further, from the results given in (8.14), the *isoparametric derivatives* defined by (8.20) are expressed as

$$\mathbf{a}(v) = n\sum_{j=0}^{n-1} B^{n-1}_j(v)\mathbf{a}_j \tag{8.22-a}$$

$$\mathbf{b}(v) = n\sum_{j=0}^{n-1} B^{n-1}_j(v)\mathbf{b}_j \tag{8.22-b}$$

$$\mathbf{c}(v) = n\sum_{j=0}^{n-1} B^{n-1}_j(v)\mathbf{c}_j \tag{8.22-c}$$

where $B^{n-1}_j(v)$ are *univariate Bernstein polynomils.* The **control vectors** $\mathbf{a}_j, \mathbf{b}_j, \mathbf{c}_j$ for $j = 0, 1, ..., n-1$ are defined as

$$\begin{aligned} \mathbf{a}_j &= \{\mathbf{p}_{1,j,n-j-1} - \mathbf{p}_{0,j,n-j}\}; \quad \mathbf{b}_j = \{\mathbf{q}_{1,j,n-j-1} - \mathbf{q}_{0,j,n-j}\} \\ \mathbf{c}_j &= \{\mathbf{p}_{0,j+1,n-j-1} - \mathbf{p}_{0,j,n-j}\} \equiv \{\mathbf{q}_{0,j+1,n-j-1} - \mathbf{q}_{0,j,n-j}\} \end{aligned} \tag{8.23}$$

An illustration of the above *control vectors* with $n = 4$ is given in Fig. 8.18. In the figure, the common boundary curve $A(0,v) \equiv B(0,v)$ is a quartic Bezier curve. The control vertices V_i of the common boundary curve are then given by

$$V_0 = \mathbf{p}_{004}; \quad V_1 = V_0 + \mathbf{c}_0; \quad \ldots; \quad V_4 = V_0 + \mathbf{c}_0 + \mathbf{c}_1 + \mathbf{c}_2 + \mathbf{c}_3 = \mathbf{p}_{040}.$$

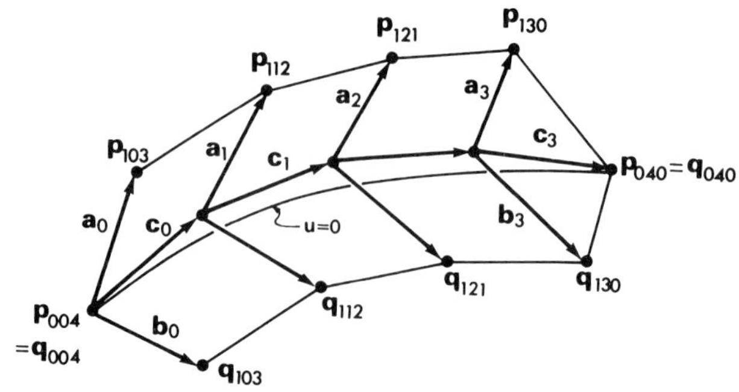

Figure 8.18 Construction of Control Vectors at Common Boundary

The scalar functions $p(v), q(v), r(v)$ in (8.21) are defined in the same Bezier form as in (7.38). That is, for some scalar coefficients p_j, q_j, r_j, we have

$$p(v) = \sum_{j=0}^{k} p_j B_j^k(v) \tag{8.24 - a}$$

$$q(v) = \sum_{j=0}^{k} q_j B_j^k(v) \tag{8.24 - b}$$

$$r(v) = \sum_{j=0}^{k} r_j B_j^k(v) \tag{8.24 - c}$$

Further, the following condition is imposed on $p(v)$ and $q(v)$:

$$p(v) + q(v) = 1 \quad and \quad p(v) > 0,\ q(v) > 0. \tag{8.25}$$

An implication of the condition (8.25) is that the two tangent vectors $\mathbf{a}(v)$ and $\mathbf{b}(v)$ be on strictly opposite sides of $\mathbf{c}(v)$. On substituting the results of (8.22) and (8.24), the G^1-condition (8.21) is evaluated to be (Shin and Choi, 1990)

$$\frac{n!\ k!}{(n+k-1)!} \sum_{j=0}^{n+k-1} \sum_{i=0}^{k} (p_i \mathbf{a}_{j-i} + q_i \mathbf{b}_{j-i} + r_i \mathbf{c}_{j-i}) \binom{j}{k} \binom{n+k-1-j}{k-i} B_j^{n+k-1}(v) = 0.$$

In obtaining the above relation, the *Bernstein polynomial identity relation* (7.32) was utilized. In order for the above equation to be identically zero for all $v \in [0, 1]$, the coefficient terms should all be zero. Namely, we obtain the following G^1-**condition for triangular Bezier patches:**

$$\sum_{i=0}^{k}(p_i\mathbf{a}_{j-i}+q_i\mathbf{b}_{j-i}+r_i\mathbf{c}_{j-i})\binom{j}{i}\binom{n+k-1-j}{k-i}=0 \quad for \quad j=0,1,...,n+k-1. \quad (8.26)$$

If $n = 4$ and $k = 1$, the G^1-condition (8.26) reduces to the following set of linear equations:

$$p_0\mathbf{a}_0 + q_0\mathbf{b}_0 + r_0\mathbf{c}_0 = \mathbf{0} \qquad (8.27-a)$$

$$3(p_0\mathbf{a}_1 + q_0\mathbf{b}_1 + r_0\mathbf{c}_1) + (p_1\mathbf{a}_0 + q_1\mathbf{b}_0 + r_1\mathbf{c}_0) = \mathbf{0} \qquad (8.27-b)$$

$$(p_0\mathbf{a}_2 + q_0\mathbf{b}_2 + r_0\mathbf{c}_2) + (p_1\mathbf{a}_1 + q_1\mathbf{b}_1 + r_1\mathbf{c}_1) = \mathbf{0} \qquad (8.27-c)$$

$$(p_0\mathbf{a}_3 + q_0\mathbf{b}_3 + r_0\mathbf{c}_3) + 3(p_1\mathbf{a}_2 + q_1\mathbf{b}_2 + r_1\mathbf{c}_2) = \mathbf{0} \qquad (8.27-d)$$

$$p_1\mathbf{a}_3 + q_1\mathbf{b}_3 + r_1\mathbf{c}_3 = \mathbf{0} \qquad (8.27-e)$$

And the constraint $p(v) + q(v) = 1$ in (8.25) becomes

$$p_0 + q_0 = 1 \quad and \quad p_1 + q_1 = 1. \qquad (8.28)$$

The degree n in the G^1-condition should be at least 4 when the degrees of the two TBPs, $A(u, v)$ and $B(u, v)$, are three. See Shin and Choi (1990) for more details. This means that the cubic TBPs should be degree-elevated in order to make them G^1-continuous across their common boundary curve.

8.5.2 Construction of Cubic Bezier Patches

Since each triangular patch in the "curve-net" is bounded by three cubic Bezier curves, all the "boundary control vertices" (of a cubic TBP) have already been fixed, as indicated by "solid dots" in Fig. 8.19. The only unknown control vertex is the "center control vertex" $\mathbf{p}_{111}$ of the cubic TBP (triangular Bezier patch). It should be noted that the choice of $\mathbf{p}_{111}$ does not affect

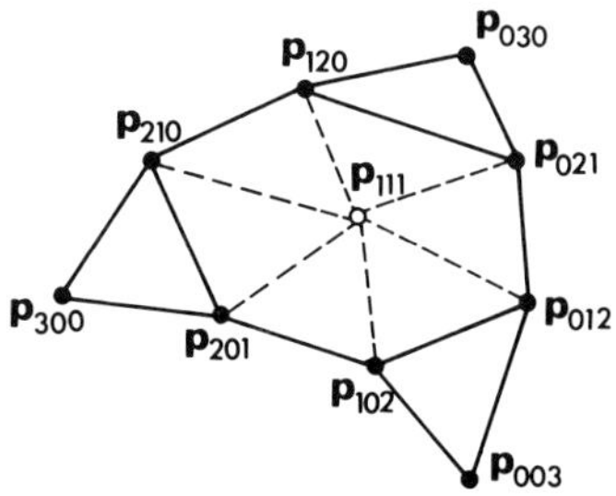

Figure 8.19 Determination of Center Control Vertex

the "visual continuity" (across patch boundaries), but it has a great influence on the "aesthetic smoothness" of the resulting composite surface.

A popular method for determining the center control vertex from the boundary control vertices is the so called **nine parameter cubic method** (Farin, 1983). In this method, the center control vertex is given by

$$\mathbf{p}_{111} = (\mathbf{p}_{210} + \mathbf{p}_{120} + \mathbf{p}_{012} + \mathbf{p}_{021} + \mathbf{p}_{201} + \mathbf{p}_{102})/4 - (\mathbf{p}_{300} + \mathbf{p}_{030} + \mathbf{p}_{003})/6. \quad (8.29-a)$$

This choice gives a **quadratic precision** patch meaning that it is an exact choice when the degree of the TBP is indeed quadratic. One may easily verify (8.29-a) by using the surface degree-elevation formula (8.15) for $\mathbf{p}_{111}$ and the curve degree-elevation formula (7.29) for the other control vertices.

Other choice for $\mathbf{p}_{111}$ is also proposed in the literature (Piper, 1987). The center control vertex is defined so that the resulting cubic patch becomes a ruled surface if one of the sets of boundary control vertices lie on a straight line. That is,

$$\mathbf{p}_{111} = (\mathbf{p}_{210} + \mathbf{p}_{120} + \mathbf{p}_{012} + \mathbf{p}_{021} + \mathbf{p}_{201} + \mathbf{p}_{102})/6. \quad (8.29-b)$$

In either case, a modification may be necessary in order to improve global smoothness (see Piper, 1987 for an approach).

8.5.3 Preparation for G^1-Correction

Each patch in the triangular composite surface is a cubic TBP (triangular Bezier patch) defined by nine control vertices $\{\mathbf{p}_{ijk} : i + j + k = 3\}$. Since it is not possible to make the composite surface visually smooth with the cubic patches (Shin and Choi, 1990), each cubic TBP is subdivided at its centroid and then the degree of sub-patches are elevated by one. The result of degree elevation followed by subdivision is depicted in Fig. 8.20.

The control vertices resulting from the subdivision at the centroid are given by (from Eqn. (8.17)):

$$\mathbf{p}^r_{ijk} = \{\mathbf{p}^{r-1}_{i+1,j,k} + \mathbf{p}^{r-1}_{i,j+1,k} + \mathbf{p}^{r-1}_{i,j,k+1}\}/3 : \quad i + j + k = 3 - r \quad (8.30)$$

with $\mathbf{p}^0_{ijk} \equiv \mathbf{p}_{ijk}$ for $r = 0, 1, 2, 3$. Then, the control vertices of the three subpatches are given by (from Eqn. (8.19))

$\{\mathbf{p}^i_{0jk} : i + j + k = 3\}$ *for the subpatch at u=0 boundary.*

$\{\mathbf{p}^j_{i0k} : i + j + k = 3\}$ *for the subpatch at v=0 boundary.*

$\{\mathbf{p}^k_{ij0} : i + j + k = 3\}$ *for the subpatch at w=0 boundary.*

Finally, the degree of each cubic subpatch is elevated by using the degree elevation formula

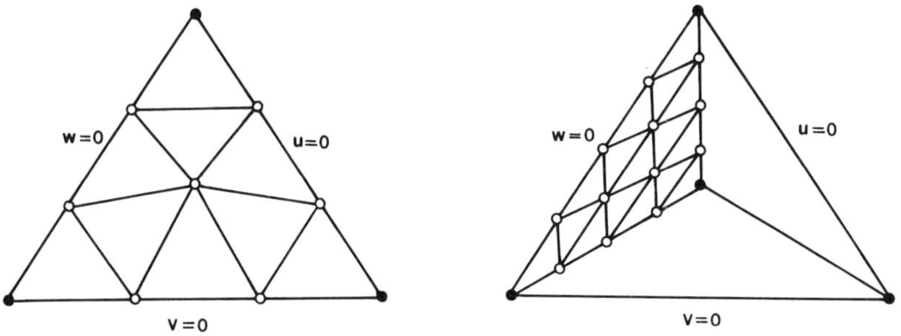

Figure 8.20 Subdivision and Degree Elevation of a Cubic Patch

(8.15). For example, the control vertices $\{\hat{\mathbf{p}}_{ijk}\}$ of the "$w = 0$ boundary" subpatch shown in the right figure of Fig. 8.20 are given by

$$\hat{\mathbf{p}}_{ijk} = \frac{i}{4}\mathbf{p}^k_{i-1,j,0} + \frac{j}{4}\mathbf{p}^k_{i,j-1,0} + \frac{k}{4}\mathbf{p}^{k-1}_{i,j,0} : \quad i+j+k=4 \tag{8.31}$$

where $\mathbf{p}^k_{ij0}$ are as defined in (8.30).

8.5.4 G^1-Correction of Off-boundary Control Vertices

Now the two adjacent patches, $A(u,v)$ and $B(u,v)$, shown earlier in Fig. 8.17 have been subdivided and degree-elevated. Let us define

$A^u(u,v)$: *quartic subpatch of* $A(u,v)$ *sharing* $u = 0$ *common boundary, and*

$B^u(u,v)$: *quartic subpatch of* $B(u,v)$ *sharing* $u = 0$ *common boundary*

as depicted in Fig. 8.21. We want to "correct" the *off-boundary control vertices* $\mathbf{p}_{112}$, $\mathbf{p}_{121}$, $\mathbf{q}_{112}$, and $\mathbf{q}_{121}$, leaving the control vertices on the (subpatch) boundaries untouched, in order to make the two subpatches G^1-continuous. Our task here is to determine the amounts of corrections for the *boundary control vectors* $\mathbf{a}_1, \mathbf{a}_2, \mathbf{b}_1, \mathbf{b}_2$ so that the G^1-conditions (8.27) are satisfied.

The first step is to determine the scalar coefficients p_i, q_i, r_i. Rewriting the boundary G^1-equations (8.27-a) and (8.27-e), we have (note $p_i + q_i = 1$)

$$p_0(\mathbf{a}_0 - \mathbf{b}_0) + r_0\mathbf{c}_0 = -\mathbf{b}_0 \tag{8.32-a}$$

$$p_1(\mathbf{a}_3 - \mathbf{b}_3) + r_1\mathbf{c}_3 = -\mathbf{b}_3. \tag{8.32-b}$$

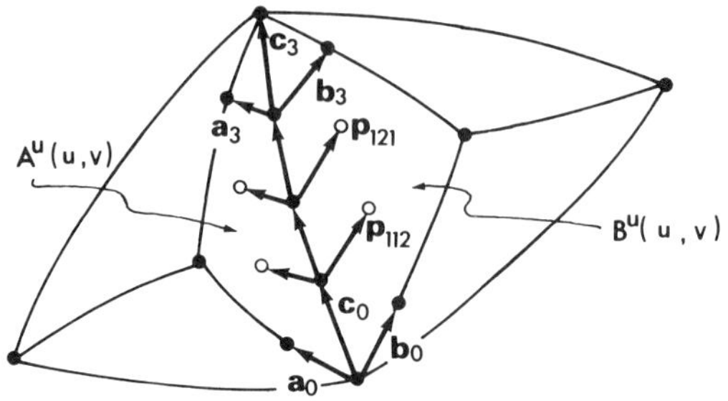

Figure 8.21 G^1-Correction of Off-boundary Control Vertices

Each of the vector equations in (8.32) is in the form

$$\alpha \mathbf{f} + \beta \mathbf{g} = \mathbf{h}$$

which is trivially solved for α, β if converted to two scalar equations:

$$(\mathbf{f} \cdot \mathbf{f})\alpha + (\mathbf{f} \cdot \mathbf{g})\beta = (\mathbf{f} \cdot \mathbf{h}), \quad and$$
$$(\mathbf{f} \cdot \mathbf{g})\alpha + (\mathbf{g} \cdot \mathbf{g})\beta = (\mathbf{g} \cdot \mathbf{h}).$$

Further, it can be shown that (8.32) always gives a valid solution as long as the two vector $\mathbf{a}_i$ and $\mathbf{b}_i$ are on strictly opposite sides of $\mathbf{c}_i$.

Having determined the scalar coefficients from (8.32), the next step is to "correct" the *boundary control vectors* $\mathbf{a}_1, \mathbf{a}_2, \mathbf{b}_1, \mathbf{b}_2$. Rewriting the three (middle) equations in (8.27), we have

$$3(p_0\mathbf{a}_1 + q_0\mathbf{b}_1) = -(p_1\mathbf{a}_0 + q_1\mathbf{b}_0 + r_1\mathbf{c}_0) - 3r_0\mathbf{c}_1 \qquad (8.33-a)$$
$$(p_0\mathbf{a}_2 + q_0\mathbf{b}_2) + (p_1\mathbf{a}_1 + q_1\mathbf{b}_1) = -(r_0\mathbf{c}_2 + r_1\mathbf{c}_1) \qquad (8.33-b)$$
$$3(p_1\mathbf{a}_2 + q_1\mathbf{b}_2) = -(p_0\mathbf{a}_3 + q_0\mathbf{b}_3 + r_0\mathbf{c}_3) - 3r_1\mathbf{c}_2 \qquad (8.33-c)$$

which may be expressed in a matrix form $\mathbf{Ax} = \mathbf{d}$:

$$\begin{bmatrix} 3p_0 & 3q_0 & 0 & 0 \\ p_1 & q_1 & p_0 & q_0 \\ 0 & 0 & 3p_1 & 3q_1 \end{bmatrix} \begin{bmatrix} \mathbf{a}_1 \\ \mathbf{b}_1 \\ \mathbf{a}_2 \\ \mathbf{b}_2 \end{bmatrix} = \begin{bmatrix} -(p_1\mathbf{a}_0 + q_1\mathbf{b}_0 + r_1\mathbf{c}_0) - 3r_0\mathbf{c}_1 \\ -(r_0\mathbf{c}_2 + r_1\mathbf{c}_1) \\ -(p_0\mathbf{a}_3 + q_0\mathbf{b}_3 + r_0\mathbf{c}_3) - 3r_1\mathbf{c}_2 \end{bmatrix}. \qquad (8.34)$$

Note that the above G^1-condition does not hold with the "current" control vectors $\mathbf{a}_i, \mathbf{b}_i$ which were obtained from the Bezier curve-net.

In order to satisfy the G^1-equation (8.34), the initial *boundary control vectors (ie, off-boundary control points)* need to be "corrected". Let $\hat{\mathbf{a}}_i, \hat{\mathbf{b}}_i$ denote "corrected" control vectors satisfying (8.34) defined as

$$\hat{\mathbf{a}}_1 = \mathbf{a}_1 + \mathbf{e}_1 \qquad (8.35-a)$$

$$\hat{\mathbf{b}}_1 = \mathbf{b}_1 + \mathbf{e}_2 \qquad (8.35-b)$$

$$\hat{\mathbf{a}}_2 = \mathbf{a}_2 + \mathbf{e}_3 \qquad (8.35-c)$$

$$\hat{\mathbf{b}}_2 = \mathbf{b}_2 + \mathbf{e}_4 \qquad (8.35-d)$$

where $\mathbf{e}_i$ are called **correction vectors**. Thus, the G^1-equation (8.34) should be expressed as

$$\mathbf{A}(\mathbf{x} + \mathbf{e}) = \mathbf{d} \qquad (8.36)$$

$$\textit{where,} \quad \mathbf{x} = [\mathbf{a}_1\ \mathbf{b}_1\ \mathbf{a}_2\ \mathbf{b}_2]^T : \textit{initial control vectors},$$

$$\mathbf{e} = [\mathbf{e}_1\ \mathbf{e}_2\ \mathbf{e}_3\ \mathbf{e}_4]^T : \textit{correction vectors},$$

$$\mathbf{A} = \begin{bmatrix} \mathbf{A}_1 \\ \mathbf{A}_2 \\ \mathbf{A}_3 \end{bmatrix} = \begin{bmatrix} 3p_0 & 3q_0 & 0 & 0 \\ p_1 & q_1 & p_0 & q_0 \\ 0 & 0 & 3p_1 & 3q_1 \end{bmatrix},$$

$$\mathbf{d} = \begin{bmatrix} \mathbf{d}_1 \\ \mathbf{d}_2 \\ \mathbf{d}_3 \end{bmatrix} = \begin{bmatrix} -(p_1\mathbf{a}_0 + q_1\mathbf{b}_0 + r_1\mathbf{c}_0) - 3r_0\mathbf{c}_1 \\ -(r_0\mathbf{c}_2 + r_1\mathbf{c}_1) \\ -(p_0\mathbf{a}_3 + q_0\mathbf{b}_3 + r_0\mathbf{c}_3) - 3r_1\mathbf{c}_2 \end{bmatrix}.$$

There are four unknown vectors $\mathbf{e}_i$ in the three vector equations (8.36), which makes it an under-determined linear system. Since the *correction vectors* $\mathbf{e}_i$ represent "deviations" from the original composite surface it would be desirable to make the amount of correction as small as possible (Recall that the original composite surface obtained in §8.5.3 is the best possible interpolant for the input data except the fact that it is a C^0-surface). Thus, we may formulate a least square problem as:

$$\textit{Minimize} \quad \mathbf{e}^T\mathbf{e}(= \textstyle\sum \mathbf{e}_i \cdot \mathbf{e}_i) \qquad (8.37)$$

$$\textit{Subject to} \quad \mathbf{Ae} = \mathbf{d} - \mathbf{Ax} \quad (\textit{See Eqn.}\ 8.36).$$

Before trying to solve the above *quadratic programming* (QP) *problem*, it is necessary to check the *rank* of the coefficient matrix $\mathbf{A}$.

It can be shown that the *rank* of $\mathbf{A}$ in (8.37) is 3 if $p_0q_1 \neq p_1q_0$. In this case, the solution for the *QP-problem* can be expressed in terms of the **Kuhn-Tucker condition** (see for example,

Avriel, 1976, pp.185-188) as

$$\mathbf{e} = \mathbf{A}^{\mathbf{T}}\mathbf{m} \quad and \quad \mathbf{A}\mathbf{e} = \mathbf{f} \quad \Rightarrow \quad \mathbf{m} = (\mathbf{A}\mathbf{A}^{\mathbf{T}})^{-1}\mathbf{f} \tag{8.38}$$

where, $\mathbf{m} = [\mathbf{m}_1\ \mathbf{m}_2\ \mathbf{m}_3]^T$*: Lagrange multipliers,*

$$\mathbf{f} = \mathbf{d} - \mathbf{A}\mathbf{x} = \begin{bmatrix} \mathbf{f}_1 \\ \mathbf{f}_2 \\ \mathbf{f}_3 \end{bmatrix} = \begin{bmatrix} -3(p_0\mathbf{a}_1 + q_0\mathbf{b}_1 + r_0\mathbf{c}_1) - (p_1\mathbf{a}_0 + q_1\mathbf{b}_0 + r_1\mathbf{c}_0) \\ -3p_0\mathbf{a}_2 + q_0\mathbf{b}_2 + r_0\mathbf{c}_2) - (p_1\mathbf{a}_1 + q_1\mathbf{b}_1 + r_1\mathbf{c}_1) \\ -3(p_1\mathbf{a}_2 + q_1\mathbf{b}_2 + r_1\mathbf{c}_2) - (p_0\mathbf{a}_3 + q_0\mathbf{b}_3 + r_0\mathbf{c}_3) \end{bmatrix}.$$

On evaluating $(\mathbf{A}\mathbf{A}^{\mathbf{T}})$, we have

$$\mathbf{A}\mathbf{A}^{\mathbf{T}} = \begin{bmatrix} 3p_0 & 3q_0 & 0 & 0 \\ p_1 & q_1 & p_0 & q_0 \\ 0 & 0 & 3p_1 & 3q_1 \end{bmatrix}, \begin{bmatrix} 3p_0 & p_1 & 0 \\ 3q_0 & q_1 & 0 \\ 0 & p_0 & 3q_1 \\ 0 & q_0 & 3q_1 \end{bmatrix}$$

$$= \begin{bmatrix} 9\alpha & \beta & 0 \\ \beta & \alpha + \beta & \beta \\ 0 & \beta & 9\gamma \end{bmatrix}.$$

$$where, \quad \alpha = (p_0)^2 + (q_0)^2,$$

$$\beta = 3(p_0 p_1 + q_0 q_1),$$

$$\gamma = (p_1)^2 + (q_1)^2.$$

Using **Cramer's rule** (see for example, Kreyszig, 1983, pp.318-321), the *Lagrange multipliers* are expressed as ($where\ \delta = det(\mathbf{A}\mathbf{A}^{\mathbf{T}})$):

$$\mathbf{m}_1 = \frac{1}{\delta}\begin{bmatrix} \mathbf{f}_1 & \beta & 0 \\ \mathbf{f}_2 & \alpha + \gamma & \beta \\ \mathbf{f}_3 & \beta & 9\gamma \end{bmatrix},$$

$$\mathbf{m}_2 = \frac{1}{\delta}\begin{bmatrix} 9\alpha & \mathbf{f}_1 & 0 \\ \beta & \mathbf{f}_2 & \beta \\ 0 & \mathbf{f}_3 & 9\gamma \end{bmatrix},$$

$$\mathbf{m}_3 = \frac{1}{\delta}\begin{bmatrix} 9\alpha & \beta & \mathbf{f}_1 \\ \beta & \alpha + \gamma & \mathbf{f}_2 \\ 0 & \beta & \mathbf{f}_3 \end{bmatrix},$$

Thus, when $p_0q_1 \neq p_1q_0$, a solution to the least square (ie, *quadratic programming*) problem (8.37) is given by

$$\mathbf{e}^T = [\mathbf{e}_1\ \mathbf{e}_2\ \mathbf{e}_3\ \mathbf{e}_4] = \mathbf{m}^\mathbf{T}\mathbf{A} \tag{8.39}$$

$$\begin{aligned}
\textit{where,}\quad & \mathbf{m}^T = [\mathbf{m}_1\ \mathbf{m}_2\ \mathbf{m}_3] : \textit{Lagrange multipliers},\\
& \mathbf{m}_1 = \{(9\gamma(\alpha+\gamma)-\beta^2)\mathbf{f}_1 - 9\beta\gamma\mathbf{f}_2 + \beta^2\mathbf{f}_3\}/\delta,\\
& \mathbf{m}_2 = 9\{-\beta\gamma\mathbf{f}_1 + 9\alpha\gamma\mathbf{f}_2 - \alpha\beta\mathbf{f}_3\}/\delta,\\
& \mathbf{m}_3 = \{\beta^2\mathbf{f}_1 - 9\alpha\beta\mathbf{f}_2 + (9\alpha(\alpha+\beta)-\beta^2)\mathbf{f}_3\}/\delta,\\
& \delta = det(\mathbf{A}\mathbf{A}^T) = 9(\alpha+\beta)(9\alpha\gamma-\beta^2),\\
& \alpha = (p_0)^2 + (q_0)^2;\ \beta = 3(p_0p_1 + q_0q_1);\ \gamma = (p_1)^2 + (q_1)^2\\
& \mathbf{f}_1, \mathbf{f}_2, \mathbf{f}_3 : \ \textit{as defined in}\ (8.38),\\
& \mathbf{A} :\ (3\times 4)\ \textit{matrix as defined in}\ (8.36).
\end{aligned}$$

If $p_0q_1 = q_1q_0$, we will have $p_0 = p_1$ and $q_0 = q_1$ (since $p_1 + q_1 = 1$ from (8.28)). In this case the rank of the coefficient matrix **A** in (8.36) is less than three because $det(\mathbf{A}\mathbf{A}^\mathbf{T}) = 0$. Further, the middle equation of (8.36) may become redundant since

$$\mathbf{A}_1 + \mathbf{A}_3 = 3\mathbf{A}_2 \tag{8.40 – a}$$

holds if $p_0 = p_1$ and $q_0 = q_1$. However, in that case, the linear system (8.36) would not have a solution unless the following also holds

$$\mathbf{d}_1 + \mathbf{d}_3 = 3\mathbf{d}_2 \tag{8.40 – b}$$

Under the condition of "$p_0 = p_1$ and $q_0 = q_1$", it can be shown that the above relation always holds if the "precision" of the quartic common boundary curve (See Fig. 8.21) is cubic (Shin and Choi, 1990). Since the *quartic* common boundary curve was obtained by elevating the degree of a cubic curve, the linear system (8.36) under the condition of $p_0 = p_1$ and $q_0 = q_1$ has a solution. Thus, we can formulate the following QP-problem which is similar to (8.37):

$$\textit{Minimize}\ \mathbf{e}^T\mathbf{e}(=\Sigma\ \mathbf{e}_i\cdot\mathbf{e}_i)\quad s.t.\ \bar{\mathbf{A}}\mathbf{e} = \bar{\mathbf{d}} - \bar{\mathbf{A}}\mathbf{x}\ (\textit{See Eqn.}\ 8.36) \tag{8.41}$$

$$\begin{aligned}
\textit{where,}\quad & \mathbf{x} = [\mathbf{a}_1\ \mathbf{b}_1\ \mathbf{a}_2\ \mathbf{b}_2]^T,\\
& \mathbf{e} = [\mathbf{e}_1\ \mathbf{e}_2\ \mathbf{e}_3\ \mathbf{e}_4]^T,\\
& \bar{\mathbf{A}} = \begin{bmatrix} 3p_0 & 3q_0 & 0 & 0\\ 0 & 0 & 3p_0 & 3q_0 \end{bmatrix};\quad \bar{\mathbf{d}} = \begin{bmatrix} -(p_0\mathbf{a}_0 + q_0\mathbf{b}_0 + r_1\mathbf{c}_0) - 3r_0\mathbf{c}_1\\ -(p_0\mathbf{a}_3 + q_0\mathbf{b}_3 + r_0\mathbf{c}_3) - 3r_1\mathbf{c}_2 \end{bmatrix}.
\end{aligned}$$

The *Kuhn-Tucker* condition for the above *QP*-problem is again given by

$$\mathbf{e} = \bar{\mathbf{A}}^T\mathbf{m} \quad and \quad \bar{\mathbf{A}}\mathbf{e} = \bar{\mathbf{f}} \quad \Rightarrow \quad \mathbf{m} = (\bar{\mathbf{A}}\bar{\mathbf{A}}^T)^{-1}\bar{\mathbf{f}} \tag{8.42}$$

where, $\mathbf{m} = [\mathbf{m}_1 \ \mathbf{m}_2]^T$*: Lagrange multipliers,*

$$\bar{\mathbf{f}} = \begin{bmatrix} \mathbf{f}_1 \\ \mathbf{f}_3 \end{bmatrix} = \begin{bmatrix} -3(p_0\mathbf{a}_1 + q_0\mathbf{b}_1 + r_0\mathbf{c}_1) - (p_0\mathbf{a}_0 + q_0\mathbf{b}_0 + r_1\mathbf{c}_0) \\ -3(p_0\mathbf{a}_2 + q_0\mathbf{b}_2 + r_1\mathbf{c}_2) - (p_0\mathbf{a}_3 + q_0\mathbf{b}_3 + r_0\mathbf{c}_3) \end{bmatrix}.$$

By following the same steps of the $p_0q_1 \neq q_1q_0$ case, the *correction vectors* for the case of $p_0q_1 = q_1q_0$ are obtained from (8.42) as

$$\mathbf{e}^T = [\mathbf{e}_1 \ \mathbf{e}_2 \ \mathbf{e}_3 \ \mathbf{e}_4] = \frac{1}{3(p_0^2 + q_0^2)}[p_0\mathbf{f}_1 \ q_0\mathbf{f}_1 \ p_0\mathbf{f}_3 \ q_0\mathbf{f}_3] \tag{8.43}$$

where $\mathbf{f}_1, \mathbf{f}_3$ are as defined in (8.42).

Having determined the correction vectors $\mathbf{e}_i$, the *off-boundary control vertices* **p,q** for the common boundary between $A(u,v)$ and $B(u,v)$ – or more exactly between the subdivided triangular patches $A^u(u,v)$ and $B^u(u,v)$ in Fig. 8.21 – are corrected as follows:

$$\hat{\mathbf{p}}_{112} = \mathbf{p}_{112} + \mathbf{e}_1 \tag{8.44 – a}$$

$$\hat{\mathbf{q}}_{112} = \mathbf{q}_{112} + \mathbf{e}_2 \tag{8.44 – b}$$

$$\hat{\mathbf{p}}_{121} = \mathbf{p}_{121} + \mathbf{e}_3 \tag{8.44 – c}$$

$$\hat{\mathbf{q}}_{121} = \mathbf{q}_{121} + \mathbf{e}_4 \tag{8.44 – d}$$

The *correction vectors* $\mathbf{e}_i'$*s* in (8.44) are given by (8.39) when $p_0q_1 \neq q_1q_0$ or by (8.43) when $p_0q_1 = p_1q_0$. The above *off-boundary control vertices* are indicated by small circles in Fig. 8.21.

8.5.5 Modification of Subpatch-boundary Control Vertices

These new (corrected) control vertices given by (8.44) define a composite surface with G^1-continuous joins between adjacent patches. However, the *corrections* given by (8.44) may cause *gradient discontinuity* across the subpatch boundaries. In order to restore smoothness across the subpatch boundaries, the *control vertices on the subpatch boundaries* need to be modified. The corrected control vertices are labeled as "L" in Fig. 8.22, and the control vertices to be modified at this stage are the ones labeled "M", "N" and "O" in Fig. 8.22.

The modification process may be carried out as follows:

a) determine the control vertices "M" as the centroid of the three vertices surrounding "M",

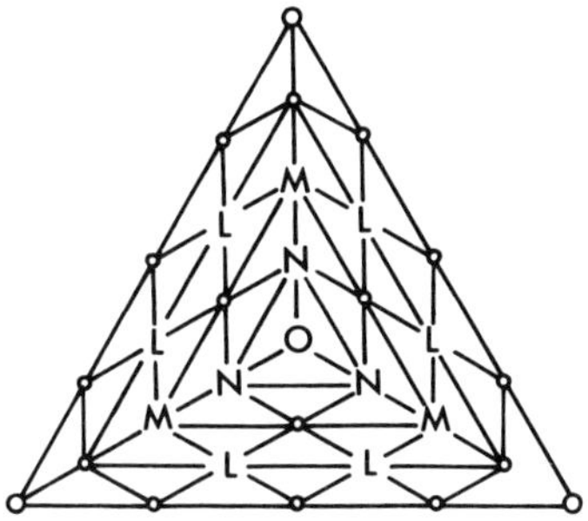

Figure 8.22 Correction of Control Vertices on Subpatch Boundaries

b) determine the control vertices "N" as the centroid of the three vertices surrounding "N", and

c) determine "O" as the centroid of the three "N's".

This correction produces C^1-continuous (ie, parametrically smooth) joins between the subpatches in each triangle (Piper, 1987). Since this correction does not affect the G^1-condition across the common boundary of adjacent triangles, the resulting surface is visually smooth every where.

8.6 SUMMARY AND DISCUSSIONS

The overall procedure of constructing a composite (triangular) surface interpolating to scattered point data are summarized in Fig. 8.23. The scattered data interpolation scheme introduced in this chapter has some distinct advantages over the rectangular surface interpolation schemes given in the previous chapter. When the input data are obtained from a physical model (by a coordinate measuring machine), it is not easy to have a regularly arranged data set. The triangular surface interpolation scheme can take any data set and then process it almost automatically.

The major drawback of the triangular scheme is that the resulting composite surface has a *scattered topology* which makes it rather difficult to be processed (eg, cutter path generation for NC machining, surface/surface intersection, blending, and rendering). In this respect, further researches are needed in these application areas.

The *triangulation* method has its own applications in addition to the scattered data interpolation. Examples include FEM mesh generation, cutter interference handling (Choi & Jun, 1989), and CAPP for sculptured surface machining.

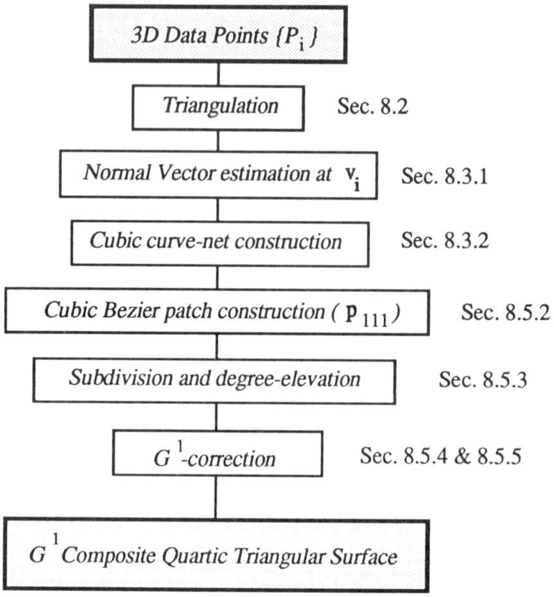

Figure 8.23 Overall Procedure of Constructing Triangular Surface

The main complication in the proposed surface construction scheme is in the G^1-correction stage for the *off-boundary control vertices* denoted by "L" in Fig. 8.22. Especially, the case distinction for the determination of the correction vectors $\mathbf{e}_i$ may cause numerical instabilities. The G^1-surface construction scheme discussed in §8.5 is in fact a *curve-net* interpolation method (to be discussed in Chapter 10). An alternative construction scheme is proposed in Chapter 10 (See §10.5.4).

CHAPTER 9

SURFACE CONSTRUCTION FROM 2D CROSS-SECTIONS

9.1 INTRODUCTION

Among the surface description methods discussed in Chapter 1 (See Fig. 1.9), the *section curve sweeping* method is the most popular one in conventional engineering drawings. A wide variety of curved objects are designed with *sweep surfaces*. A smooth sculptured surface that is best described as a trajectory of "cross-section curves" swept along "profile curves" is called a **sweep surface**. The popularity of sweep surface seems to be due to 1) it is easy to describe the sculptured surface by specifying 2D cross-sections, and 2) the resulting surface is aesthetically appealing. The designer may envision the surface as a "blended trajectory" of cross-section curves swept along profile curves.

Presented in this chapter are various approaches to constructing mathematical representations $\mathbf{r}(u, v)$ for the type of sweep surfaces illustrated in Figs. 9.1 through 9.3. The topics to be discussed include:

- **sweeping**: coordinate transformation of cross section curves,
- **lofting**: *Hermite* blending of cross section curves, and
- **skinning**: *B-spline* blending of cross section curves.

Shown in Fig. 9.1a is the transition surface (called a *front mask*) in between the front face and the screen of a commercial TV set. The "front mask" surface is defined by two *boundary* curves and a *section* curve (in this case, a straight line) as depicted in Fig. 9.1b.

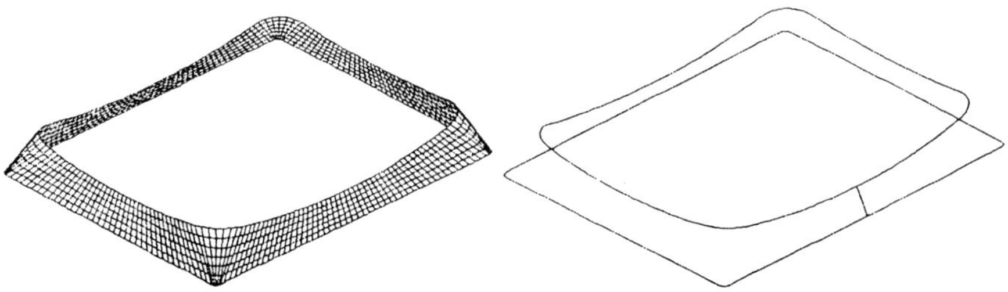

(a) (b)

Figure 9.1 Sweep Surface Example (Front Mask)

9.2.2 Overall Procedure of Sweep Surface Modeling

Depicted in Fig. 9.5 is a 3D view of the sweep surface patch of Fig. 9.4. Also shown in the figure are:

- **base plane**: an xy-plane corresponding to the *main drawing*,
- **section planes**: planes on which *section curves* are drawn.

Imagine that the section plane is a solid plate on which the section curve is drawn from point P_0 to point P_1. Now we make a through hole in the plate at P_0 where the first boundary curve C, which is now denoted as $\mathbf{b}_0(v)$, initially intersects with the plate. The boundary curves are also regarded as physical wires in 3D space. And then the plate is moved in space so that the wire $\mathbf{b}_0(v)$ passes through the hole P_0 while keeping the plate in a vertical position.

To obtain a unique motion of the moving plate (on which a section curve is drawn) a sweeping rule has to be specified. For the sweep surface of Fig. 9.5, a natural choice would be to move the plate so that the "flow rates" of the two boundary curves are *synchronized*. In other words, if an *intermediate section plane* (ie, the moving plate at an intermediate location) intersects the first boundary curve $\mathbf{b}_0(v)$ at, say, 30% of the curve length, it should intersect the second boundary curve $\mathbf{b}_1(v)$ also at 30% of its length.

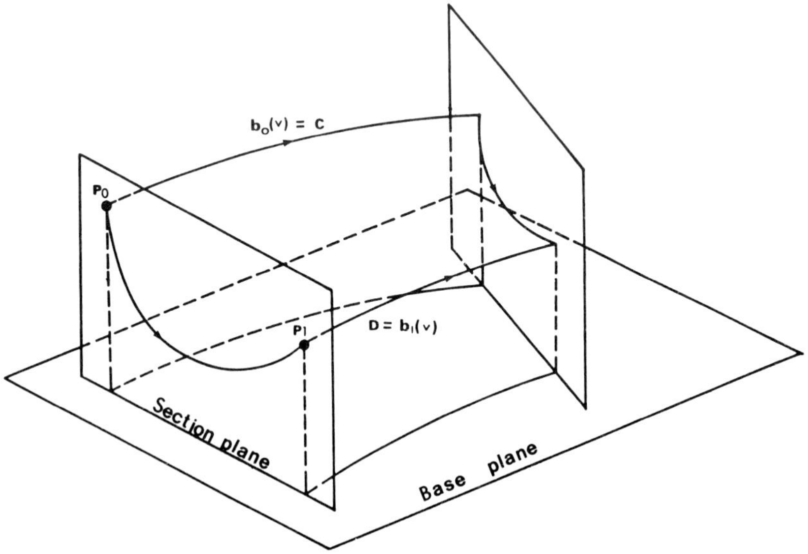

Figure 9.5 3D View of the Sweep surface in Figure 9.4

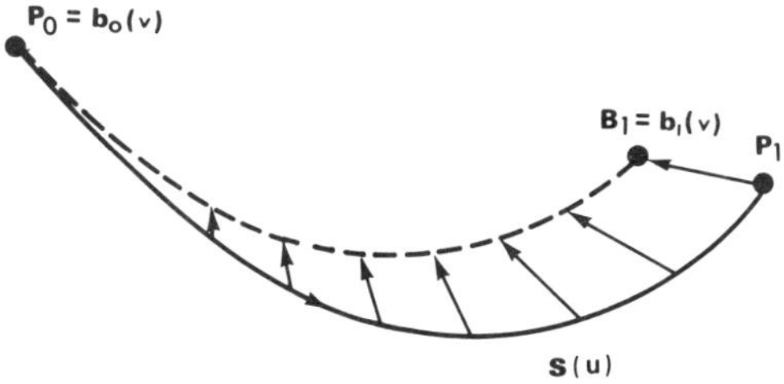

Figure 9.6 Correction of Intermediate Section Curve

Our sweeping rule guarantees that the first boundary curve $\mathbf{b}_0(v)$ always passes through P_0 on the moving plate but the second boundary curve $\mathbf{b}_1(v)$ does not in general intersect the plate at P_1 as depicted in Fig. 9.6, where B_1 is the current intersection point of $\mathbf{b}_1(v)$ on the plate. Thus, it is necessary to "correct" the original section curve so that the "initial intersection point" P_1 is brought to the "current intersection point" B_1. The "corrected" trajectory of the first section curve (Section A-A') defines a sweep surface. By repeating the same steps for the second cross section curve (Section B-B'), another sweep surface is obtained. Finally, the two sweep surfaces are blended, according to a suitable blending rule, to obtain the final sweep surface.

A mathematical model of a surface is a mapping from a 2D domain (of parameters u, v) to an E^3-space. Our goal here is to obtain the mapping function $\mathbf{r}(u, v)$ from the descriptions of the two boundary curves and two section curves as given in Fig. 9.4. The proposed modeling procedure consists of the following four steps which are nothing but a mathematical replica of the descriptive procedure presented above:

- *Definition of coordinate frames and curve equations.*
- *Sweep transformation.*
- *Correction of section curves for boundary conditions.*
- *Blending.*

In the subsections that follow, each of the above steps is described in detail using the (synchronized) sweep surface of Fig. 9.4 as an example. As will be seen in §9.2.6, the result for the *synchronized sweeping* rule is also applicable to other sweeping rules.

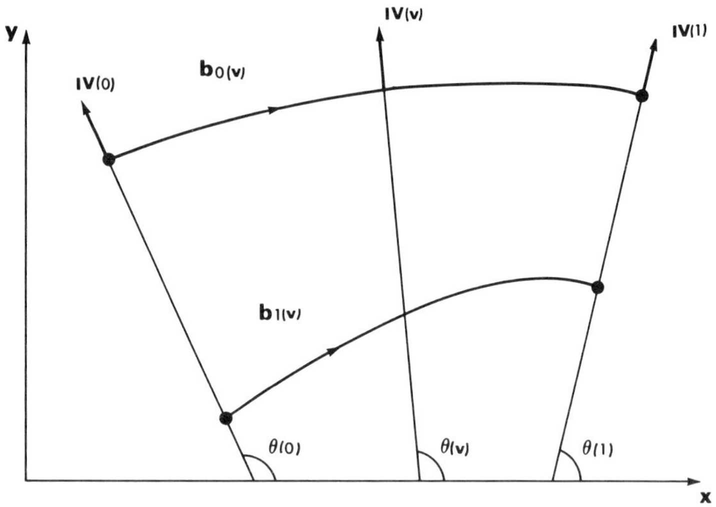

Figure 9.7 Intersection Vectors and Intersection Angles on Base Plane

9.2.3 Definition of Coordinate Frames and Curve Equations

An xy-coordinate frame can be defined on the main drawing (ie, base plane) as shown in Fig. 9.7. And then, the two boundary curves are represented as parametric curve equations with respect to a *profile parameter* $v \in [0, 1]$:

$$\mathbf{b}_0(v) = [x_0(v) \;\; y_0(v) \;\; z_0(v)] \; : \; \textit{first boundary curve.}$$
$$\mathbf{b}_1(v) = [x_1(v) \;\; y_1(v) \;\; z_1(v)] \; : \; \textit{second boundary curve.}$$

Depicted in Fig. 9.7 are 2D vectors, called **intersection vector** IV, denoting intersection lines between "intermediate section planes" and the base plane, and **intersection angles** θ representing angles between the x-axis and "intersection vectors". The *intersection vector* IV is a function of the *profile parameter* v and is expressed as

$$IV(v) = [(x_0(v) - x_1(v)) \;\; (y_0(v) - y_1(v)) \;\; 0]. \qquad (9.1 - a)$$

The *intersection angle* θ is also a function of v and is given by

$$\theta(v) = atan2(y_0(v) - y_1(v), \quad x_0(v) - x_1(v)). \qquad (9.1 - b)$$

In order to have a well-defined intersection angle $\theta(v)$, the projected images of the two boundary curves, $\mathbf{b}_0(v)$ and $\mathbf{b}_1(v)$, are not allowed to intersect with each other.

The parameterization of the boundary curves is not unique: it could be a B-spline curve, a composite Bezier curve, or a "non-functional curve" (meaning that the curve does not have an explicit form). The only requirement here is that the mapping from $v \in [0,1]$ to $\mathbf{b}(v)$ is regular and non-singular. That is, a one-to-one mapping is required. At any rate, we simply assume that *two boundary curves* $\mathbf{b}_0(v)$ and $\mathbf{b}_1(v)$ are given.

On each of the section-view drawings, an xy-coordinate frame is defined as shown in Fig. 9.8, and then each section curve is represented as a parametric curve $\mathrm{s}(u)$ with $0 \le u \le 1$. Also specified in the section plane are

- **guide point** (G): *a point on the section plane through which* $\mathbf{b}_0(v)$ *passes,*
- *intersection vector IV' appearing on the section plane, and*
- *intersection angle (ϕ): angle between x-axis and IV' (on the section plane).*

Notice in Fig. 9.8 that the guide point is given by

$$G = (g_x,\ g_y,\ 0) = \mathrm{s}(0).$$

Once the *section coordinate frame* is chosen, the *intersection angle* ϕ and the guide point G become fixed values. In summary, we have defined the following geometric entities:

$\mathbf{b}_0(v), \mathbf{b}_1(v)$ *for* $v \in [0,1]$: *3D boundary curves,*
$\mathrm{s}(u)$ *for* $u \in [0,1]$: *section curve on the xy-plane of section coordinate,*
$IV(v)$: intersection vector on the base plane,
$\theta(v)$: intersection angle on the base plane,
G : guide point on the section plane,
IV' : intersection vector on the section plane,
ϕ : intersection angle on the section plane.

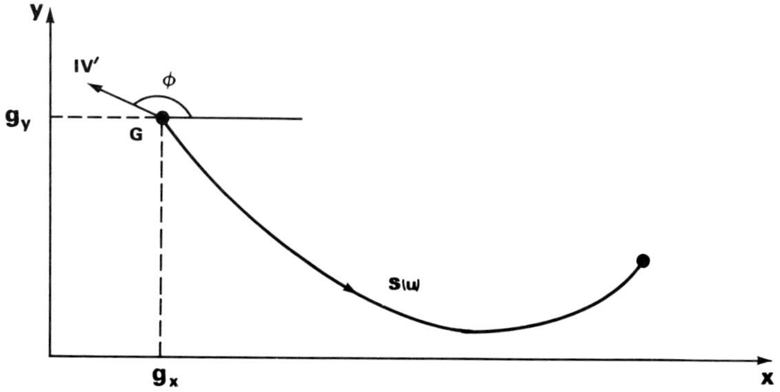

Figure 9.8 Section Plane Coordinate Frame for Synchronized Sweeping

9.2.4 Sweep Transformation of Section Curves

The first *boundary curve* $\mathrm{b}_0(v)$ and the *intersection vector* $IV(v)$ are shown in Fig. 9.9a, and the *section coordinate frame* is depicted in Fig. 9.9b. Notice that Fig. 9.9a represents the *base coordinate frame* and Fig. 9.9b is the xy-plane of a section coordinate frame.

The objective here is to "bring" the section plane to the desired position in the base coordinate frame so that the intersection vector IV' on the section plane (See Fig. 9.8) coincides with the intersection vector $IV(v)$ attached to $\mathrm{b}_0(v)$ in Fig. 9.9a. For a given value of v, the 3D location of the section curve, $\mathrm{p}(u, v)$ in Fig. 9.9c, may be obtained by performing the following sequence of coordinate transformations (Imagine that the section plane of Fig. 9.9b is originally located on the xy-plane of the 3D coordinate in Fig. 9.9a):

a) Translate so that the guide point G is moved to the origin;
b) Rotate ϕ degree around the z-axis;
c) Rotate -90 degrees around the x-axis;
d) Rotate $\theta(v)$ degrees around the z-axis;
e) Translate so that G is moved back to $\mathrm{b}_0(v)$.

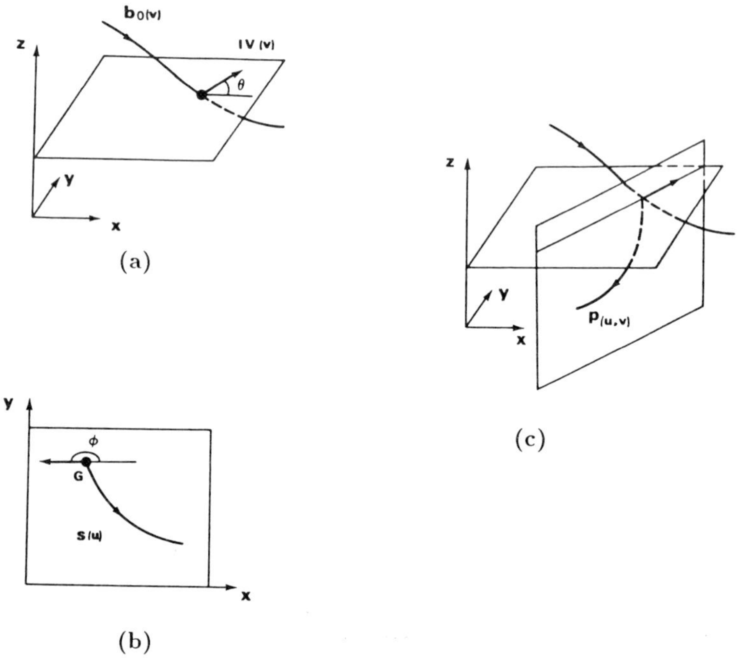

Figure 9.9 Sweep Transformation

Among the five transformations the three rotations form the core of the sweep transformation, and can conveniently be expressed in a single matrix form:

$$
\begin{aligned}
SWEEP(\phi,\theta(v)) &= ROT(z,\phi)\ ROT(x,-90)\ ROT(z,\theta(v)) \\
&= \begin{bmatrix} \cos\phi & \sin\phi & 0 \\ -\sin\phi & \cos\phi & 0 \\ 0 & 0 & 1 \end{bmatrix} \begin{bmatrix} 1 & 0 & 0 \\ 0 & 0 & -1 \\ 0 & 1 & 0 \end{bmatrix} \begin{bmatrix} \cos\theta(v) & \sin\theta(v) & 0 \\ -\sin\theta(v) & \cos\theta(v) & 0 \\ 0 & 0 & 1 \end{bmatrix} \\
&= \begin{bmatrix} \cos\phi\ \cos\theta(v) & \cos\phi\ \sin\theta(v) & -\sin\phi \\ -\sin\phi\ \cos\theta(v) & -\sin\phi\ \sin\theta(v) & -\cos\phi \\ -\sin\theta(v) & \cos\theta(v) & 0 \end{bmatrix}
\end{aligned}
\tag{9.2}
$$

where ϕ and $\theta(v)$ are *intersection angles* on the section plane and on the base plane, respectively. Thus, the sweep surface $\mathbf{p}(u,v)$ obtained by sweeping a section curve $\mathbf{s}(u)$ along a boundary curve $\mathbf{b}_0(v)$ is expressed as the following **sweep transformation**:

$$\mathbf{p}(u,v) = (\mathbf{s}(u) - G)\ SWEEP(\phi,\theta(v)) + \mathbf{b}_0(v) :\ 0 \le u, v \le 1 \tag{9.3}$$

where, $\mathbf{s}(u)$ *for* $u \in [0,1]$: *section curve on the section plane,*

G : *guide point on the section plane* (*Fig.* 9.8),

$SWEEP(\phi,\theta(v))$: 3×3 *sweep transformation matrix* (9.2),

ϕ : *intersection angle on the section plane* (*Fig.* 9.8),

$\theta(v)$: *intersection angle on the base plane* (*Fig.* 9.7),

$\mathbf{b}_0(v)$ *for* $v \in [0,1]$: *3D boundary curve.*

9.2.5 Correction and Blending of Section Curves

The sweep transformation given by (9.3) will bring the initial section plane to an intermediate location as depicted in Fig. 9.9c. Since the end point of the intermediate section curve $\mathbf{p}(u,v)$ is in general away from the second boundary curve $\mathbf{b}_1(v)$, the end point $\mathbf{p}(1,v)$ of the curve has to be brought to $\mathbf{b}_1(v)$. One way of making this correction is to

a) scale the section curve $\mathbf{p}(u,v)$ *so that the distance between its "start point"* $\mathbf{p}(0,v)$ *and "end point"* $\mathbf{p}(1,v)$ *becomes equal to the distance between the two boundary curves, and then*

b) blend the two scaled section curves by using a blending function.

The scaling factor $f(v)$ to be used in step a) is given by

$$f(v) = |\mathbf{b}_0(v) - \mathbf{b}_1(v)| / |\mathbf{s}(0) - \mathbf{s}(1)|. \tag{9.4}$$

Thus, the corrected *intermediate section curve* $\mathbf{q}(u,v)$ may be expressed as a blend of the two scaled (and translated) intermediate section curves:

$$\begin{aligned}\mathbf{q}(u,v) &= \alpha(u)\{f(v)[\mathbf{p}(u,v)-\mathbf{p}(0,v)]+\mathbf{b}_0(v)\}\\ &+\beta(u)\{f(v)[\mathbf{p}(u,v)-\mathbf{p}(1,v)]+\mathbf{b}_1(v)\}\\ &= f(v)[\mathbf{s}(u)-\alpha(u)\mathbf{s}(0)-\beta(u)\mathbf{s}(1)]SWEEP(\phi,\theta(v))\\ &+[\alpha(u)\mathbf{b}_0(v)+\beta(u)\mathbf{b}_1(v)],\end{aligned} \tag{9.5}$$

where $\alpha(u), \beta(u)$ are blending functions.

Since there are two sweep surfaces, one from each section curve (Sections A-A' and B-B' in Fig. 9.4), the two sweep surfaces $\mathbf{q}_i(u,v)\ :\ i=0,1$ of (9.5) have to be blended. The equation of the **blended sweep surface** is expressed as:

$$\begin{aligned}\mathbf{r}(u,v) &= \alpha(v)\mathbf{q}_0(u,v)+\beta(v)\mathbf{q}_1(u,v)\\ &= \alpha(v)f_0(v)[\mathbf{s}_0(u)-\alpha(u)\mathbf{s}_0(0)-\beta(u)\mathbf{s}_0(1)]\ SWEEP(\phi_0,\theta(v))\\ &+\beta(v)f_1(v)[\mathbf{s}_1(u)-\alpha(u)\mathbf{s}_1(0)-\beta(u)\mathbf{s}_1(1)]\ SWEEP(\phi_1,\theta(v))\\ &+[\alpha(u)\mathbf{b}_0(v)+\beta(u)\mathbf{b}_1(v)]\ ; \qquad 0\le u,v\le 1,\end{aligned} \tag{9.6}$$

where, $\mathbf{s}_i(u), \mathbf{b}_j(v)$: *section curve and boundary curve equations,*

$f_i(v)$: *scaling factor given by* (9.4),

ϕ_i : *intersection angle on section plane* i (*Fig.* 9.8),

$\theta(v)$: *intersection angle on base plane given by* (9.1),

$SWEEP(\phi_i,\theta(v))$: *sweep transformation matrix* (9.2),

$\alpha(*), \beta(*)$: *blending functions.*

9.2.6 Representation of General Sweep Surfaces

In all, there are four sweeping rules considered as depicted in Fig. 9.10. They are:

- **Parallel sweep**: section planes are parallel with each other (Fig. 9.10a).
- **Rotational sweep**: section planes intersect with each other at a fixed vertical line which becomes the axis of rotation (Fig. 9.10b).
- **Spined sweep**: section planes are perpendicular to the tangent of the spine curve projected on the base plane (Fig. 9.10c).
- **Synchronized sweep**: each section plane intersects the two boundary curves at the same parameter value v (Fig. 9.10d).

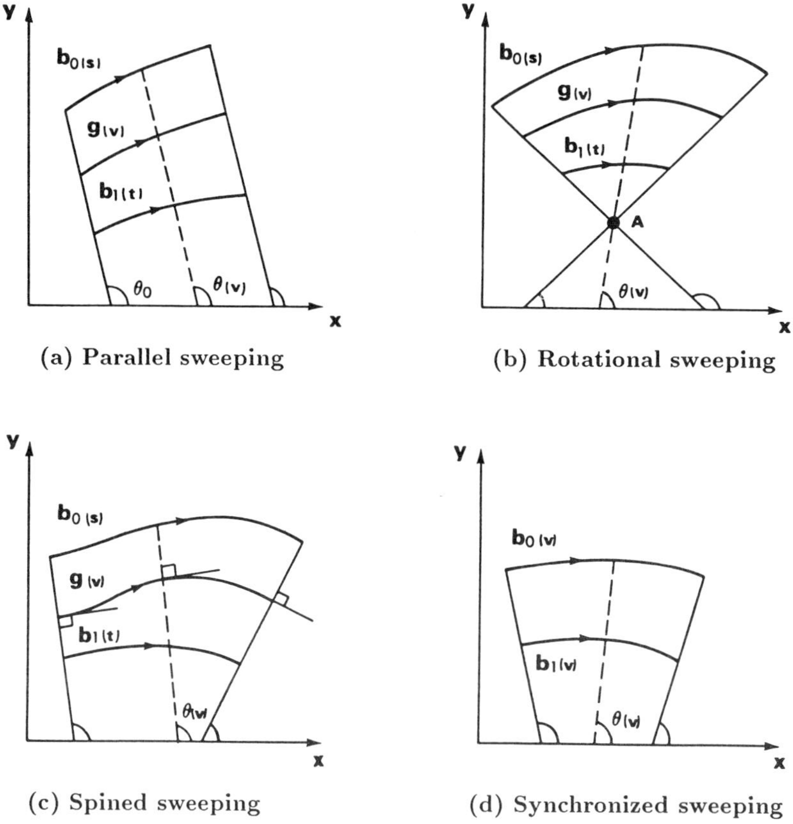

Figure 9.10 Sweeping Rules

For the construction of general sweep surfaces, the concept of a **guide curve** is introduced. In each of the above four sweeping rules, the sweeping action is regulated by the flow rate of the guide curve $\mathbf{g}(v)$.

A *guide curve* is either a *spine curve* (ie, a design axis) or a boundary curve. In the previous section, the first boundary curve $\mathbf{b}_0(v)$ was treated as a guide curve. Under the synchronized sweeping rule, either boundary curve can be used as a guide curve. That is, $\mathbf{g}(v) \equiv \mathbf{b}_0(v)$ and $\mathbf{g}(v) \equiv \mathbf{b}_1(v)$ would give the same result. A boundary curve is required to be an actual boundary of the sweep surface, while the spine curve is not necessarily located on the surface.

The geometric entities appearing in Fig. 9.10 may be formally defined as follows:

$$\mathbf{b}_0(s) = (x_0(s),\ y_0(s),\ z_0(s)), \quad 0 \le s \le 1\ ;\ (\textit{first boundary curve})$$

$$\mathbf{b}_1(t) = (x_1(t),\ y_1(t),\ z_1(t)), \quad 0 \le t \le 1\ ;\ (\textit{second boundary curve})$$

$$\mathbf{g}(v) = (x_g(v),\ y_g(v),\ z_g(v)), \quad 0 \le v \le 1\ ;\ (guide\ curve)$$

$$d\mathbf{g}(v)/dv = (\dot{x}_g(v),\ \dot{y}_g(v),\ \dot{z}_g(v));\ (derivative\ of\ \mathbf{g}(v))$$

$$A = (a_x,\ a_y,\ 0)\ ;\ (coordinates\ of\ the\ axis\ of\ rotation\ in\ Fig.\ 9.10b)$$

Then the **intersection angle** $\theta(v)$ under each sweeping rule is given by:

$$\begin{aligned}
\theta(v) &= \theta_0;\ (parallel\ sweep)\\
\theta(v) &= atan2(y_g(v) - a_y,\ x_g(v) - a_x);\ (rotational\ sweep)\\
\theta(v) &= atan2(\dot{x}_g(v), -\dot{y}_g(v));\ (spined\ sweep)\\
\theta(v) &= atan2(y_0(v) - y_1(v),\ x_0(v) - x_1(v));\ (synchronized\ sweep).
\end{aligned} \tag{9.7}$$

Shown in Fig. 9.11 is an initial section plane defined on its local coordinate system. The guide point G, intersection vector IV', and intersection angle ϕ are defined the same as before (Fig. 9.8). The only difference between the synchronized case of Fig. 9.8 and the general case is that the guide point G in Fig. 9.11 is allowed to depart from the section curve $\mathbf{s}(u)$. For a given value v of the profile parameter, the intermediate section curve $\mathbf{p}(u, v)$ is expressed as the following sweep transformation:

$$\mathbf{p}(u, v) = (\mathbf{s}(u) - G)\ SWEEP(\phi, \theta(v)) + \mathbf{g}(v). \tag{9.8}$$

Notice the similarity between (9.8) above and (9.3) in §9.2.4. We could have obtained the same result by simply replacing $\mathbf{b}_0(v)$ by $\mathbf{g}(v)$.

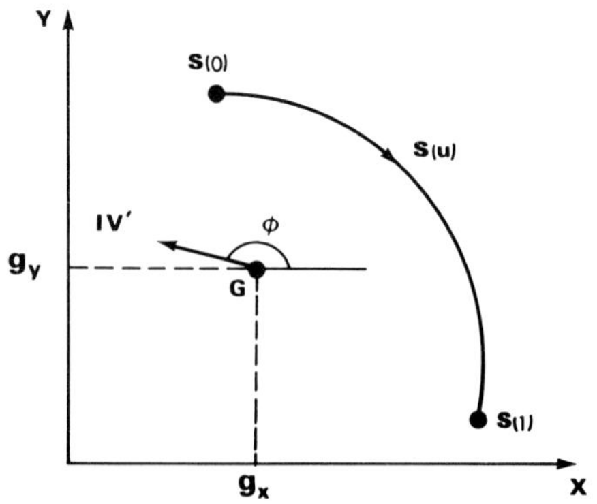

Figure 9.11 Section Plane Coordinate Frame for General Sweeping Rules

In order to define a "correction transformation", we need to find the points on the intermediate section plane at which the two boundary curves intersect with it. The intersection points are easily obtained by numerically solving the following equations for s and t

$$\mathbf{b}_0(s) \cdot \mathbf{n}(v) = \mathbf{g}(v) \cdot \mathbf{n}(v) \qquad (9.9-a)$$

$$\mathbf{b}_1(t) \cdot \mathbf{n}(v) = \mathbf{g}(v) \cdot \mathbf{n}(v) \qquad (9.9-b)$$

where $\mathbf{n}(v) = (sin\theta(v),\ -cos\theta(v),\ 0)$ is the *normal vector* of the *intermediate section plane.* Let s^* and t^* be the solutions to (9.9-a) and (9.9-b), respectively, then the scaling factor $f(v)$ for a general sweeping rule is given by:

$$f(v) = |\mathbf{b}_0(s^*) - \mathbf{b}_1(t^*)| / |\mathbf{s}(0) - \mathbf{s}(1)|. \qquad (9.10)$$

The "corrected" intermediate section curve $\mathbf{q}(u, v)$ is then expressed as the following correction transformation (compare it with Eqn. (9.5)):

$$\begin{aligned} \mathbf{q}(u, v) = {} & f(v)[\mathbf{s}(u) - \alpha(u)\mathbf{s}(0) - \beta(u)\mathbf{s}(1)] SWEEP(\phi, \theta(v)) \\ & + [\alpha(u)\mathbf{b}_0(s^*) + \beta(u)\mathbf{b}_1(t^*)]. \end{aligned} \qquad (9.11)$$

Finally the desired sweep surface $\mathbf{r}(u, v)$ is obtained by blending the two corrected intermediate section curves $\mathbf{q}_i(u, v)$ for $i = 0, 1$:

$$\begin{aligned} \mathbf{r}(u, v) = {} & \alpha(v) f_0(v)[\mathbf{s}_0(u) - \alpha(u)\mathbf{s}_0(0) - \beta(u)\mathbf{s}_0(1)]\ SWEEP(\phi_0, \theta(v)) \\ & + \beta(v) f_1(v)[\mathbf{s}_1(u) - \alpha(u)\mathbf{s}_1(0) - \beta(u)\mathbf{s}_1(1)]\ SWEEP(\phi_1, \theta(v)) \\ & + [\alpha(u)\mathbf{b}_0(s^*) + \beta(u)\mathbf{b}_1(t^*)]\ ; \quad 0 \le u, v \le 1, \end{aligned} \qquad (9.12)$$

where, $f_i(v)$: *scaling factor i given by* (9.10),
s^*, t^*: *intersection points between section plane and* $\mathbf{b}_j$ (*Eqn.* 9.9),
$\mathbf{s}_i(u), \mathbf{b}_j(v)$: *section curve and boundary curve equations,*
ϕ_i: *intersection angle on section plane i* (*Fig.* 9.8),
$\theta(v)$: *intersection angle on base plane given by* (9.7),
$SWEEP(\phi_i, \theta(v))$: *sweep transformation matrix* (9.2),
$\alpha(*), \beta(*)$: *blending functions.*

Note that the sweep surface equation (9.6) for a synchronized sweeping is in fact a special case of (9.12) where $s^* = t^* = v$.

In summary, a sweep surface patch should have a) one and only one *guide curve,* b) up to two *boundary curves,* and c) one or two *section curves* as listed in *Table 9.1.* When there is no boundary curve, a spine curve is necessary. A spine curve or a boundary curve acts as

Table 9.1 Cases of Sweep Surfaces

cases	number of boundary cvs	number of section cvs	applicable sweeping rules			
			paral	rotat	spine	synch
1	0	1	O	O	O	X
2	0	2	O	O	O	X
3	1	1	O	O	O	X
4	1	2	O	O	O	X
5	2	1	O	O	O	O
6	2	2	O	O	O	O

a guide curve. The sweep surface expression (9.12) corresponds to the "case 6" in the table where BC=2/SC=2 which represents the most general sweep surface patch. The cases 1 and 2 are given below (the remaining cases are left to the reader as an exercise):

1) $BC = 0/SC = 1:$

$$\mathbf{r}(u,v) = (\mathbf{s}_0(u) - G_0)SWEEP(\phi_0, \theta(v)) + \mathbf{g}(v)$$

2) $BC = 0/SC = 2:$

$$\mathbf{r}(u,v) = \{(\mathbf{s}_0(u) - G_0)SWEEP(\phi_0, \theta(v)) + \mathbf{g}(v)\}\alpha(v) + \{(\mathbf{s}_1(u) - G_1)SWEEP(\phi_1, \theta(v)) + \mathbf{g}(v)\}\beta(v).$$

The choice of the blending function $\alpha(*)$ and $\beta(*)$ in (9.12) affects the "quality" of the surface. The simplest choice is a *linear blending function:*

$$\alpha(u) = 1 - u\ ; \quad \beta(u) = u.$$

When an array of sweep surface patches is to be constructed from a network of boundary curves and section curves, the following choice which is the *Hermite blending function* introduced in §3.3.2 (see Eqn. (3.9)) can provide tangent continuity across patch boundaries:

$$\alpha(u) = 1 - 3u^2 + 2u^3,$$

$$\beta(u) = 3u^2 - 2u^3.$$

Note that $\alpha(u) \equiv H_0^3(u)$ and $\beta(u) \equiv H_3^3(u)$.

9.3 HERMITE BLENDED CROSS-SECTIONAL SURFACES

This section presents a different approach to constructing smooth surfaces from a sequence of cross-sectional curves: the method is known as **lofting** where the *surface* is modeled by an *interpolating function* (instead of sweep transformation). Input data for the sweep surface now consists of 3D section curves (without profile curves). We use the term "section curve" in place of "cross-sectional curve" for brevity. Methods of constructing smooth surfaces from cross-sectional data have long been utilized, especially in ship-yards, and are given a special name *"lofting"*.

In CAGD community, however, the same lofting is given different names depending on how the interpolating surface is constructed. If the section curves are blended with *Hermite functions*, the resulting surface is called a *Hermite blended cross-sectional surface.* If the section curves are B-spline curves and they are blended with *B-spline basis functions*, the lofting method is called *skinning* (Woodward, 1988) which is the subject of the next section.

9.3.1 Construction of 3D Parametric Curves from 2D Cross-sections

Let us consider the 3D section curves $\{\mathbf{p}_j(u) : j = 0, 1, ..., n\}$ as shown in Fig. 9.12. The basic idea for constructing a Hermite blended surface is to fit cubic polynomial curves along the longitudinal direction. When the section curves $\mathbf{p}_j(u)$'s are general parametric curves, a *Hermite lofted surface* (See Eqn. (5.18)) can be constructed from each pair of adjoining section curves. We will call the resulting surface a **Hermite blended surface.** If the section curves are conic sections, a (composite) *polyconic* surface is obtained by fitting longitudinal cubic spline curves through the control vertices.

Before presenting methods of constructing *blended cross-sectional* surfaces from 3D section curves $\mathbf{p}_j(u)$, methods of obtaining a 3D section curve $\mathbf{p}(u)$ from a 2D section curve $\mathbf{s}(u)$ will be introduced first. As it is not easy to define a 3D curve, each of the section curves in Fig. 9.12 is usually specified as a 2D curve $\mathbf{s}_j(u)$ on its local xy-coordinate system as depicted in Fig. 9.13b.

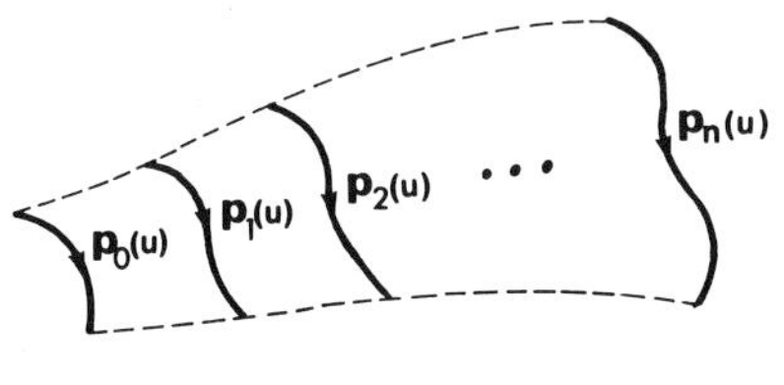

Figure 9.12 3D Cross-sectional Curves to be Blended

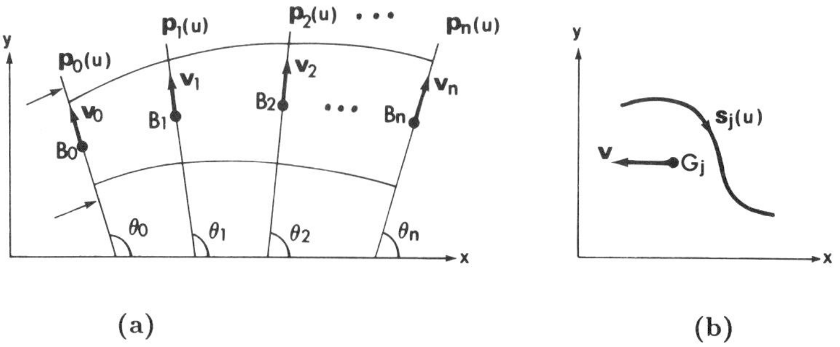

Figure 9.13 Description of Cross-section Curves

As discussed in §9.2, the section planes appearing on the *base plane* of Fig. 9.13a have to be specified first in order to fix the 3D location of a section curve. Depicted in Fig. 9.13 are

B_j : *reference point for section j on the base plane,*
$\mathbf{v}_j$: *intersection vector on the base plane,*
θ_j : *intersection angle on the base plane,*
$\mathbf{p}_j(u)$: *3D section curve j,*
G_j : *reference point on section plane j,*
$\mathbf{v}$: *intersection vector on section plane j,*
$\mathbf{s}_j(u)$: *2D section curve on its local coordinate frame.*

Assuming all the cross-sections are vertical planes, each section plane (it is a straight on the base plane) defines an *intersection angle* θ_j with the x-axis. In order to establish a unique transformation between the *section coordinate frame* of Fig. 9.13b and the *base coordinate frame* of Fig. 9.13a, "reference points" and "intersection vectors" have to be specified. B_j's in Fig. 9.13a denote reference points whose 3D coordinates are known. The same reference point B_j appears as G_j on the local xy-plane (Fig. 9.13b) of jth cross-section. The local section coordinate frame is defined such that the x-axis is a horizontal line in the base coordinate system (this means that the *intersection vector* $\mathbf{v}$ in Fig. 9.13b is parallel to the minus x direction).

Then, from the results given in §9.2 (See Eqns. 9.2 and 9.3), the 3D section curve $\mathbf{p}_j(u)$ is expressed as a *sweep transformation* of $\mathbf{s}_j(u)$. That is, we have

$$\mathbf{p}_j(u) = [\mathbf{s}_j(u) - G_j]\ SWEEP(\theta_j) + B_j, \tag{9.13}$$

where both B_j and G_j denote the same reference point defined in the base and local coordinate systems, respectively. In (9.13), the sweep transformation matrix SWEEP(θ) is given by

$\mathbf{t}_{0j}$: tangent vectors at V_{0j} for $j = 0, 1, ..., n$,
$\mathbf{t}_{1j}$: tangent vectors at V_{1j} for $j = 0, 1, ..., n$,
$\mathbf{t}_{2j}$: tangent vectors at V_{2j} for $j = 0, 1, ..., n$, and
τ_j: tangent vectors at ω_j for $j = 0, 1, ..., n$.

These *longitudinal tangents* may be determined by using any one of the "curve fitting" methods mentioned in §9.3.2: the FMILL method given by Eqn. (9.20), *Ferguson fitting* of Eqn. (9.21) or (9.22), and *chord-length spline fitting* of Eqn. (4.27). When the physical spacing of data points is relatively even, the Ferguson fitting method is recommended.

The procedure of constructing a polyconic surface based on the Ferguson fitting method is as follows: Assuming a "free end condition", $\mathbf{t}_{0j}$'s are obtained by solving the following linear equation system (See Eqn. (9.22)).

$$\begin{bmatrix} 2 & 1 & 0 & 0 & & & \\ 1 & 4 & 1 & 0 & & & \\ 0 & 1 & 4 & 1 & & & \\ & & & \cdot & & & \\ & & & & \cdot & & \\ & & & & 1 & 4 & 1 \\ & & & & 0 & 1 & 2 \end{bmatrix} \begin{bmatrix} \mathbf{t}_{00} \\ \mathbf{t}_{01} \\ \mathbf{t}_{02} \\ \cdot \\ \cdot \\ \cdot \\ \mathbf{t}_{0n} \end{bmatrix} = \begin{bmatrix} 3(V_{01} - V_{00}) \\ 3(V_{02} - V_{00}) \\ 3(V_{03} - V_{01}) \\ \cdot \\ \cdot \\ 3(V_{0n} - V_{0,n-2}) \\ 3(V_{0n} - V_{0,n-1}) \end{bmatrix}. \tag{9.23}$$

When "end tangents" $\hat{\mathbf{t}}_{00}, \hat{\mathbf{t}}_{0n}$ are given, a linear equation system of the form (9.21) is solved for internal tangents $\{\mathbf{t}_{0j} : j = 1, ..., n-1\}$. In this case, a *composite curve* $\{\mathbf{c}_{0j}(v) : j = 1, 2, ..., n\}$ passing through the first "row" of control vertices $\{V_{0j} : j = 0, 1, ..., n\}$ is constructed as follows:

$$\begin{aligned} \mathbf{c}_{0j}(v) &= H_0^3(v)V_{0,j-1} + H_1^3(v)\mathbf{t}_{0,j-1} + H_2^3(v)\mathbf{t}_{0,j} + H_3^3(v)V_{0,j} \\ &= \mathbf{V}\ \mathbf{C}\ \mathbf{S}^j\ ; \quad j = 1, 2, ..., n \end{aligned} \tag{9.24}$$

$$\begin{aligned} \textit{where,}\quad \mathbf{V} &= [\,1 \quad v \quad v^2 \quad v^3\,], \\ \mathbf{S}^j &= [V_{0,j-1} \quad V_{0,j} \quad \mathbf{t}_{0,j-1} \quad \mathbf{t}_{0,j}]^T, \\ \mathbf{C} &= \begin{bmatrix} 1 & 0 & 0 & 0 \\ 0 & 0 & 1 & 0 \\ -3 & 3 & -2 & -1 \\ 2 & -2 & 1 & 1 \end{bmatrix}. \end{aligned}$$

Composite curves $\mathbf{c}_{1j}(v), \mathbf{c}_{2j}(v)$ for the remaining control vertices may be obtained in the same way. A cubic spline function $\{w_j(v) : j = 1, ..., n\}$ is also constructed from $\{\omega_j, \tau_j : j = 0, 1, ..., n\}$. Then, the **polyconic patch** $\mathbf{r}_j(u, v)$ bounded by $\mathbf{p}_{j-1}(u)$ and $\mathbf{p}_j(u)$ is given by

$$\mathbf{r}_j(u,v) = \frac{(1-u)^2\mathbf{c}_{0j}(v) + 2u(1-u)w_j(v)\mathbf{c}_{1j}(v) + u^2\mathbf{c}_{2j}(v)}{(1-u)^2 + 2u(1-u)w_j(v) + u^2}. \tag{9.25}$$

When the 2D cross-sectional curves $\mathbf{s}_j(u)$ in Fig. 9.13b are conic section curves, so are the 3D section curves $\mathbf{p}_j(u)$. Since a rational quadratic Bezier curve (ie, conic section curve) is invariant under a linear transformation, it is enough to transform only the control vertices into 3D space by using (9.13) or (9.18). The same is true for non-rational Bezier curves and B-spline curves. If cross-sectional curves $\mathbf{p}_j(u)$'s are B-spline curves, they may better be blended by using B-spline functions. This last case is called *skinning.* In fact, the terms "skinning" and "lofting" both may mean the same operation of constructing a surface by "blending" cross-sectional data. But, in this book, we use the term "skinned surface" to mean a *B-spline blended surface constructed from B-spline cross-section curves.*

The composite surface given by (9.25) is very useful in constructing *edge blending surfaces* as will be seen in Chapter 10. Metrical properties of a polyconic surface are more easily evaluated than those of a Hermite blended surface (See §10.5). In general, it is better to blend control vertices of section curves than to blend section curves themselves.

9.4 B-SPLINE BLENDED SURFACES (SKINNING)

Now we consider the case where each of the section curves in Fig. 9.12 is a composite (cubic) B-spline curve composed of m curve segments as depicted in Fig. 9.15a. As shown in the figure, the jth section curve $\mathbf{p}_j(t)$ is composed of m "NUB curve segments" $\{\mathbf{p}_j^i(u) : i = 0, ..., m-1;\ u \in [0,1]\}$. It is defined by the following "control polygon" $\mathbf{C}_j$ consisting of $m+3$ control points:

$$\mathbf{C}_j = \{C_{ij} : i = 0, 1, ..., m+2\} \ for\ j = 0, 1, ..., n.$$

It is further assumed that each of the section curves is supported by the same *knot span vector* $\mathbf{D}_u$ (See §4.3.3):

$$\mathbf{D}_u = \{\Delta_i : i = -2, -1, ..., m, m+1\}. \tag{9.26}$$

The reader who is not comfortable with the treatment of non-uniform B-spline (NUB) is advised to review §3.4, §4.2.3, and §7.5.

The fist step in *skinning* is to fit a NUB curve through each "row" i of B-spline control vertices, $C_{i0}, C_{i1}, ..., C_{in}$. For this, another knot span vector in the longitudinal direction has to be determined. One way of determining *longitudinal knot spans* $\{\nabla_j\}$ is to use the average distances between control vertices on adjacent section curves. That is, the *supporting knot spans* are determined from the following:

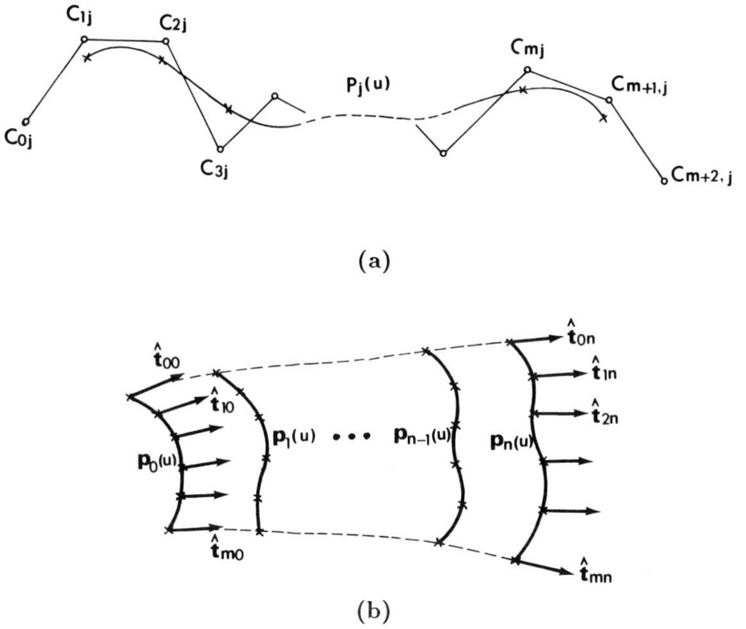

Figure 9.15 NUB Cross-sectional Curves for Skinning

$$\nabla_j = \frac{1}{m+3} \sum_{i=0}^{m+2} |C_{i,j+1} - C_{i,j}| \quad for\ j = 0, 1, ..., n-1 \tag{9.27}$$

The *extended knot spans* are simply set to zero. Thus, the **longitudinal knot span** *vector* $\mathbf{D}_v = \{\nabla_j : j = -2, -1, ..., n+1\}$ is given by

$$\mathbf{D}_v = \{0,\ 0,\ \nabla_0,\ \nabla_1, ..., \nabla_{n-1},\ 0,\ 0\}. \tag{9.28}$$

Having determined the knot span vectors, the next step is to define B-spline control vertices $\{V_{ij}\}$ of the skinned surface. In order to determine the control vertices, *end tangent vectors* $\{\hat{\mathbf{t}}_{i0}, \hat{\mathbf{t}}_{in} : i = 0, 1, ..., m\}$ should be known (in Fig. 9.15b). If they are not given they can be estimated by applying a suitable end condition (see §4.2.2). Note that $\mathbf{t}_{i0}$ and $\mathbf{t}_{in}$ are *end tangents* of a spline curve passing through $\mathbf{p}_0^i(0), ..., \mathbf{p}_n^i(0)$. Thus, if a free end condition is desired, for example, a linear equation system of the type (9.23) may be solved to determine the end tangents. Now the remaining skinning problem becomes the problem of fitting a composite NUB surface discussed in §7.5.

Thus, from the results of §7.5, the control vertices may be obtained by solving the following linear equation *(for* $i=0,1,\ldots,m+2$*)*:

$$\begin{bmatrix} -3 & 3 & 0 & 0 & & & \\ f_0 & (1-f_0-g_0) & g_0 & 0 & & & \\ & & & \cdot & & & \\ & & & & \cdot & & \\ & & & & f_n & (1-f_n-g_n) & g_m \\ & & & & 0 & -3 & 3 \end{bmatrix} \begin{bmatrix} V_{i,0} \\ V_{i,1} \\ \cdot \\ \cdot \\ V_{i,n+1} \\ V_{i,n+2} \end{bmatrix} = \begin{bmatrix} \mathbf{d}_i \\ C_{i,0} \\ \cdot \\ \cdot \\ V_{i,n} \\ \mathbf{e}_i \end{bmatrix} \tag{9.29}$$

$$\begin{aligned} \textit{where,}\quad & f_j = (\nabla_j)^2/(\nabla_{j-1}^2\nabla_{j-2}^3) \\ & g_j = (\nabla_{j-1})^2/(\nabla_{j-1}^2\nabla_{j-1}^3) \\ & \nabla_j^{k+1} = \nabla_j + \nabla_{j+1} + \ldots + \nabla_{j+k}, \\ & \nabla_j : \textit{ v-direction knot spans in } (9.28). \end{aligned}$$

As discussed in §7.5, the "boundary vectors" $\mathbf{d}_i$ in (9.29) are determined by solving the following linear equations:

$$\begin{aligned} & 3(\mathbf{d}_1 - \mathbf{d}_0) = \hat{\mathbf{x}}_{00} \\ & 3(\mathbf{d}_{m+2} - \mathbf{d}_{m+1}) = \hat{\mathbf{x}}_{m0} \\ & f_i\mathbf{d}_i + (1-f_i-g_i)\mathbf{d}_{i+1} + g_i\mathbf{d}_{i+2} = \hat{\mathbf{t}}_{i0} \quad \textit{for } i = 0,\ldots,m \end{aligned} \tag{9.30}$$

$$\begin{aligned} \textit{where,}\quad & f_i = (\Delta_i)^2/(\Delta_{i-1}^2\Delta_{i-2}^3) \\ & g_i = (\Delta_{i-1})^2/(\Delta_{i-1}^2\Delta_{i-1}^3) \\ & \Delta_i : \ u - \textit{direction knot spans.} \end{aligned}$$

Similarly, $\mathbf{e}_i$'s *(for* $i=0,\ldots,m$*)* in (9.29) are obtained from the following:

$$\begin{aligned} & 3(\mathbf{e}_1 - \mathbf{e}_0) = \hat{\mathbf{x}}_{0n} \\ & 3(\mathbf{e}_{m+2} - \mathbf{e}_{m+1}) = \hat{\mathbf{x}}_{mn} \\ & f_i\mathbf{e}_i + (1-f_i-g_i)\mathbf{e}_{i+1} + g_i\mathbf{e}_{i+2} = \hat{\mathbf{t}}_{in} \end{aligned} \tag{9.31}$$

In (9.30) and (9.31), the *corner twist vectors* $\hat{\mathbf{x}}_{00}, \hat{\mathbf{x}}_{m0}, \hat{\mathbf{x}}_{0n}, \hat{\mathbf{x}}_{mn}$ are usually set to zero. Once the control vertices are determined, the *skinned surface* is represented as a composite (bicubic) NUB surface defined by $(m+2)\times(n+2)$ control vertices $\{V_{ij}\}$ and two orthogonal *knot span vectors* given by (9.26) and (9.28). Some details of skinning techniques for NUB surface interpolation may be found in Woodward (1988).

The skinning method discussed above is applicable to Bezier curves as well. In theory, it is not necessary to treat the Bezier case separately because Bezier surface is a special case of NUB surface. But, the Bezier surface interpolation method has some advantages over the NUB method as will be seen shortly. One may think of two cases where the NUB-based skinning method gives some undesirable results. The first case is when the distances between two adjoining section curves, $\mathbf{p}_{j+1}(t)$ and $\mathbf{p}_j(t)$, vary considerably. In other words, the maximum distance between the two curves is much larger than the minimum distance between them. That is,

$$\max_t |\mathbf{p}_{j+1}(t) - \mathbf{p}_j(t)| >> \min_t |\mathbf{p}_{j+1}(t) - \mathbf{p}_j(t)|.$$

An example of such a situation was given in Fig. 7.17 (See §7.6).

The second case is when the sweep surface is defined by two longitudinal *boundary curves*, $\mathbf{b}_0(v)$ and $\mathbf{b}_1(v)$, as well as a series of section curves $\mathbf{p}_j(t)$. If all the input curves are composite Bezier curves, the NUB-based skinned surface may not be visually smooth. This situation arises more frequently than one might expect because most of the curves in engineering drawing are composite curves consisting of lines and circular arcs.

Recall from §4.4.4 that a circular arc can be modeled exactly as a quadratic rational Bezier curve or as an approximate cubic (non-rational) Bezier curve. In the latter case, the skinning problem becomes the "composite Bezier surface fitting" problem treated in §7.6. In a sense, Bezier methods are more general than B-spline methods because B-spline curves can always be converted to Bezier curves by inserting (multiple) knots (Boehm, 1980) as discussed in Chapter 6.

9.5 DISCUSSIONS

The surface construction schemes based on the blending of cross-sectional data have a definite advantage over the point data interpolation methods (Chapts 7 & 8) as far as shape control is concerned. They are also superior to the curve-net interpolation schemes (to be discussed in Chap. 10) in terms of "description power". In other words, it is easier to *describe* a shape (on blue prints or on a graphics screen) by using cross-sectional views than by using 3D curve-net. Further research is needed to combine the two so that the advantages of each can be fully realized.

CHAPTER 10

SURFACE CONSTRUCTION FROM 3D CURVE-NETS

10.1 INTRODUCTION

This chapter considers the problem of constructing a smooth surface interpolating to 3D curve-net data. Even though it is our goal to develop a unified interpolation scheme which can be used for any type of curve-net data, it is desirable to exploit inherent characteristics of input data. Curve-net data may be characterized in terms of its *geometry* and *topology*. **Mesh curves** in the curve-net may be *smooth (composite) curves* as in Fig. 10.1a, or they may be *segmented curves* as in Fig. 10.1c. The curve-net may have a *rectangular topology* as in Fig. 10.1a or have a *irregular topology* as in Fig. 10.1b. Thus, one may distinguish three types of curve-net as depicted in Fig. 10.1. They are:

- *smooth mesh curves with rectangular topology (Fig.* 10.1*a),*
- *smooth mesh curves with irregular topology (Fig.* 10.1*b), and*
- *segmented mesh curves (Fig.* 10.1*c).*

Mesh curves in a curve-net may intersect with each other at **mesh points** (more exactly *mesh intersection points*). A portion of a mesh curve between two mesh points is called a **mesh segment**. A mesh curve may be a general parametric curve or a *cubic polynomial curve*. In fact, the *composite Bezier scheme* of §7.6 and the *triangular Bezier scheme* of §8.5 are *curve-net interpolation* schemes for *cubic mesh curves*.

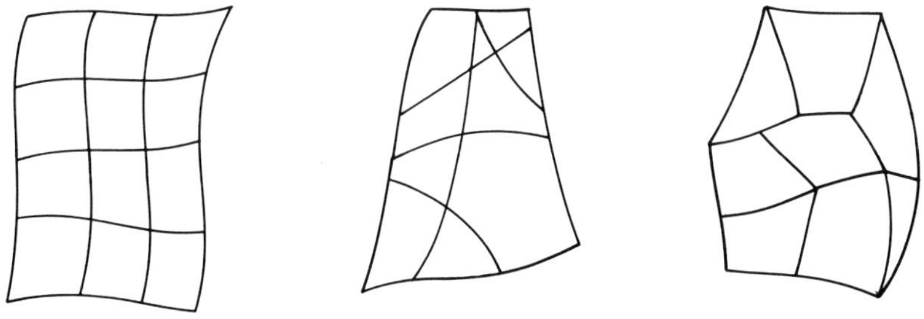

(a) (b) (c)

Figure 10.1 Three Types of Curve-net Data

Curve-net interpolation schemes for "smooth mesh curves having rectangular topology" will be presented in the next section, and interpolation schemes for "smooth mesh curves having irregular topology" in §10.3. Interpolation schemes for "segmented mesh curves" are discussed in §10.4. A *rectangular patch* in the curve-net is modeled as a *Coons patch* (see §5.3.3), and we use the *n-sided Gregory patch* of §5.3.4 in representing an *nonrectangular* patch. Cubic mesh curve interpolation schemes are introduced in §10.5.

10.2 INTERPOLATION OF SMOOTH RECTANGULAR CURVE-NET

Let us consider a curve-net of *smooth mesh curves having rectangular topology* (See Fig. 10.1a). The curve-net is composed of

a) $\{\mathbf{a}_{i*}(v) : i = 0, ..., m;\ v \in [0,1]\}$*: a set of "horizontal" mesh curves, and*
b) $\{\mathbf{b}_{*j}(u) : j = 0, ..., n;\ u \in [0,1]\}$*: a set of "vertical" mesh curves.*

The vertical mesh curves $\mathbf{b}_{*j}(u)$ and horizontal mesh curves $\mathbf{a}_{i*}(v)$ intersect with each other at *mesh points* $\{\mathbf{p}_{ij}\}$ such that

$$\mathbf{p}_{ij} = \mathbf{a}_{ij}(0) = \mathbf{a}_{i,j-1}(1) = \mathbf{b}_{ij}(0) = \mathbf{b}_{i-1,j}(1)$$

From these curve-net data, we want to construct a surface patch $\mathbf{r}^{ij}(u,v)$ *filling in* the rectangular region bounded by four mesh curves

$$\mathbf{a}_{ij}(v),\ \mathbf{b}_{ij}(u),\ \mathbf{a}_{i+1,j}(v),\ \textit{and}\ \mathbf{b}_{i,j+1}(u)$$

as depicted in Fig. 10.2.

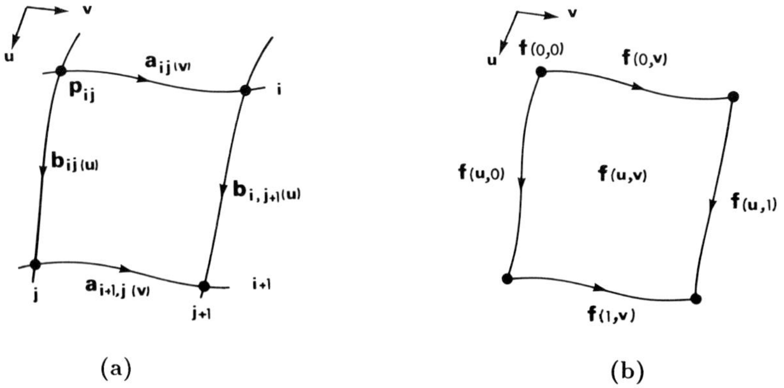

Figure 10.2 A Patch in a Rectangular Curve-net

10.2.1 Hermite-Blended Sweep-Surface Patch

An intuitive scheme to interpolate to the curve data of Fig. 10.2a is to use the *sweeping idea* of §5.4 and §9.2. A major requirement here is that the resulting composite surface $\{\mathbf{r}^{ij}(u,v)\}$ maintain visual continuity across patch boundaries. For a notational simplicity, let us define (See Fig. 10.2b)

$$\mathbf{f}(0,v) \equiv \mathbf{a}_{ij}(v) \tag{10.1 – a}$$

$$\mathbf{f}(1,v) \equiv \mathbf{a}_{i+1,j}(v) \tag{10.1 – b}$$

$$\mathbf{f}(u,0) \equiv \mathbf{b}_{ij}(u) \tag{10.1 – c}$$

$$\mathbf{f}(u,1) \equiv \mathbf{b}_{i,j+1}(u) \tag{10.1 – d}$$

Thus, $\mathbf{f}(u,v)$ may be treated as an imaginary surface patch whose boundary matches to the input curve-net data.

Let $\mathbf{r}_i(u,v)$ denote *sweep surfaces* obtained by sweeping $\mathbf{f}(u,0)$ along $\mathbf{f}(i,v)$ for $i = 0,1$. Then, the *translational sweep surfaces* are expressed as (See Eqn. (5.33))

$$\mathbf{r}_i(u,v) = \mathbf{f}(u,0) + \mathbf{f}(i,v) - \mathbf{f}(i,0) \quad for \quad i = 0,1 \tag{10.2}$$

By employing suitable blending functions, the two translational sweep surfaces in (10.2) may be blended to obtain a *blended sweep surface* $\mathbf{p}_0(u,v)$:

$$\begin{aligned}\mathbf{p}_0(u,v) &= \alpha(u)\ \mathbf{r}_0(u,v) + \beta(u)\ \mathbf{r}_1(u,v)\\ &= \mathbf{f}(u,0) + \alpha(u)[\mathbf{f}(0,v) - \mathbf{f}(0,0)] + \beta(u)[\mathbf{f}(1,v) - \mathbf{f}(1,0)]\end{aligned} \tag{10.3}$$

where, $\alpha(u), \beta(u)$: *blending functions,*
$\mathbf{r}_i(u,v)$: *translational sweep surfaces in* (10.2).

Since there is another blended sweep surface $\mathbf{p}_1(u,v)$ of the type (10.3) obtained by sweeping $\mathbf{f}(u,1)$, $\mathbf{p}_0(u,v)$ and $\mathbf{p}_1(u,v)$ are blended again to form a "doubly blended" sweep surface patch.

$$\begin{aligned}\mathbf{r}(u,v) &= \alpha(v)\ \mathbf{p}_0(u,v) + \beta(v)\ \mathbf{p}_1(u,v)\\ &= \alpha(v)\{\mathbf{f}(u,0) + \alpha(u)[\mathbf{f}(0,v) - \mathbf{f}(0,0)] + \beta(u)[\mathbf{f}(1,v) - \mathbf{f}(1,0)]\}\\ &\quad + \beta(v)\{\mathbf{f}(u,1) + \alpha(u)[\mathbf{f}(0,v) - \mathbf{f}(0,1)] + \beta(u)[\mathbf{f}(1,v) - \mathbf{f}(1,1)]\}\\ &= [\alpha(u)\ \beta(u)]\begin{bmatrix}\mathbf{f}(0,v)\\ \mathbf{f}(1,v)\end{bmatrix} + [\mathbf{f}(u,0)\ \mathbf{f}(u,1)]\begin{bmatrix}\alpha(v)\\ \beta(v)\end{bmatrix}\\ &\quad - [\alpha(u)\ \beta(u)]\begin{bmatrix}\mathbf{f}(0,0) & \mathbf{f}(0,1)\\ \mathbf{f}(1,0) & \mathbf{f}(1,1)\end{bmatrix}\begin{bmatrix}\alpha(v)\\ \beta(v)\end{bmatrix} \quad for\ u,v \in [0,1]\end{aligned} \tag{10.4}$$

where $\mathbf{f}(i,\ j)$ are as given in (10.1).

If the blending functions $\alpha(t), \beta(t)$ are linear, namely, if

$$\alpha(t) = (1-t)\ ;\ \beta(t) = t,$$

then the blended sweep surface $\mathbf{r}(u,v)$ in (10.4) becomes the *bilinear Coons patch* introduced in §5.3.3 (See Eqn. (5.24)). If the blending functions are in *Hermite forms* such that

$$\alpha(t) \equiv H_0^3(t) = 1 - 3t^2 + 2t^3 \tag{10.5 - a}$$

$$\beta(t) \equiv H_3^3(t) = 3t^2 - 2t^3, \tag{10.5 - b}$$

then the blended sweep surface (10.4) becomes a **Hermite-blended sweep surface patch.** An important property of the Hermite blending functions (10.5) is that all the *end derivatives* are zero, namely,

$$\dot{\alpha}(0) = \dot{\alpha}(1) = \dot{\beta}(0) = \dot{\beta}(1) = 0. \tag{10.6}$$

By using the results in (10.6), the *v-direction tangent* along $v = 0$ boundary, for example, is evaluated as

$$\begin{aligned} \mathbf{r}_v(u,0) &\equiv \partial\mathbf{r}(u,v)/\partial v|_{v=0} \\ &= \alpha(u)\mathbf{f}_v(0,0) + [1-\alpha(u)]\mathbf{f}_v(1,0). \end{aligned} \tag{10.7}$$

In terms of the original notations of Fig. 10.2a, the *cross boundary tangent* (10.7) can be rewritten as

$$\mathbf{r}_v^{ij}(u,0) = \alpha(u)\dot{\mathbf{a}}_{ij}(0) + \beta(u)\dot{\mathbf{a}}_{i+1,j}(0) \tag{10.8 - a}$$

$$where, \quad \dot{\mathbf{a}}_{ij}(0) \equiv \partial\mathbf{a}_{ij}(v)/\partial v|_{v=0}.$$

Similarly, the v-direction derivative along the same boundary curve on the"left" patch $\mathbf{r}^{i,j-1}(u,v)$ may be expressed as

$$\mathbf{r}_v^{i,j-1}(u,1) = \alpha(u)\dot{\mathbf{a}}_{i,j-1}(1) + \beta(u)\dot{\mathbf{a}}_{i+1,j-1}(1). \tag{10.8 - b}$$

If the two cross boundary tangents in (10.8-a) and (10.8-b) are collinear, the two surface patches would be *parametrically smooth* along their common boundary curve $\mathbf{b}_{ij}(u)$. That is, for some scalar λ_j, we should have

$$\mathbf{r}_v^{ij}(u,0) = \lambda_j \mathbf{r}_v^{i,j-1}(u,1)$$

which in turn is satisfied if the following holds

$$\dot{\mathbf{a}}_{ij}(0) = \lambda_j \dot{\mathbf{a}}_{i,j-1}(1) \quad for \quad all \quad i,j \tag{10.9 - a}$$

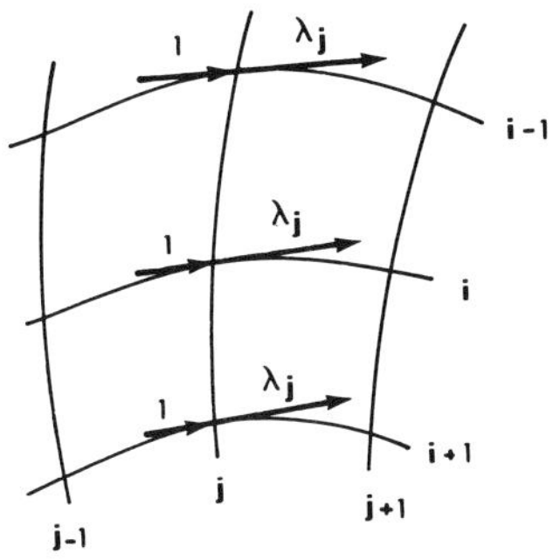

Figure 10.3 Collinear G^1-Condition for Blended Sweep Surface

The G^1*-condition* (10.9-a) assures a first order continuity in *v-direction.* A similar G^1-condition for a *u-direction* continuity is given by (*for some* μ_i)

$$\dot{\mathbf{b}}_{ij}(0) = \mu_i \dot{\mathbf{b}}_{i-1,j}(1) \quad for \quad all \quad i,j \qquad (10.9-b)$$

The *collinear* G^1*-condition* (10.9-a) is depicted in Fig. 10.3.

If the G^1-conditions in (10.9) are not met, *the Hermite blended sweep surfaces* (10.4) may not be smooth at the input mesh curves. In other words, the sweep surface scheme does not produce a smooth composite surface unless the *collinear* G^1*-conditions* in (10.9) are met, which is a severe restriction in most practical applications.

10.2.2 Hermite-Blended Coons Patch

If the rectangular region in Fig. 10.2a is filled with a *bicubic Coons patch* instead of the Hermite blended sweep surface (10.4), the restrictive G^1-condition (10.9) is no longer necessary. In order to use the *bicubic Coons patch* model, however, cross boundary tangents have to be specified in advance. Let $\mathbf{s}_{ij}$ and $\mathbf{t}_{ij}$, respectively, denote *u-direction* and *v-direction tangents* at mesh points $\mathbf{p}_{ij}$. One way of estimating the **mesh point tangents** is to take an average of *end tangents* of the mesh curves at $\mathbf{p}_{ij}$. Namely,

$$\mathbf{s}_{ij} = \{\dot{\mathbf{b}}_{ij}(0) + \dot{\mathbf{b}}_{i-1,j}(1)\}/2 \qquad (10.10-a)$$

$$\mathbf{t}_{ij} = \{\dot{\mathbf{a}}_{ij}(0) + \dot{\mathbf{a}}_{i,j-1}(1)\}/2 \qquad (10.10-b)$$

where $\dot{\mathbf{a}}_{ij}(0) \equiv \partial \mathbf{a}_{ij}(v)/\partial v|_{v=0}$ and $\dot{\mathbf{b}}_{ij}(0) \equiv \partial \mathbf{b}_{ij}(u)/\partial u|_{u=0}$.

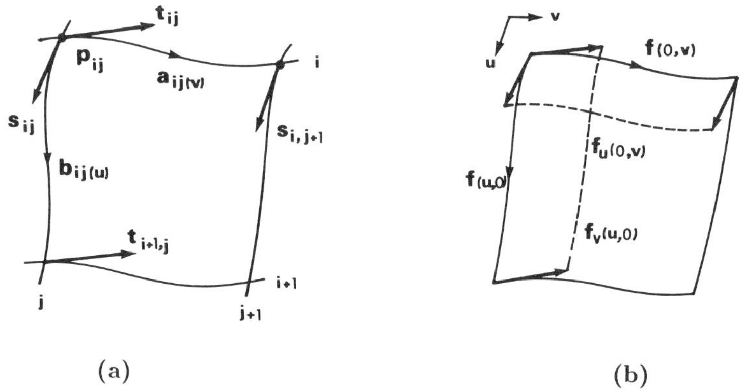

Figure 10.4 Estimation of Cross-Boundary Tangents Functions

The *mesh point tangents* $\mathbf{s}_{ij}, \mathbf{t}_{ij}$ are depicted in Fig. 10.4a.

Having estimated *mesh point tangents* $\mathbf{s}_{ij}$ and $\mathbf{t}_{ij}$, the next step is to construct *cross-boundary tangents* $\mathbf{s}_{ij}(v)$ and $\mathbf{t}_{ij}(u)$, along the mesh segments $\mathbf{a}_{ij}(v)$ and $\mathbf{b}_{ij}(u)$, respectively. One way of determining the **cross boundary tangents** is to take a *Hermite blending* of two *mesh point tangents*. That is,

$$\begin{aligned}\mathbf{s}_{ij}(v) &= \alpha(v)\mathbf{s}_{ij} + \beta(v)\mathbf{s}_{i,j+1} \\ \mathbf{t}_{ij}(u) &= \alpha(u)\mathbf{t}_{ij} + \beta(u)\mathbf{t}_{i+1,j}\end{aligned} \tag{10.11}$$

where $\alpha(t)$, $\beta(t)$ are Hermite blending functions given by (10.5).

As depicted in Fig. 10.4b, cross boundary tangents along the four boundary curves of the patch are denoted as

$$\begin{aligned}\mathbf{f}_u(0,v) &\equiv \mathbf{s}_{ij}(v) \\ \mathbf{f}_u(1,v) &\equiv \mathbf{s}_{i+1,j}(v) \\ \mathbf{f}_v(u,0) &\equiv \mathbf{t}_{ij}(u) \\ \mathbf{f}_v(u,1) &\equiv \mathbf{t}_{i,j+1}(u).\end{aligned} \tag{10.12}$$

From the property of Hermite blending functions given by (10.6), one may easily see that *twist vectors* at mesh intersection points are all zero:

$$\begin{aligned}\mathbf{f}_{uv}(0,0) &= \partial \mathbf{f}_u(0,v)/\partial v|_{v=0} \\ &= \dot{\alpha}(0)\mathbf{s}_{ij} + \dot{\beta}(0)\mathbf{s}_{i,j+1} = \mathbf{0}, \quad etc.\end{aligned} \tag{10.13}$$

Finally, from the result of §5.3.3 (See Eqn. (5.29)), the **Hermite blended Coons patch** for the rectangular region in Fig. 10.4b is expressed as follows:

$$\mathbf{r}(u,v) = \mathbf{r}_1(u,v) + \mathbf{r}_2(u,v) - \mathbf{r}_3(u,v) \tag{10.14}$$

with

$$\mathbf{r}_1(u,v) = [\alpha(u)\ \beta(u)\ \gamma(u)\ \delta(u)] \begin{bmatrix} \mathbf{f}(0,v) \\ \mathbf{f}(1,v) \\ \mathbf{f}_u(0,v) \\ \mathbf{f}_u(1,v) \end{bmatrix} \tag{10.15 - a}$$

$$\mathbf{r}_2(u,v) = [\mathbf{f}(u,0)\ \mathbf{f}(u,1)\ \mathbf{f}_v(u,0)\ \mathbf{f}_v(u,1)] \begin{bmatrix} \alpha(v) \\ \beta(v) \\ \gamma(v) \\ \delta(v) \end{bmatrix} \tag{10.15 - b}$$

$$\mathbf{r}_3(u,v) = [\alpha(u)\beta(u)\gamma(u)\delta(u)] \begin{bmatrix} \mathbf{f}(0,0) & \mathbf{f}(0,1) & \mathbf{f}_v(0,0) & \mathbf{f}_v(0,1) \\ \mathbf{f}(1,0) & \mathbf{f}(1,1) & \mathbf{f}_v(1,0) & \mathbf{f}_v(1,1) \\ \mathbf{f}_u(0,0) & \mathbf{f}_u(0,1) & \mathbf{0} & \mathbf{0} \\ \mathbf{f}_u(1,0) & \mathbf{f}_u(1,1) & \mathbf{0} & \mathbf{0} \end{bmatrix} \begin{bmatrix} \alpha(v) \\ \beta(v) \\ \gamma(v) \\ \delta(v) \end{bmatrix} \tag{10.16}$$

$$\begin{aligned} where,\quad & \alpha(u) = H_0^3(u) = (1 - 3u^2 + 2u^3), \\ & \beta(u) = H_3^3(u) = (3u^2 - 2u^3), \\ & \gamma(u) = H_1^3(u) = (u - 2u^2 + u^3), \\ & \delta(u) = H_2^3(u) = (-u^2 + u^3), \\ & \mathbf{f}(i,v), \mathbf{f}(u,j) \quad for\ i,j = 0,1:\ boundary\ curves (10.1), \\ & \mathbf{f}_u(i,v), \mathbf{f}_v(u,j) \quad for\ i,j = 0,1:\ cross\text{-}boundary\ tangents\ (10.12). \end{aligned}$$

The curve-net interpolation scheme based on Hermite blended Coons patches gives a visually smooth composite surface interpolating over the 3D curve-net:

$$\{\mathbf{a}_{i*}(v) \times \mathbf{b}_{*j}(u): \quad i = 0, ..., m; \quad j = 0, ..., n\}.$$

The resulting surface has "pseudo flat spots" at mesh intersection points because all the *twist vectors* are set to zero.

$\mathbf{C}$: *Ferguson coefficient matrix as in* (5.5),

$$\mathbf{Q} = \begin{bmatrix} \mathbf{p}_{00} & \mathbf{p}_{01} & \nu\mathbf{t}_{00} & \nu\mathbf{t}_{01} \\ \mathbf{p}_{10} & \mathbf{p}_{11} & \nu\mathbf{t}_{10} & \nu\mathbf{t}_{11} \\ \mu\mathbf{s}_{00} & \mu\mathbf{s}_{01} & 0 & 0 \\ \mu\mathbf{s}_{10} & \mu\mathbf{s}_{11} & 0 & 0 \end{bmatrix},$$

$\mu = \{|\mathbf{p}_{10} - \mathbf{p}_{00}| + |\mathbf{p}_{11} - \mathbf{p}_{01}|\}/2$: *u-direction expansion factor,*

$\nu = \{|\mathbf{p}_{01} - \mathbf{p}_{00}| + |\mathbf{p}_{11} - \mathbf{p}_{10}|\}/2$: *v-direction expansion factor,*

$\mathbf{s}_i(v), \mathbf{t}_j(u)$: *cross-boundary tangents* (10.18).

Note in (10.19) that *expansion factors* are introduced to make the magnitudes of *cross boundary tangents* close to respective "chord lengths".

10.3.2 Triangular Interpolant (Convex Combination)

The triangular region *CDE* of the curve-net in Fig. 10.5 is shown in Fig. 10.7a. The boundary curves of the triangular patch are ordered anti-clockwise, and are denoted as

$\mathbf{e}_1(s_1)$ *for* $s_1 \in [0, 1]$: *mesh segment EC,*
$\mathbf{e}_2(s_2)$ *for* $s_2 \in [0, 1]$: *mesh segment CD, and*
$\mathbf{e}_3(s_3)$ *for* $s_3 \in [0, 1]$: *mesh segment DE.*

And the three corner points are denoted as

$\mathbf{p}_1 = \mathbf{e}_2(1) = \mathbf{e}_3(0)$: *mesh point D,*
$\mathbf{p}_2 = \mathbf{e}_3(1) = \mathbf{e}_1(0)$: *mesh point E, and*
$\mathbf{p}_3 = \mathbf{e}_1(1) = \mathbf{e}_2(0)$: *mesh point C.*

The domain of the triangular patch in Fig. 10.7a is defined in Fig. 10.7b as an *equilateral triangle of height unity* with its boundary data:

$\{V_i : i = 1, 2, 3\}$: *domain corner points (vertices) for* $\mathbf{p}_i$, *and*
$\{E_i : i = 1, 2, 3\}$: *domain edges (sides) for* $\mathbf{e}_i(s_i)$.

The perpendicular distances λ_i from an inside point V to the side E_i of the domain triangle form the barycentric coordinates of the domain point V.

We want to construct a triangular surface patch interpolating to the boundary curve data $\mathbf{e}_i(s_i)$ by using the *convex combination* scheme introduced earlier in §5.3.4. The first step is to establish a relation between the *curve parameters* s_i and the *barycentric coordinates* λ_i. A point $E_1(s_1)$ on the edge V_2V_3 of the domain triangle may be expressed as a linear blending of the two vertices. That is,

$$E_1(s_1) = (1 - s_1)\ V_2 + s_1\ V_3.$$

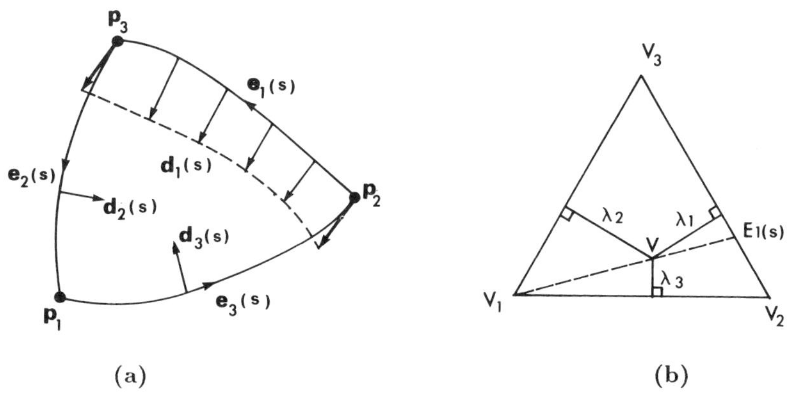

Figure 10.7 Construction of a Triangular Interpolant

For a given domain point V whose barycentric coordinates are $(\lambda_1,\ \lambda_2,\ \lambda_3)$, the intersection of the radial line V_1V with side 1 is given by (See Fig. 10.7b)

$$E_1(V) = [\lambda_2/(\lambda_2+\lambda_3)]\ V_2 + [\lambda_3/(\lambda_2+\lambda_3)]\ V_3.$$

From the above two expressions, the curve parameter s_1 is obtained as

$$s_1 = \lambda_3/(\lambda_2+\lambda_3). \qquad (10.20-a)$$

Other curve parameters are also expressed in terms of λ_i as (See Eqn. (5.30))

$$s_2 = \lambda_1/(\lambda_3+\lambda_1). \qquad (10.20-b)$$

$$s_3 = \lambda_2/(\lambda_1+\lambda_2). \qquad (10.20-c)$$

The next step is to estimate *cross-boundary tangents* $\mathbf{d}_i(s_i)$ along the boundary curves $\mathbf{e}_i(s_i)$. By following the same steps in the rectangular case, the (unit) *cross boundary tangents* $\mathbf{d}_i(s_i)$ may be obtained as

$$\mathbf{d}_1(s_1) = \alpha(s_1)\{-\dot{\mathbf{e}}_3(1)/|\dot{\mathbf{e}}_3(1)|\} + \beta(s_1)\{\dot{\mathbf{e}}_2(0)/|\dot{\mathbf{e}}_2(0)|\} \qquad (10.21-a)$$

$$\mathbf{d}_2(s_2) = \alpha(s_2)\{-\dot{\mathbf{e}}_1(1)/|\dot{\mathbf{e}}_1(1)|\} + \beta(s_2)\{\dot{\mathbf{e}}_3(0)/|\dot{\mathbf{e}}_3(0)|\} \qquad (10.21-b)$$

$$\mathbf{d}_3(s_3) = \alpha(s_3)\{-\dot{\mathbf{e}}_2(1)/|\dot{\mathbf{e}}_2(1)|\} + \beta(s_3)\{\dot{\mathbf{e}}_1(0)/|\dot{\mathbf{e}}_1(0)|\} \qquad (10.21-c)$$

where $\alpha(t), \beta(t)$ are Hermite blending functions in (10.5). The cross-boundary tangents are pointing inwards as depicted in Fig. 10.7a.

From the results of §5.3.4, a **triangular convex combination patch** $\mathbf{r}(V)$ interpolating to the three boundary curves $\mathbf{e}_1(s_1), \mathbf{e}_2(s_2), \mathbf{e}_3(s_3)$ is expressed as follows (See Eqn. (5.32)):

$$\mathbf{r}(V) = \sum_{i=1}^{3} \gamma_i(V)\{\mathbf{e}_i(s_i) + \lambda_i \nu_i \mathbf{d}_i(s_i)\} \tag{10.22}$$

$$\begin{aligned} \textit{where,}\quad & V = (\lambda_1, \lambda_2, \lambda_3): \quad \textit{barycentric coordinates}\,(\Sigma\lambda_i = 1), \\ & s_i = \lambda_{i-1}/(1-\lambda_i) \quad \textit{as defined in}(10.20), \\ & \gamma_i(V) = (\lambda_{i-1}\lambda_{i+1})^2/\{(\lambda_1\lambda_2)^2 + (\lambda_2\lambda_3)^2 + (\lambda_3\lambda_1)^2\}, \\ & \nu_i = \{|\dot{\mathbf{e}}_{i-1}(1)| + |\dot{\mathbf{e}}_{i+1}(0)|\}/2: \quad \textit{expansion factor}. \end{aligned}$$

The role of the *expansion factors* ν_i is to make the magnitude of the cross boundary tangent, $\nu_i \mathbf{d}_i$, roughly equal to the distance from the edge $\mathbf{e}_i$ to the vertex $\mathbf{p}_i$. The index i is to be interpreted as "$(i-1)$ *mod* $3+1$".

10.3.3 Pentagonal Interpolant (Convex Combination)

Finally, we condiser the problem of constructing a pentagonal surface patch for the pentagonal region *BCEFH* in Fig. 10.5. The pentagonal region is bounded by five edges

$$\mathbf{e}_i(s_i): \; s_i \in [0,1] \quad \textit{for} \quad i = 1, ..., 5$$

as depicted in Fig. 10.8a. The edges have anti-clockwise directions and are ordered accordingly. An edge $\mathbf{e}_i$ starts at $\mathbf{p}_i$ and ends at $\mathbf{p}_{i+1}$. That is,

$$\mathbf{p}_i = \mathbf{e}_i(0) = \mathbf{e}_{i-1}(1). \tag{10.23}$$

The index i is interpreted as "$(i-1)$ *mod* $5+1$". The domain of the surface patch is a *regular pentagon* of height unity having vertices V_i *for* $i = 1, ..., 5$. The interpolation scheme here is almost identical to the one for a triangular patch discussed in the previous subsection.

It is first necessary to define a coordinate system on the domain pentagon. Let λ_i denote the perpendicular distance of an interior point V to *side* i (joining the two vertices V_i and V_{i+1}) as depicted in Fig. 10.8b. There are five of them, but they are linearly dependent such that (Charrot and Gregory, 1984)

$$\lambda_{i-1} = 2\kappa(1 + \lambda_{i-2}) - \lambda_{i+2} \tag{10.24 - a}$$

$$\lambda_i = 1 - 2\kappa(\lambda_{i-2} + \lambda_{i+2}) \tag{10.24 - b}$$

$$\lambda_{i+1} = 2\kappa(1 + \lambda_{i+2}) - \lambda_{i-2} \tag{10.24 - c}$$

$$\textit{where} \quad \kappa = (\sqrt{5} - 1)/4.$$

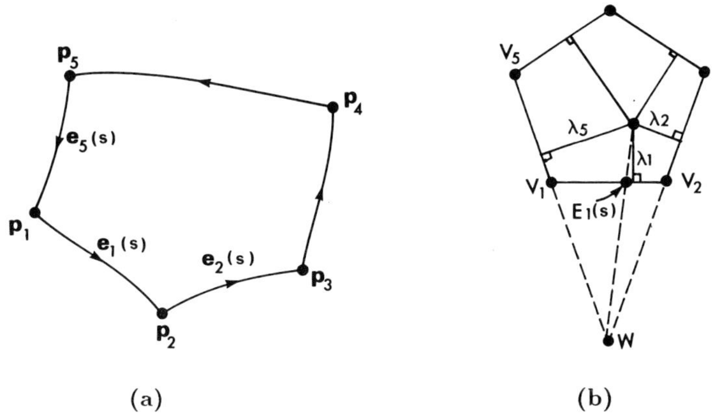

Figure 10.8 Construction of a Pentagonal Interpolant

Further, one may verify that the following holds:

$$\Sigma\lambda_i = \sqrt{5}.$$

Recall that the index i is interpreted as "$(i-1)\ mod\ 5+1$".

The next step is to establish a relation between the *curve parameters* s_i and the *coordinates* λ_i of the pentagonal domain. An edge $E_i(s_i)$ of the domain pentagon may be defined as a linear blending of its end vertices:

$$E_i(s_i) = (1-s_i)\ V_i + s_i\ V_{i+1}. \tag{10.25}$$

Let W denote the intersection point of the two edges E_{i-1} and E_{i+1} (See Fig. 10.8b). Then, the intersection of the edge E_1 with the line joining W and an interior point $V=(\lambda_1,\lambda_2,\lambda_3,\lambda_4,\lambda_5)$ is given by (see the figure)

$$E_i(V) = [\lambda_{i+1}/(\lambda_{i-1}+\lambda_{i+1})]\ V_i + [\lambda_{i-1}/(\lambda_{i-1}+\lambda_{i+1})]\ V_{i+1}. \tag{10.26}$$

On comparing (10.25) and (10.26), we get

$$s_i = \lambda_{i-1}/(\lambda_{i-1}+\lambda_{i+1}) \quad for\ i=1,...,5 \tag{10.27}$$

The *cross-boundary tangents* $\mathbf{d}_i(s_i)$ along the boundary curve $\mathbf{e}_i(s_i)$ are then defined as a Hermite blending of *normalized* tangents of adjacent curves. Namely,

$$\mathbf{d}_i(s_i) = \alpha(s_i)\frac{-\dot{\mathbf{e}}_{i-1}(1)}{|\dot{\mathbf{e}}_{i-1}(1)|} + \beta(s_i)\frac{\dot{\mathbf{e}}_{i+1}(0)}{|\dot{\mathbf{e}}_{i+1}(0)|} \tag{10.28}$$

where $\alpha(t), \beta(t)$ are Hermite blending functions in (10.5). In fact, the above expression is the same as the one given in (10.21) for the triangular case.

Finally, the **pentagonal convex combination patch** $\mathbf{r}(V)$ interpolating to the boundary curves $\{\mathbf{e}_i(s_i) : i = 1, ..., 5\}$ is expressed as (Gregory, 1986):

$$\mathbf{r}(V) = \sum_{i=1}^{5} \gamma_i(V)\{\mathbf{e}_i(s_i) + \lambda_i \nu_i \mathbf{d}_i(s_i)\} \tag{10.29}$$

where, $V = (\lambda_1, \lambda_2, \lambda_3, \lambda_4, \lambda_5)$ *with dependency given by* (10.24),

$s_i = \lambda_{i-1}/(\lambda_{i-1} + \lambda_{i+1})$ *for* $i = 1, ..., 5$ *as in* (10.27),

$$\gamma_i(V) = \Big(\prod_{j\neq i}^{5} \lambda_j\Big)^2 \Big/ \Big\{\sum_{k=1}^{5}\big(\prod_{j\neq k}^{5} \lambda_j\big)^2\Big\},$$

$\nu_i = \{|\dot{\mathbf{e}}_{i-1}(1)| + |\dot{\mathbf{e}}_{i+1}(0)|\}/2 :$ *expansion factor,*

$\mathbf{d}_i(s_i) :$ *cross-boundary tangent function* (10.28).

"i" *is interpreted as* "$(i-1)$ *mod* $5+1$".

In this section, we have introduced methods of constructing rectangular, triangular, and pentagonal patches in a smooth but irregular curve-net. The basic approach here is

a) to determine unit tangent vectors (called "mesh point tangent" or "corner tangent") at each mesh intersection point,
b) to estimate cross-boundary tangent functions by blending a pair of corner tangents, and
c) to construct a surface patch by making a Boolean sum (for rectangles) or a convex combination (for triangles and pentagons) of the boundary data.

A key point in the above construction schemes is that the cross-boundary tangent of each *mesh segment* is determined by blending a pair of mesh point tangent vectors, which is based on the assumption that all the mesh curves are smooth at mesh intersection points.

10.4 INTERPOLATION SCHEMES FOR SEGMENTED CURVE-NET

This section presents a method for constructing smooth composite surface interpolating to a "segmented" curve-net data of the type depicted earlier in Fig. 10.1c. The basic strategy is the same as that for the "smooth" mesh curve case. The main difference is in the estimation of the common cross boundary tangent $\mathbf{d}_i(s)$. For a segmented curve-net interpolation, cross-boundary tangents $\mathbf{d}_i(s)$ are estimated from "mesh point normals", while they are determined by blending "mesh point tangents" in a smooth curve-net interpolation. The curve-net interpolation scheme is

1) to determine "mesh point tangents" at both ends of a mesh segment,

2) to compute "mesh point normals" from the mesh point tangents,

3) to estimate a "cross boundary tangent" using the mesh point normals, and

4) to construct a surface patch from the boundary data.

Shown in Fig. 10.9 is a pentagonal patch in a segmented curve-net. The boundary curves (ie, mesh segments) of the patch are denoted as

$$..., \mathbf{e}_{i-1}(s),\ \mathbf{e}_i(s),\ \mathbf{e}_{i+1}(s), ... \quad for\ s \in [0,1]. \tag{10.30}$$

And, **mesh point tangents** at both ends of $\mathbf{e}_i(s)$ are defined as follows:

$$\begin{aligned} \mathbf{s}_0 &= d\mathbf{e}_i(s)/ds|_{s=0} \equiv \dot{\mathbf{e}}_i(0) \\ \mathbf{t}_0 &= -\dot{\mathbf{e}}_{i-1}(1) \\ \mathbf{s}_1 &= \dot{\mathbf{e}}_i(1) \\ \mathbf{t}_1 &= \dot{\mathbf{e}}_{i+1}(0) \end{aligned} \tag{10.31}$$

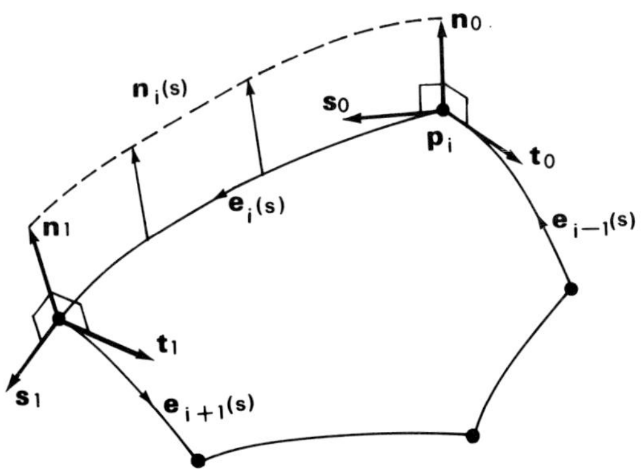

Figure 10.9 Boundary Normal Vector Estimation

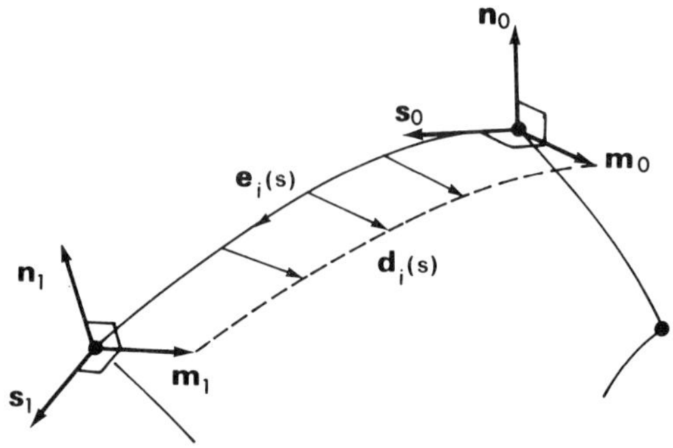

Figure 10.10 Determination of Cross Boundary Tangent

The **mesh point normals** at the ends of the boundary curve $\mathbf{e}_i(s)$ are computed as

$$\mathbf{n}_j = (\mathbf{s}_j \times \mathbf{t}_j)/|\mathbf{s}_j \times \mathbf{t}_j| \quad for \quad j = 0, 1 \tag{10.32}$$

As depicted in Fig. 10.10, a *"virtual" mesh point tangent* may be constructed at each end of the boundary curve so that it is perpendicular to both the mesh point normal and the boundary curve. That is, **virtual mesh point tangents** are determined as

$$\mathbf{m}_j = (\mathbf{n}_j \times \mathbf{s}_j)/|\mathbf{n}_j \times \mathbf{s}_j| \quad for \quad j = 0, 1 \tag{10.33}$$

Finally, by blending the two *virtual mesh point tangents*, we obtain a **virtual cross boundary tangent** $\mathbf{d}_i(s)$ along the boundary curve $\mathbf{e}_i(s)$:

$$\mathbf{d}_i(s) = \alpha(s)\mathbf{m}_0 + \beta(s)\mathbf{m}_1 \; s \in [0, 1] \tag{10.34}$$

where $\alpha(s), \beta(s)$ are blending functions.

When the *virtual cross boundary tangent of* (10.34) is used as the cross boundary tangent of the boundary curve $\mathbf{e}_i$, the resulting surface patch may possess some undesirable "directional properties". Thus, if we want to improve "directional properties of the surface patch" at its boundaries, the *cross boundary tangent* $\mathbf{d}_i(s)$ may have to be "corrected" so that it satisfies the following end conditions (See Figure 10.11):

$$\begin{aligned} \mathbf{d}_i(0) &= \mathbf{t}_0/|\mathbf{t}_0| \\ \mathbf{d}_i(1) &= \mathbf{t}_1/|\mathbf{t}_1|. \end{aligned} \tag{10.35}$$

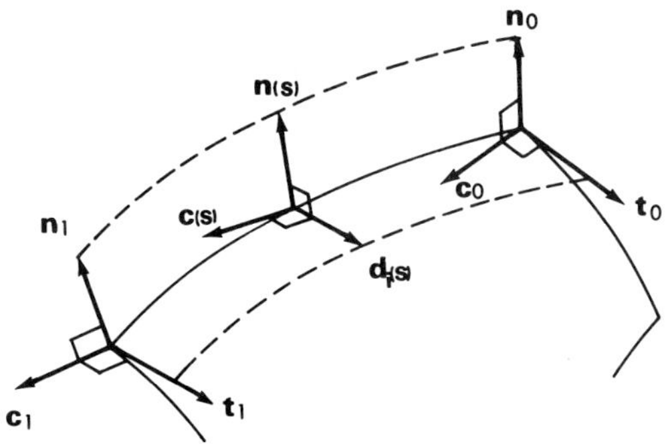

Figure 10.11 Correction of Cross Boundary Tangents

The following construction scheme for $\mathbf{d}_i(s)$ meets the end condition (10.35):

$$\mathbf{d}_i(s) = \{\mathbf{n}(s) \times \mathbf{c}(s)\}/|\mathbf{n}(s) \times \mathbf{c}(s)| \quad for\ s \in [0,1] \tag{10.36}$$

where, $\mathbf{n}(s) = \alpha(s)\mathbf{n}_0 + \beta(s)\mathbf{n}_1$: *boundary normal,*

$\mathbf{c}(s) = \alpha(s)\mathbf{c}_0 + \beta(s)\mathbf{c}_1$,

$\mathbf{c}_j = \mathbf{t}_j \times \mathbf{n}_j : \quad j = 0, 1,$

$\mathbf{n}_0, \mathbf{n}_1$: *mesh point normal* (10.32),

$\alpha(s), \beta(s)$: *blending functions.*

The "corrected" *cross boundary tangent* (10.36) is shown in Fig. 10.11.

Having determined all the *boundary data* of the patch, an n-sided surface patch $\mathbf{r}(V)$ interpolating to the boundary curves $\{\mathbf{e}_i(s_i)\}$ may be modeled as a *convex combination surface* (Gregory, 1986):

$$\mathbf{r}(V) = \sum_{i=1}^{n} \gamma_i(V)\{\mathbf{e}_i(s_i) + \lambda_i \nu_i \mathbf{d}_i(s_i)\} \quad s_i \in [0,1] \tag{10.37}$$

where, $V = \{\lambda_i : i = 1, ...n\}$,

$s_i = \lambda_{i-1}/(\lambda_{i-1} + \lambda_{i+1})$,

$\gamma_i(V) = \left(\prod_{j \neq i}^{n} \lambda_j\right)^2 \Big/ \left\{\sum_{k=1}^{n}(\prod_{j \neq k}^{n} \lambda_j)^2\right\}$: *convex combination factor,*

ν_i : *expansion factor,*

$i = 1, ..., n$ *with* "$(i-1)$ *mod* $n + 1$".

In (10.37), the boundary curves $\mathbf{e}_i(s)$ and cross boundary tangents $\mathbf{d}_i(s)$ are ordered in an anti-clockwise order. The **expansion factor** ν_i for $\mathbf{d}_i(s)$ may simply be set to

$$\nu_i = \{|\mathbf{t}_i| + |\mathbf{t}_{i+1}|\}/2. \tag{10.38}$$

The choice for the *cross boundary tangent* $\mathbf{d}_i(s)$ is a trade-off between surface quality and computational efficiency. The virtual cross boundary tangent (10.34) would give a computational efficiency (at an expense of surface quality).

10.5 CUBIC CURVE-NET INTERPOLATION

When all the *mesh segments* $\mathbf{e}_i(s)$ in a curve-net are cubic polynomial curves, the patches in the curve-net can be represented by the *polynomial patch models* introduced in §5.2 (instead of the *Boolean sum or convex combination schemes*). As with the general curve-net case, cubic curve-nets may also be classified into the following three types:

- *smooth curve-net having rectangular topology,*
- *smooth curve-net having irregular topology, and*
- *segmented curve-net.*

In this section, however, we will limit our discussion to the interpolation of the third type cubic curve-net. Note that the first two types are special cases of the third. In fact, a cubic curve-net interpolation scheme for the first type has already been presented in §7.6. The cubic curve-net interpolation scheme to be introduced is due to Chiyokura and Kimura (1984).

10.5.1 A Non-Standard Rational Bicubic Bezier Patch

Before presenting the bicubic curve-net interpolation scheme, a non-standard rational bicubic Bezier patch model will be introduced first. As discussed in §5.2.3, a bicubic Bezier patch is given by

$$\mathbf{r}(u,v) = \sum_{i=0}^{3}\sum_{j=0}^{3} V_{ij} B_i^3(u) B_j^3(v) \qquad for \quad u,v \in [0,1], \tag{10.39}$$

where V_{ij} are control vertices and $B_i^3(u)$ are *Bernstein polynomial.* The control vertices $\{V_{ij}\}$ with indices $i,j = 1,2$ are called *interior control vertices* and the rest are *boundary control vertices.*

Now the "interior control vertex" V_{ij} is represented as a blending of a pair of 3D points, P_{ij} and Q_{ij}, as follows:

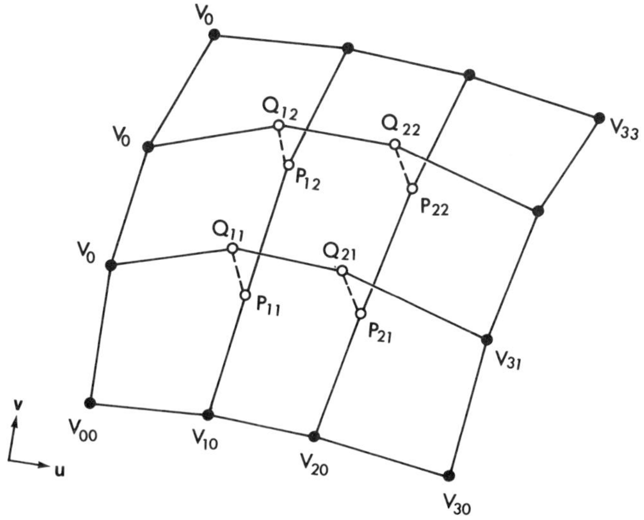

Figure 10.12 "Gregory Patch"

$$
\begin{aligned}
V_{11}(u,v) &= \{u\ P_{11} + v\ Q_{11}\}/(u+v) \\
V_{12}(u,v) &= \{u\ P_{12} + (1-v)Q_{12}\}/\{u+(1-v)\} \\
V_{21}(u,v) &= \{(1-u)\ P_{21} + v\ Q_{21}\}/\{(1-u)+v\} \\
V_{22}(u,v) &= \{(1-u)\ P_{22} + (1-v)\ Q_{22}\}/\{(1-u)+(1-v)\}.
\end{aligned} \qquad (10.40-a)
$$

The *boundary control vertices* are given as usual, namely,

$$V_{ij}(u,v) = V_{ij} = P_{ij} = Q_{ij} \quad for \quad i=0,3 \quad or \quad j=0,3. \qquad (10.40-b)$$

The rational Bezier patch with its control vertices given by (10.40) is shown in Fig. 10.12. With these control vertices, the Bezier patch (10.39) is expressed as

$$\mathbf{r}(u,v) = \sum_{i=0}^{3}\sum_{j=0}^{3} V_{ij}(u,v)B_i^3(u)B_j^3(v) \qquad for \quad u,v \in [0,1]. \qquad (10.41)$$

The above rational bicubic Bezier patch is named a **Gregory patch** by Chiyokura and Kimura (not to be confused with the *Gregory patches* of §5.3). The 3D points P_{ij} and Q_{ij} are called Gregory points.

Derivatives of the *Gregory patch* (10.41) are easily evaluated. For example, the *u-direction derivative* is expressed as

$$\mathbf{r}_u(u,v) = \sum_{j=0}^{3}\left(\sum_{i=0}^{3} \frac{\partial}{\partial u} V_{ij}(u,v) B_i^3(u) + V_{ij}(u,v) \frac{d}{du} B_i^3(u)\right) B_j^3(v). \qquad (10.42)$$

Derivatives of Bernstein polynomials in (10.42) are given by (See Eqn. (6.5)):

$$\frac{d}{du} B_i^3(u) = 3\{B_{i-1}^2(u) - B_i^2(u)\}. \qquad (10.43)$$

And the partials of the control vertices are given by

$$\frac{\partial}{\partial u} V_{11}(u,v) = v(P_{11} - Q_{11})/(u+v)^2 \qquad (10.44-a)$$

$$\frac{\partial}{\partial u} V_{12}(u,v) = (1-v)(P_{12} - Q_{12})/(1+u-v)^2 \qquad (10.44-b)$$

$$\frac{\partial}{\partial u} V_{21}(u,v) = v(Q_{21} - P_{21})/(1-u+v)^2 \qquad (10.44-c)$$

$$\frac{\partial}{\partial u} V_{22}(u,v) = (1-v)(Q_{22} - P_{22})/(2-u-v)^2 \qquad (10.44-d)$$

$$\frac{\partial}{\partial u} V_{ij}(u,v) = 0 \quad for \quad i = 0,3 \ or \ j = 0,3. \qquad (10.44-e)$$

Thus, on evaluating (10.42) at $u = 0, 1$, the u-direction *cross boundary tangents* of the Gregory patch are expressed as

$$\mathbf{r}_u(0,v) = 3\sum_{j=0}^{3}\left(Q_{1j} - V_{0j}\right) B_j^3(v). \qquad (10.45-a)$$

$$\mathbf{r}_u(1,v) = 3\sum_{j=0}^{3}\left(V_{3j} - Q_{2j}\right) B_j^3(v). \qquad (10.45-b)$$

The v-direction cross boundary tangents are also similarly obtained as

$$\mathbf{r}_v(u,0) = 3\sum_{i=0}^{3}\left(P_{i1} - V_{i0}\right) B_i^3(u). \qquad (10.46-a)$$

$$\mathbf{r}_v(u,1) = 3\sum_{i=0}^{3}\left(V_{i3} - P_{i2}\right) B_i^3(u). \qquad (10.46-b)$$

And the tangent of the $u = 0$ boundary curve, for example, is given by

$$\mathbf{r}_v(0,v) = 3\sum_{j=0}^{2}\left(V_{0,j+1} - V_{0,j}\right) B_j^2(v). \qquad (10.47)$$

which is the same as that of a bicubic Bezier patch (10.39).

From the above result, one may see that *u-direction cross boundary tangents* of the Gregory patch (10.41) are influenced by Q_{ij} only and v-direction cross boundary tangents by P_{ij}. This is an important property of the *Gregory patch* as will be seen shortly.

10.5.2 Local Interpolation Scheme for Rectangular Patches

The overall procedure of cubic curve-net interpolation is similar to that of the *segmented (general) curve-net interpolation* discussed in the previous section. In this subsection, a method for constructing rectangular surface patch is presented. The rectangular surface patch will be represented as a *Gregory patch* depicted in Fig. 10.12. Surface interpolation methods for non-rectangular patches will be presented later.

Shown in Fig. 10.13 is a rectangular patch bounded by four *cubic (Bezier) curves*. The control vertices defining the patch boundaries will become the *boundary control points* of the Gregory patch. Thus, it remains to determine the *Gregory points* Q_{ij}, P_{ij} *for* $i, j = 1, 2$. These eight *Gregory points* are to be determined as follows:

Q_{11}, Q_{12} *are obtained from* $u = 0$ *boundary data,*
Q_{21}, Q_{22} *are obtained from* $u = 1$ *boundary data,*
P_{11}, P_{21} *are obtained from* $v = 0$ *boundary data, and*
P_{12}, P_{22} *are obtained from* $v = 1$ *boundary data.*

We will take the $u = 0$ *boundary curve* as an example. As depicted in Fig. 10.13, each boundary curve is defined by four *Bezier points*.

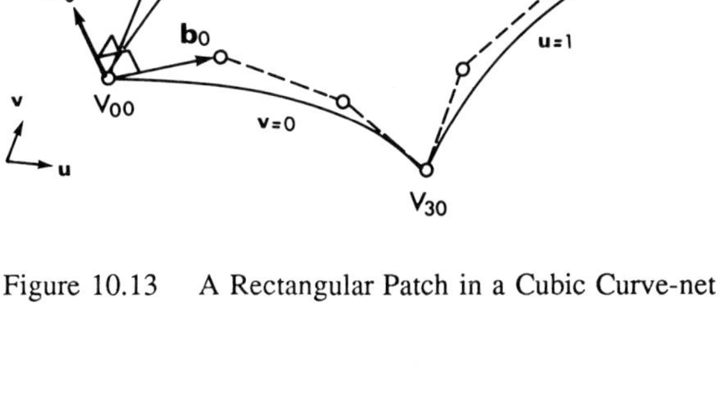

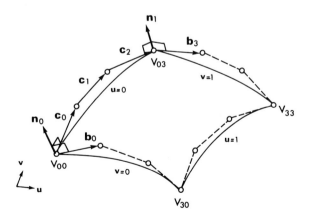

Figure 10.13 A Rectangular Patch in a Cubic Curve-net

The four *Bezier points* on the $u = 0$ boundary curve are $V_{00}, V_{01}, V_{02}, V_{03}$. Compare Fig. 10.13 with Fig. 10.12. As depicted in Fig. 10.13, let us define the following **control vectors** along the $u = 0$ boundary curve:

$$\mathbf{c}_j = V_{0,j+1} - V_{0,j} \quad for \quad j = 0, 1, 2. \tag{10.48}$$

And the first *control vectors* on the $v = 0, 1$ boundary curves are defined as

$$\begin{aligned} \mathbf{b}_0 &= V_{1,0} - V_{0,0} \\ \mathbf{b}_3 &= V_{1,3} - V_{0,3}. \end{aligned} \tag{10.49}$$

Then, the **mesh point normals** (also called corner normals) $\mathbf{n}_0, \mathbf{n}_1$ at the ends of the boundary curve are given by (See Eqn. (10.32))

$$\begin{aligned} \mathbf{n}_0 &= (\mathbf{b}_0 \times \mathbf{c}_0)/|\mathbf{b}_0 \times \mathbf{c}_0| \\ \mathbf{n}_1 &= (\mathbf{b}_3 \times \mathbf{c}_2)/|\mathbf{b}_3 \times \mathbf{c}_2|. \end{aligned} \tag{10.50}$$

Now, as shown in Fig. 10.14, we construct **virtual corner tangents** as follows (See Eqn. (10.33)):

$$\begin{aligned} \mathbf{m}_0 &= (\mathbf{n}_0 \times \mathbf{c}_0)/|\mathbf{n}_0 \times \mathbf{c}_0| \\ \mathbf{m}_1 &= (\mathbf{n}_1 \times \mathbf{c}_2)/|\mathbf{n}_1 \times \mathbf{c}_2|. \end{aligned} \tag{10.51}$$

From the above *virtual corner tangents*, a **virtual cross boundary tangent** for the $u = 0$ boundary curve may be defined as

$$\mathbf{t}(v) = \alpha(v)\mathbf{m}_0 + \beta(v)\mathbf{m}_1, \quad v \in [0, 1] \tag{10.52}$$

which is identical in form to the $\mathbf{d}_i(s)$ given by (10.34). If $\alpha(v), \beta(v)$ are linear blending functions, (10.52) represents a Bezier curve of degree 1 (in fact it is a *hodograph* not a curve, see Chapter 6).

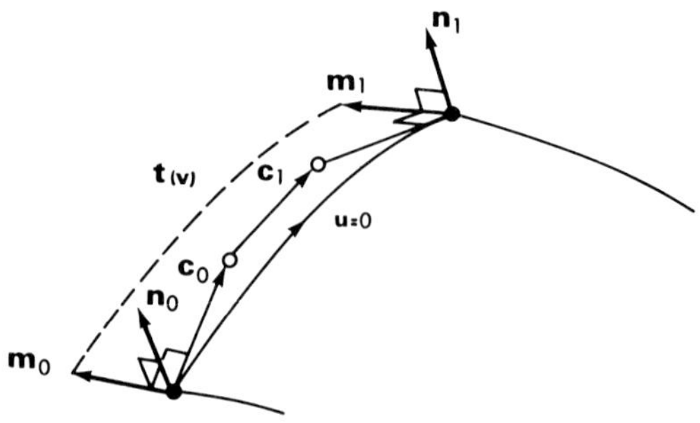

Figure 10.14 Construction of Virtual Cross Boundary Tangent

That is, we have (See §6.4 for degree elevation)

$$\begin{aligned}
\mathbf{t}(v) &= (1-v)\mathbf{m}_0 + v\mathbf{m}_1 \\
&= \sum_{i=0}^{1} \mathbf{m}_i B_i^1(v) \\
&= \sum_{i=0}^{2} \mathbf{t}_i B_i^2(v) \quad \textit{for } v \in [0,1]
\end{aligned} \tag{10.53}$$

$$\textit{where,} \quad \mathbf{t}_0 = \mathbf{m}_0\ ; \quad \mathbf{t}_2 = \mathbf{m}_1,$$
$$\mathbf{t}_1 = (\mathbf{m}_0 + \mathbf{m}_1)/2.$$

The reason for this degree elevation will be explained shortly.

Now we are ready to construct a *Gregory patch* (ie, determine Q_{11}, Q_{12}) so that it is G^1-continuous with the *virtual cross boundary tangent* (10.53) across the $u = 0$ boundary curve. The first step is to construct an expression for the "cross boundary tangent" $\mathbf{b}(v)$ of the $u = 0$ boundary curve. From the result of (10.45-a), we have

$$\mathbf{b}(v) = 3\sum_{j=0}^{3} \mathbf{b}_j B_j^3(v) \tag{10.54}$$

$$\textit{where,} \quad \mathbf{b}_0 = V_{10} - V_{00}\ ; \quad \mathbf{b}_3 = V_{13} - V_{03}\ : \quad \textit{as defined in } (10.49),$$
$$\mathbf{b}_1 = Q_{11} - V_{01},$$
$$\mathbf{b}_2 = Q_{12} - V_{02}.$$

The control vectors $\mathbf{b}_i$ are shown in Fig. 10.15.

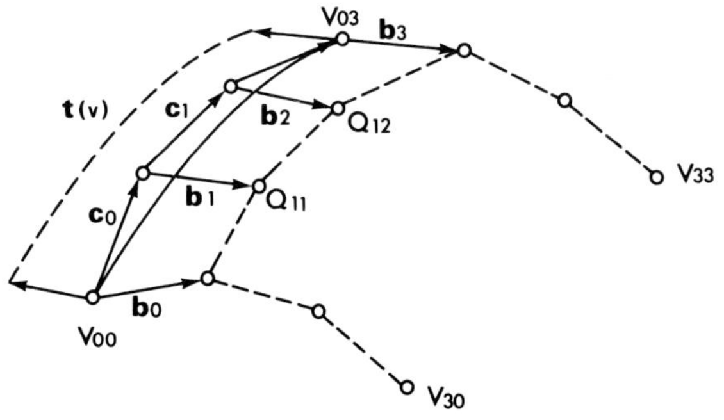

Figure 10.15 Determination of Gregory Points Q_{11} and Q_{12}

And, from (10.47), the *curve tangent* $\mathbf{c}(v)$ of the $u = 0$ boundary curve is given by

$$\mathbf{c}(v) = 3\sum_{j=0}^{3} \mathbf{c}_j B_j^2(v) \tag{10.55}$$

$$where, \quad \mathbf{c}_j = V_{0,j+1} - V_{0,j} \quad for\ j = 0, 1, 2.$$

From §7.6, one may construct the following G^1-condition which will ensure the necessary geometric continuity between the Gregory patch and the *virtual cross boundary tangent* $\mathbf{t}(v)$ of (10.54):

$$\mathbf{b}(v) = p(v)\mathbf{c}(v) + 3q(v)\mathbf{t}(v). \tag{10.56}$$

In (10.56), "3" is multiplied to $q(v)$ to simplify subsequent computations. The polynomials $\mathbf{c}(v)$ and $\mathbf{t}(v)$ are desired to have an equal degree (their degrees are two). This is the reason for the degree elevation of $\mathbf{t}(v)$. Since the degree of $\mathbf{b}(v)$ is three, the degrees of the scalar functions $p(v), q(v)$ should be one. A convenient form for $p(v),\ q(v)$ is :

$$\begin{aligned} p(v) &= (1-v)p_0 + vp_1 \\ q(v) &= (1-v)q_0 + vq_1. \end{aligned} \tag{10.57}$$

Thus, by substituting the results of (10.53), (10.55) and (10.57) into the right hand side of the G^1-condition (10.56), we have

$$\begin{aligned} \mathbf{b}(v) &= p(v)\mathbf{c}(v) + 3q(v)\mathbf{t}(v) \\ &= 3\sum_{j=0}^{3}\left(\frac{3-j}{3}(p_0\mathbf{c}_j + q_0\mathbf{t}_j) + \frac{j}{3}(p_1\mathbf{c}_{j-1} + q_1\mathbf{t}_{j-1})\right)B_j^3(v). \end{aligned} \tag{10.58}$$

On comparing the coefficients of $B_j^3(v)$ in (10.54) and in (10.58), we get

$$\mathbf{b}_j = \frac{3-j}{3}(p_0\mathbf{c}_j + q_0\mathbf{t}_j) + \frac{j}{3}(p_1\mathbf{c}_{j-1} + q_1\mathbf{t}_{j-1}) \quad for\ j = 0, 1, 2, 3 \tag{10.59}$$

which is an evaluated form of the G^1-condition (10.56).

However, we have yet to determine the unknown constants $p_i,\ q_i$ in (10.59). For $j = 0, 3$, (10.59) are written as

$$\mathbf{b}_0 = p_0\mathbf{c}_0 + q_0\mathbf{t}_0 \tag{10.60 - a}$$

$$\mathbf{b}_3 = p_1\mathbf{c}_2 + q_1\mathbf{t}_2 \tag{10.60 - b}$$

$$\begin{aligned} where, \quad & \mathbf{b}_0 = V_{10} - V_{00}\ ; \quad \mathbf{b}_3 = V_{13} - V_{03}, \\ & \mathbf{c}_j = V_{0,j+1} - V_{0,j} \quad for \quad j = 0, 2, \\ & \mathbf{t}_0 = \mathbf{m}_0\ ; \quad \mathbf{t}_2 = \mathbf{m}_1\ :\ \mathbf{m}_j\ as\ defined\ in\ (10.51). \end{aligned}$$

Observe that $\mathbf{b}_0, \mathbf{c}_0, \mathbf{t}_0$ are *coplanar (but not collinear)*. Thus, p_0 and q_0 are uniquely determined from (10.60-a). Similarly, p_1 and q_1 are obtained from (10.60-b).

Having determined p_i, q_i from (10.60), the *evaluated G^1-conditions* in (10.59) for $j = 1, 2$ provide expressions for the remaining *control vector* $\mathbf{b}_1$ and $\mathbf{b}_2$: They are

$$\mathbf{b}_1 = \frac{2}{3}(p_0\mathbf{c}_1 + q_0\mathbf{t}_1) + \frac{1}{3}(p_1\mathbf{c}_0 + q_1\mathbf{t}_0) \qquad (10.61-a)$$

$$\mathbf{b}_2 = \frac{1}{3}(p_0\mathbf{c}_2 + q_0\mathbf{t}_2) + \frac{2}{3}(p_1\mathbf{c}_1 + q_1\mathbf{t}_1) \qquad (10.61-b)$$

$$\begin{aligned} where,\quad & \mathbf{c}_j = V_{0,j+1} - V_{0,j} \quad for \quad j = 0, 1, 2, \\ & \mathbf{t}_0 = \mathbf{m}_0, \\ & \mathbf{t}_1 = (\mathbf{m}_0 + \mathbf{m}_1)/2, \\ & \mathbf{t}_2 = \mathbf{m}_1 : \ \mathbf{m}_j \ as\ defined\ in (10.51), \\ & p_1, q_1 \quad for\ i = 0, 1\ as\ determined\ from\ (10.60). \end{aligned}$$

Finally, the *Gregory points* Q_{11}, Q_{12} in Fig. 10.15 are determined from

$$Q_{i,j} = V_{0,j} + \mathbf{b}_j \quad for \quad j = 1, 2. \qquad (10.62)$$

The remaining *Gregory points* are determined in the same way from other boundary curves.

10.5.3 Local Interpolation Schemes for Non-rectangular Patches

When the cubic curve-net has non-rectangular patches such as triangular or pentagonal patches, they are *subdivided into rectangular patches* as shown in Fig. 10.16. Shown in Fig. 10.16a is a triangular patch in which the "left" boundary curve is defined by four control vertices V_0, V_1, V_2, *and* V_3.

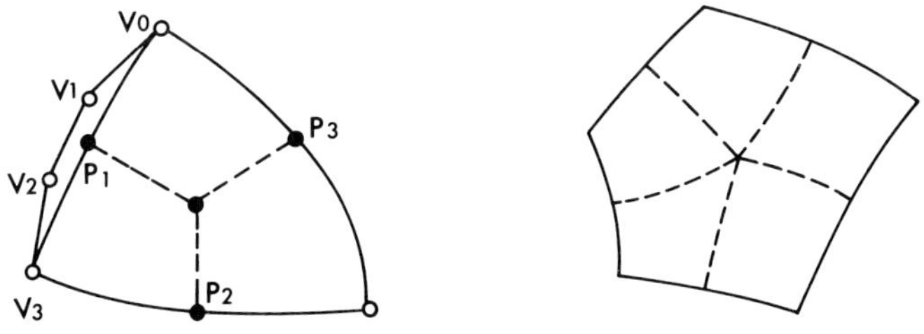

(a) (b)

Figure 10.16 Non-rectangular Patches in a Cubic Curve-net

In Fig. 10.16a, "subdivision points" on the boundsary curves are denoted as P_1, P_2, P_3 which correspond to "middle" points of the boundary curves. Namely,

$$P_i = \mathbf{r}_i(0.5) \quad for \quad i = 1, 2, 3.$$

The *subdivision point* P_1, for example, is expressed in terms of its "control vertices" as follows:

$$P_1 = (V_0 + 3V_1 + 3V_2 + V_3)/8. \tag{10.63}$$

One may easily verify the above result by applying the *de Casteljau* algorithm given in §6.3.

Boundary control vertices of the triangular patch of Fig. 10.16a are denoted by small circles in Fig. 10.17. Also depicted in Fig. 10.17 are *boundary control vectors* $\{\mathbf{c}_j : j = 0, 1, 2\}$ and *off-boundary control vectors* $\mathbf{b}_0, \mathbf{b}_3$ of the first boundary curve. The "boundary control vectors" are defined as

$$\mathbf{c}_j = V_{j+1} - V_j \quad for \quad j = 0, 1, 2. \tag{10.64}$$

As discussed in §10.5.2 (See Eqns. (10.50) and (10.51)), the cross boundary tangents, $\mathbf{t}_0$ and $\mathbf{t}_2$, are uniquely determined from $\mathbf{c}_j$'s and the "off-boundary control vectors". Further, $\mathbf{t}_1$ is defined as the average of $\mathbf{t}_0$ and $\mathbf{t}_2$. Then the *virtual cross boundary tangent* $\mathbf{t}(v)$ is expressed as (*Eqn.* (10.53))

$$\mathbf{t}(v) = \sum_{i=0}^{2} \mathbf{t}_i B_i^2(v) \quad for \quad v \in [0, 1]. \tag{10.65 - a}$$

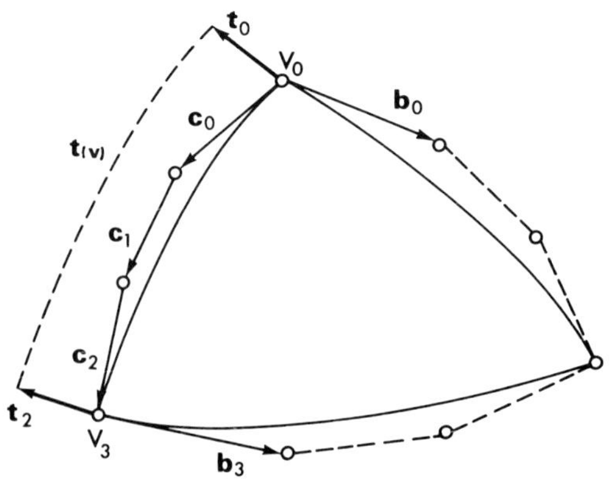

Figure 10.17 Boundary Data of a Triangular Patch

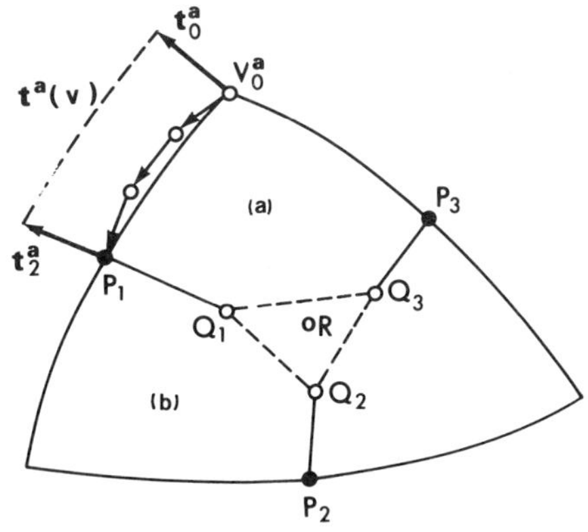

Figure 10.18 Subdivision of the Triangular Patch

Rewriting (10.54), the *cross boundary tangent* $\mathbf{b}(v)$ of the boundary curve is expressed as

$$\mathbf{b}(v) = 3\sum_{j=0}^{3} \mathbf{b}_j B_j^3(v), \qquad (10.65-b)$$

where $\mathbf{b}_1, \mathbf{b}_2$ are as given in (10.61).

Now we subdivide the triangular patch as shown in Fig. 10.18. In order to construct *internal boundary curves* (ie, common boundary curves between the subdivided patches), a *cross boundary tangent vector* is defined at each *subdivision point* P_i, and then a Bezier point Q_i is located in that vector direction. A heuristic choice for the Bezier point Q_1 for example is

$$Q_1 = P_1 + \frac{1}{3}\mathbf{b}(0.5) \qquad (10.66)$$

where, P_1 : *subdivision point*(10.63),

$\mathbf{b}(v)$: *cross boundary tangent function* (10.54).

The remaining Bezier points Q_2, Q_3 are determined in the same way. Finally a *center Bezier point* R is defined as an average of Q_i's. That is

$$R = (Q_1 + Q_2 + Q_3)/3. \qquad (10.67)$$

Thus, the three points P_1, Q_1, R become the control vertices of the *common boundary curve* between the *subpatch-(a) and subpatch-(b)* in Fig. 10.18.

The "left" boundary curve and its *virtual cross boundary tangent* are also subdivided at the center. Let $\{V_i^a : i = 0, ..., 3\}$ denote control vertices of the "left" boundary curve of *subpatch-(a)*. Then, from the *de Casteljau algorithm* of §6.3, they are given by

$$V_0^a = V_0 \qquad (10.68-a)$$

$$V_1^a = (V_0 + V_1)/2 \qquad (10.68-b)$$

$$V_2^a = (V_0 + 2V_1 + V_2)/4 \qquad (10.68-c)$$

$$V_3^a = P_1. \qquad (10.68-d)$$

The remaining boundary curves are subdivided similarly. The subdivided *virtual cross boundary tangent* $\mathbf{t}^a(v)$ of the boundary curve is expressed as

$$\mathbf{t}^a(v) = \sum_{i=0}^{2} \mathbf{t}_i^a B_i^2(v) \quad for\ v \in [0,1] \qquad (10.69)$$

$$where, \quad \mathbf{t}_0^a = \mathbf{t}_0,$$

$$\mathbf{t}_1^a = (\mathbf{t}_0 + \mathbf{t}_1)/2,$$

$$\mathbf{t}_2^a = (\mathbf{t}_0 + 2\mathbf{t}_1 + \mathbf{t}_2)/4.$$

Finally, the *internal boundary curves* are degree-elevated to make them cubic Bezier curves.

Shown in Fig. 10.19 are all the relevant boundary data for the rectangular subpatch. From the degree elevation formula of §6.4, one may find that the *off-boundary control vector* $\mathbf{b}_3^a$ is given by

$$\mathbf{b}_3^a = \frac{2}{3}(Q_1 - P_1). \qquad (10.70)$$

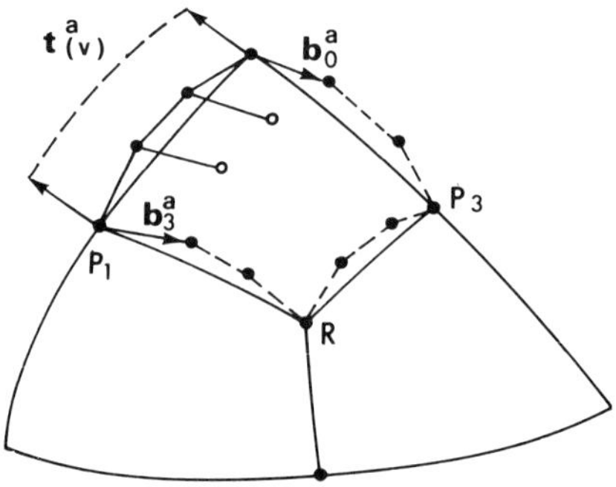

Figure 10.19 Boundary Data of Rectangular Sub-patch

The rectangular patch in Fig. 10.19 is exactly the one treated in the previous subsection. Thus, its "*Gregory points*" are given by (10.62). Local interpolation for "five-sided" and "six-sided" patches can be made the same way. In an actual implementation of the interpolation scheme, however, different strategies may be employed. See Chiyokura and Kimura (1984) for details.

The main objection to the local interpolation schemes discussed in this section is that the resulting surface patch is a *Gregory patch* which is a *non-standard rational Bezier patch* (Sarraga, 1987).

10.5.4 Local Interpolation Scheme for Triangular Patches

In this subsection, we present a different interpolation scheme for a triangular patch in a cubic curve-net (instead of the subdivision scheme of §10.5.3). In fact, if a cubic curve-net has a *triangular topology*, meaning that it consists only of triangular patches, the triangular interpolation method of Chapter 8 may be used as it is. The local interpolation scheme to be discussed is a modified version of the method presented in Chapter 8. The overall procedure of the interpolation scheme consists of the following steps:

a) construction of virtual cross boundary tangents for the boundary curves,
b) construction of a cubic triangular Bezier patch from its boundary data,
c) subdivision of the triangular Bezier patch (TBP) at its centroid,
d) degree elevation of each sub-TBP,
e) determination of off-boundary control points, and
f) modification of interior control points.

Shown in Fig. 10.20 are *boundary control vertices* of a triangular patch in a cubic curve-net with *boundary control vectors* $\mathbf{c}_0, \mathbf{c}_1, \mathbf{c}_2$ for a *boundary curve* $\mathbf{r}(v)$. As discussed in §10.5.2, the *virtual cross boundary tangent* of $\mathbf{r}(v)$ is given by (See Eqn. (10.53))

$$\mathbf{t}(v) = \sum_{i=0}^{2} \mathbf{t}_i B_i^2(v) \quad for \quad v \in [0,1]$$

and the *tangent* of the curve is given by (See Eqn. (10.55))

$$\mathbf{c}(v) = 3\sum_{i=0}^{2} \mathbf{c}_i B_i^2(v) \quad for \quad v \in [0,1].$$

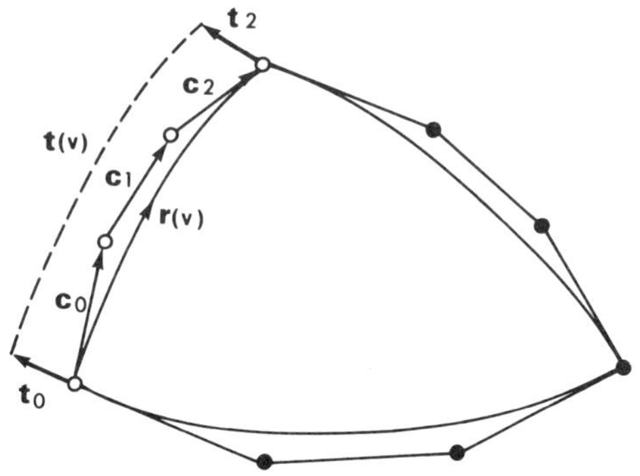

Figure 10.20 Boundary Data of a Cubic Triangular Patch

The next step is to determine the *center control point* of the patch to make it a *TBP (triangular Bezier patch).* For this, we use the method presented in §8.5.2 (See Eqn. (8.29)). Now the TBP is *subdivided* at its centroid using (8.30) and then its degree is elevated by using (8.31). The result of which is depicted in Fig. 10.21.

Also shown in Fig. 10.21 are *off boundary control vectors* $\{\mathbf{b}_i : i = 0, ..., 3\}$ (See Eqn. (8.23-b) for details). Then, from the result of (8.22-b), the *isoparametric derivative* $\mathbf{b}(v)$ is given by

$$\mathbf{b}(v) = 4\sum_{i=0}^{3} \mathbf{b}_i B_i^3(v) \quad for \quad v \in [0,1]. \tag{10.71}$$

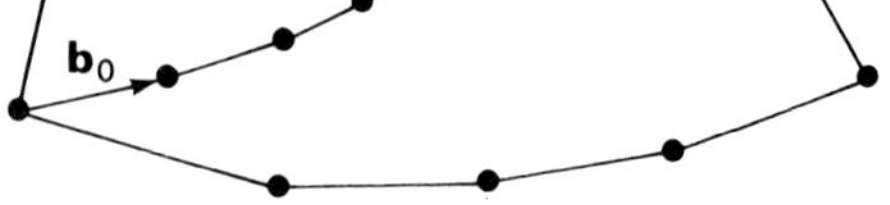

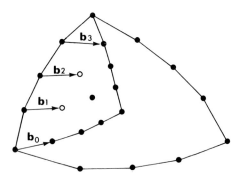

Figure 10.21 Determination of Off-boundary Control Points

We now construct a G^1-condition similar to (10.56) as follows:

$$\frac{1}{4}\mathbf{b}(v) = \frac{1}{3}p(v)\mathbf{c}(v) + q(v)\mathbf{t}(v). \tag{10.72}$$

Then the *off boundary control points* (marked as small circles in Fig. 10.21) are obtained from the procedure introduced in §10.5.2. In other words, the *off boundary control vector* $\mathbf{b}_1, \mathbf{b}_2$ are given by (10.61) as

$$\mathbf{b}_1 = \frac{2}{3}(p_0\mathbf{c}_1 + q_0\mathbf{t}_1) + \frac{1}{3}(p_1\mathbf{c}_0 + q_1\mathbf{t}_0)$$

$$\mathbf{b}_2 = \frac{1}{3}(p_0\mathbf{c}_2 + q_0\mathbf{t}_2) + \frac{2}{3}(p_1\mathbf{c}_1 + q_1\mathbf{t}_1)$$

where, p_i, q_i *for* $i = 0, 1$ *are given in* (10.60).

Having determined the off boundary control vertices, the rest of the control vertices are easily modified (See §8.5.5).

10.6 DISCUSSIONS

As far as surface interpolation is concerned, a major classification of a curve-net is whether its *mesh curves* are general parametric curves or *cubic polynomial curves*. The former type is called a *"general curve-net"* and the latter a *"cubic curve-net"*. If all the mesh curves in a curve-net are at least G^1-*continuous* at *mesh intersection points*, it is called a *"smooth curve-net"*. Otherwise, it is called a *"segmented curve-net"*.

The general curve-net interpolation schemes discussed in this chapter may be summarized as follows:

Geometry	*Topology*	*Patch-type*	*Interpolant-type*	*Reference*
smooth	*rectangular*	*4-sided*	*blended sweep, Coons*	*Sec. 10.2*
	irregular	*4-sided*	*Coons*	*Sec. 10.3*
		3-, 5-sided	*Boolean sum*	
segmented	*irregular*	*n-sided*	*Boolean sum*	*Sec. 10.4*

The main distinction between the "*smooth*" and the "*segmented*" curve-net cases is in the way the *cross-boundary tangents* are determined. In the *"smooth"* case, a pair of *mesh point (unit) tangents* are blended to construct a *"common" cross boundary tangent function,* while a *"virtual" cross boundary function* is defined from a pair of *mesh point (unit) normals* in the *"segmented"* case.

The cubic curve-net interpolation schemes introduced in this chapter (and other chapters) may be summarized as follows:

Geometry	*Topology*	*Patch-type*	*Interpolant*	*Reference*
smooth	*rectangular*	*4-sided*	*bisextic Bezier*	*Sec. 7.6*
segmented	*triangular*	*3-sided*	*triangular Bezier*	*Sec. 8.5*
	irregular	*4-sided*	*Gregory patch*	*Sec. 10.5.2*
		3-sided	*Gregory patch*	*Sec. 10.5.3*
			triangular Bezier	*Sec. 10.5.4*
		n-sided	*Gregory patch*	*Sec. 10.5.3*

For an *irregular* cubic curve-net, the concept of a *virtual cross boundary curve* is utilized, which enables a completely local interpolation. A cubic curve-net of arbitrary topology may be converted to a *triangular curve-net* (using subdivision) so that the interpolation schemes in §8.5 or §10.5.4 can be employed.

CHAPTER 11

CONSTRUCTION OF BLENDING SURFACES

11.1 INTRODUCTION

The term *blending surface* (or simply *blend*) means a surface forming a smooth localized transition between neighboring surfaces. Such a blending surface is required for various reasons, such as improving aesthetics, reducing stress concentrations, and ease of manufacture. The term "rounding" or "filleting" is more popular among engineers. However it is called, blending is a very important issue in geometric modeling and has been received quite an extensive attention. Blending surfaces in solid modeling were studied by, for example, Rossignac and Requicha (1984) and Rockwood and Owen (1987). The blending methods for solid modelers are limited to the blending of simple implicit surfaces.

The blending method introduced in this chapter, on the other hand, is for the blending of general parametric surfaces. Blending of parametric surfaces is more general than that of implicit surfaces in the sense that most implicit surfaces of practical importance (eg, conicoids and cubicoids) can be represented as rectangular parametric surfaces either in a rational polynomial form as shown in Waggenspack *et al* (1987) or in a form having trigonometric functions (see Chapter 5). This chapter presents methods for constructing *edge blends* and *corner blends* of parametrically defined surfaces. The materials in this chapter are mostly from Choi and Ju (1989) and Ju (1989).

In the proposed blending method, we construct explicit blending surfaces by simulating the action of a "rolling-ball" as depicted in Fig. 11.1. The procedure for constructing a **constant radius edge blend** consists of the following four steps:

a) construct offset surfaces of a pair of base surfaces,
b) find intersection curves of the two offset surfaces in order to determine "ball contact points" (C_0 and C_2 in Fig. 11.1b),
c) construct a conic section curve in between the two ball contact points as shown in Fig. 11.1c, and
d) construct an edge blend by "sweeping" the conic section (Fig. 11.1d).

The trajectory of the center of a ball rolling along a "valley" formed by two base surfaces corresponds to the intersection curve of the offset surfaces. If the "offset surface intersection curve" is defined on each domain of the two parametric surfaces, "ball contact curves" on the base surfaces are obtained. Thus, it is essential to have a robust and efficient algorithm for finding the intersection curves of a pair of offset surfaces. For this, we use an SSI algorithm proposed by Barnhill *et al* (1987).

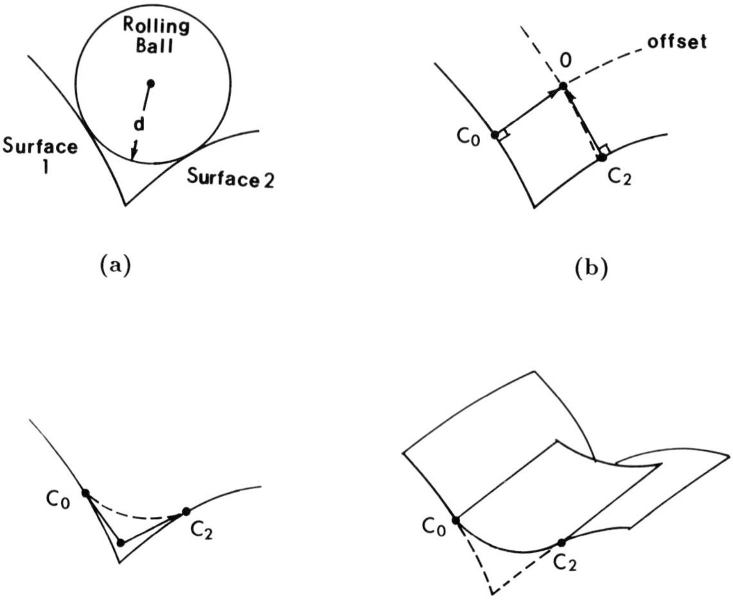

(a) (b) (c) (d)

Figure 11.1 Schematic View of Constructing an Edge Blend

The *constant radius edge blending method* is easily extended to a "variable" *radius edge blending*. The strategy for constructing a **variable radius edge blend** is as follows:

a) *"average" ball-contact points are obtained by using an average ball radius value, which is the same as the constant radius case,*
b) *"actual" ball contact points are numerically computed by applying a variable offset distance function, and*
c) *a variable radius edge blend is constructed by sweeping "variable radius" conic section curves.*

A **corner blend** is modeled as a *convex combination* of its boundary curves (See §10.4). Also introduced in this chapter is a **polyhedron rounding modification** method proposed in Chiyokura and Kimura (1983). The *rounding modification* is not exactly a *blending operation*, but it serves the same purpose.

11.2 OFFSET SURFACE INTERSECTION

This section presents a method of determining intersection curves of two offset surfaces by using the SSI (surface/surface intersection) algorithm proposed by Barnhill *et al* (1987). The SSI Algorithm works on rectangular parametric C^1 surface patches without any restriction on the functional form of the surfaces. The SSI algorithm consists of three separate stages: *Mesh generation, detection, and tracing*. Input to the SSI algorithm are:

- **f**(r,s), **g**(t,u): *two parametric base surfaces.*
- d: *offset distance which is equal to the blending radius.*
- δ: *(stepping) distance between sampled points on intersection curve.*
- ε: *tolerance value for numerical iteration.*

The SSI algorithm generates a sequence of domain values

$$\{(r_j, s_j),\ (t_j, u_j):\ j = 0, ..., n\}$$

for each *offset surface intersection curve (OSIC)*.

Before describing the SSI algorithm, we first show that *partial derivatives of offset surfaces* are easily obtained if second derivatives of the base surfaces are given. Let $\mathbf{f}(r,s)$ and $\mathbf{g}(t,u)$ be rectangular parametric patches with unit surface normal vectors $\mathbf{m}(r,s)$ and $\mathbf{n}(t,u)$, respectively. Then their **offset surfaces** are defined as:

$$\mathbf{f}^{\circ}(r,s) = \mathbf{f}(r,s) + d\ \mathbf{m}(r,s) \quad for\ r,s \in [0,1] \qquad (11.1-a)$$

$$\mathbf{g}^{\circ}(t,u) = \mathbf{g}(t,u) + d\ \mathbf{n}(t,u) \quad for\ t,u \in [0,1] \qquad (11.1-b)$$

where, d: *offset distance,*

$\mathbf{m}$: *unit surface normal of* $\mathbf{f}(r,s)$,

$\mathbf{n}$: *unit surface normal of* $\mathbf{g}(t,u)$.

The r-direction derivative of $\mathbf{f}^{\circ}(r,s)$ in (11.1-a) is given by (Farouki, 1986):

$$\begin{aligned} \mathbf{f}^{\circ}_r &\equiv \partial \mathbf{f}^{\circ}(r,s)/\partial r \\ &= \frac{\partial}{\partial r}\{\mathbf{f}(r,s) + d\ \mathbf{m}(r,s)\} \\ &= \mathbf{f}_r + d\left[\frac{E_r}{|E|} - \frac{E \cdot E_r}{|E|^3}E\right] \end{aligned} \qquad (11.2)$$

where, $E = \mathbf{f}_r \times \mathbf{f}_s$,

$E_r = \mathbf{f}_{rr} \times \mathbf{f}_s + \mathbf{f}_r \times \mathbf{f}_{rs}$,

$\mathbf{f}_r = \partial \mathbf{f}(r,s)/\partial r$, *etc.*

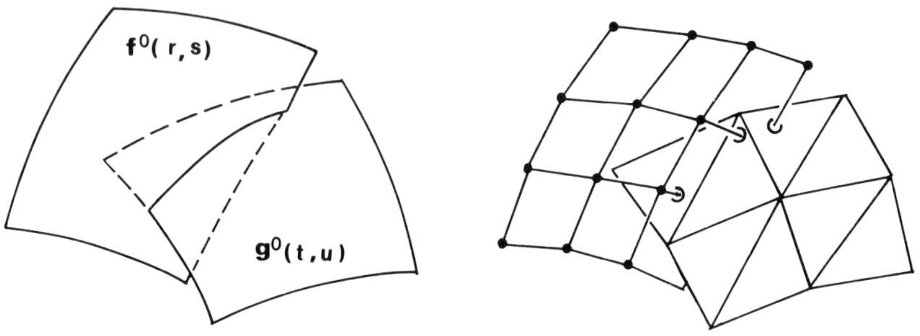

(a) (b)

Figure 11.2 (a) Two Intersecting Surfaces and (b) Mesh Generation

11.2.1 Mesh Generation Step

The first step of the SSI algorithm is to create a rectangular grid in the domain of each surface. As shown in Fig. 11.2, the first surface $\mathbf{f}^\circ(r,s)$ is approximated by a *mesh of straight line segments* joining each pair of adjacent grid points on the surface, and second surface $\mathbf{g}^\circ(t,u)$ is approximated by a *triangular polyhedron* by splitting each rectangle into two triangles. A suitable data structure is constructed for subsequent look-up operations.

11.2.2 Detection Stage

In the *detection stage*, at least one *initial intersection point* is detected should the two offset surfaces $\mathbf{f}^\circ, \mathbf{g}^\circ$ intersect with each other. The *detection* is carried out in two phases:

a) A rough intersection point is obtained from a *line/triangle intersection;*
b) It is then refined by performing a *curve/surface intersection.*

1) Line/triangle Intersection:

For each line-triangle pair – a line segment from $\mathbf{f}^\circ(r,s)$ and a triangular face from $\mathbf{g}^\circ(t,u)$ – an attempt is made to find an intersection point between the line segment and the flat triangle. If the line segment intersects with the triangle at a point, a domain tuple $\{(\hat{r},\hat{s}),\ (\hat{t},\hat{u})\}$ corresponding to that point is estimated. A method of finding the intersection between a line segment and a triangle is described below.

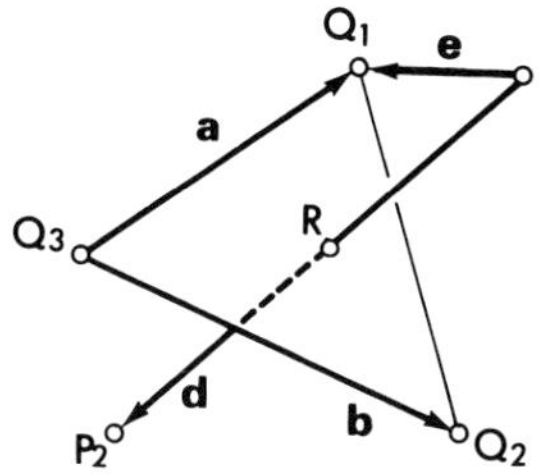

Figure 11.3 Intersection between a Line and a Triangle

Let P_1P_2 be a line segment from $\mathbf{f}^\circ(r,s)$ and $\Delta Q_1Q_2Q_3$ denote a triangle from $\mathbf{g}^\circ(t,u)$ such that

$$\begin{aligned} P_i &= \mathbf{f}^\circ(r^*, s_i) \quad for\ i = 1,2 \\ Q_i &= \mathbf{g}^\circ(t_i, u_i) \quad for\ i = 1,2,3. \end{aligned} \tag{11.3}$$

Now define the following vectors as depicted in Fig. 11.3

$$\begin{aligned} &\mathbf{a} = Q_1 - Q_3; \quad \mathbf{b} = Q_2 - Q_3 \\ &\mathbf{c} = (\mathbf{a} \times \mathbf{b})/|\mathbf{a} \times \mathbf{b}| : \quad unit\ normal\ of\ the\ plane \\ &\mathbf{d} = P_2 - P_1; \quad \mathbf{e} = Q_1 - P_1. \end{aligned} \tag{11.4}$$

Then, the implicit equation of the plane containing $\Delta Q_1Q_2Q_3$ is given by

$$\mathbf{c} \cdot \mathbf{r} = \mathbf{c} \cdot Q_1, \quad where \quad \mathbf{r} = (x, y, z).$$

Further, a parametric equation of the line P_1P_2 is given by

$$\mathbf{r}(\lambda) = P_1 + \lambda \mathbf{d}.$$

On substituting the above line equation into the plane equation, the unknown λ is determined as

$$\lambda^* = (\mathbf{c} \cdot \mathbf{e})/(\mathbf{c} \cdot \mathbf{d}) \tag{11.5 - a}$$

where **c**,**d**,**e** are as defined in (11.4).

A necessary condition to have an intersection between the line P_1P_2 and $\Delta Q_1Q_2Q_3$ is

$$0 \leq \lambda^* \leq 1. \tag{11.5 - b}$$

If this *necessary condition* is met, the domain value "s" of $\mathbf{f}^\circ(r,s)$ is estimated as (note that $\hat{r} = r^*$ is fixed):

$$\hat{s} = s_1 + \lambda^*(s_2 - s_1) \quad if\ 0 \leq \lambda^* \leq 1. \tag{11.6}$$

The intersection point R is obtained by plugging λ^* into the line equation:

$$R = P_1 + \lambda^* \mathbf{d}. \tag{11.7}$$

The domain value $(\hat{t}, \hat{u})$ of the intersection point R can be expressed in terms of barycentric coordinates (see §5.2.5) of Q_i's. That is,

$$R = \alpha\, Q_1 + \beta\, Q_2 + (1 - \alpha - \beta)\, Q_3. \tag{11.8}$$

Then the unknowns α, β in (11.8) are easily solved as

$$\alpha^* = \{\mathbf{c} \cdot (\mathbf{r} \times \mathbf{b})\}/\{\mathbf{c} \cdot (\mathbf{a} \times \mathbf{b})\} \qquad (11.9-a)$$

$$\beta^* = \{\mathbf{c} \cdot (\mathbf{r} \times \mathbf{a})\}/\{\mathbf{c} \cdot (\mathbf{b} \times \mathbf{a})\} \qquad (11.9-b)$$

$$where, \quad \mathbf{a}, \mathbf{b}, \mathbf{c} : \ as\ defined\ in\ (11.4),$$

$$\mathbf{r} = R - Q_3.$$

The line $P_1 P_2$ intersects with the triangle $\triangle Q_1 Q_2 Q_3$ if the following hold

$$\alpha^* \geq 0, \quad \beta^* \geq 0, \quad and\ \alpha^* + \beta^* \leq 1. \tag{11.10}$$

When the conditions in (11.10) are satisfied, the initial (t, u)-domain values for the intersection point R are expressed as

$$\hat{t} = \alpha^* t_1 + \beta^* t_2 + (1 - \alpha^* - \beta^*) t_3. \qquad (11.11-a)$$

$$\hat{u} = \alpha^* u_1 + \beta^* u_2 + (1 - \alpha^* - \beta^*) u_3. \qquad (11.11-b)$$

The line/triangle intersection method discussed above may be summarized as follow (assuming the line segment is on the $r = r^*$ grid of $\mathbf{f}°(r, s)$):

Procedure_LTI *(line/triangle intersection):*

0. *Input: line end points* (P_1, P_2) *and triangle vertices* (Q_1, Q_2, Q_3).
1. *Define the vectors* **a**, **b**, **c**, **d**, *and* **e** *in* (11.4).
2. *Fine* λ^* *from* $(11.5 - a)$.
3. *If the condition* $(11.5 - b)$ *is* not *met then* stop, *else compute* $\hat{s}$ *from* (11.6).
4. *Find* α^*, β^* *from* (11.9).
5. *If the condition* (11.10) *is* not *met then* stop, *else compute* $\hat{t}, \hat{u}$ *from* (11.11).
6. Return $\{\hat{s}, \hat{t}, \hat{u}\}$.

Once a domain value tuple $\{r^*, \hat{s}, \hat{t}, \hat{u}\}$ is successfully returned it is refined by finding an intersection point between the isoparametric curve $\mathbf{f}°(r^*, s)$ and the surface $\mathbf{g}°(t, u)$.

2) Curve/surface Intersection:

Domain values of a "true" curve/surface intersection point are obtained by solving the following equation (when $r = r^*$):

$$\mathbf{f}^\circ(r^*, s) - \mathbf{g}^\circ(t, u) = \mathbf{0}. \tag{11.12}$$

By using $\hat{s}, \hat{t}, \hat{u}$ as starting points, (11.12) is numerically solved for the unknowns s, t, u from the following Newton-Raphson iteration scheme which is a general algorithm for finding an intersection between an isoparametric curve and a parametric surface.

Procedure_CSI (curve/surface intersection):

0. *Input: fixed parameter value* (r^*) *and initial domain values* $(\hat{s}, \hat{t}, \hat{u})$.
1. *Find intersection "G" between the s-direction tangent line of* $\mathbf{f}^\circ(r, s)$ *at* $(r^*, \hat{s})$ *with the tangent plane of* $\mathbf{g}^\circ(t, u)$ *at* $(\hat{t}, \hat{u})$.
2. *Using G, invert the Jacobians to update domain values* $(\hat{s}, \hat{t}, \hat{u})$.
3. *If* $|\mathbf{f}^\circ(r^*, \hat{s}) - \mathbf{g}^\circ(\hat{t}, \hat{u})| \le \varepsilon$ *then stop,*
   else go to 1.

If the above iteration does not converge, the line segment and the triangle may be subdivided locally in order to recompute an intersection point from the "Procedure_LTI". If a domain value tuple is successfully obtained from the CSI iteration, it is stored in a list, called the "**initial intersection point (IIP) list**", together with the line segment number (of r, s-domain grid). This process is repeated for every pair of intersecting line/triangles. After the completion of the detection stage, we have a list of domain value tuples $\{(r_i, s_i),\ (t_i, u_i) : i = 1, ..., n\}$ stored in the IIP-list.

11.2.3 Jacobian Inversion

The procedure of finding the nearest point on a parametric surface from a 3D point G is called a **Jacobian inversion.** Since Jacobian inversion forms an essential step in SSI (surface/surface intersection) algorithms, a version of Jacobian inversion is described here. Let $\mathbf{r}(u, v)$ be a parametric surface and G be a 3D point. We want to find a point $\mathbf{r}(u^*, v^*)$ on the parametric surface that is nearest to G. Assume that we have a reasonable initial parameter value $(\hat{u}, \hat{v})$ in order to use the Newton-Raphson iteration scheme. Let us define the following as depicted in Fig. 11.4.

$$\mathbf{d} = G - \mathbf{r}(\hat{u}, \hat{v}),$$

$$\mathbf{e} = \mathbf{r}(u^*, v^*) - G,$$

$$\nabla u = (u^* - \hat{u}); \quad \nabla v = (v^* - \hat{v}).$$

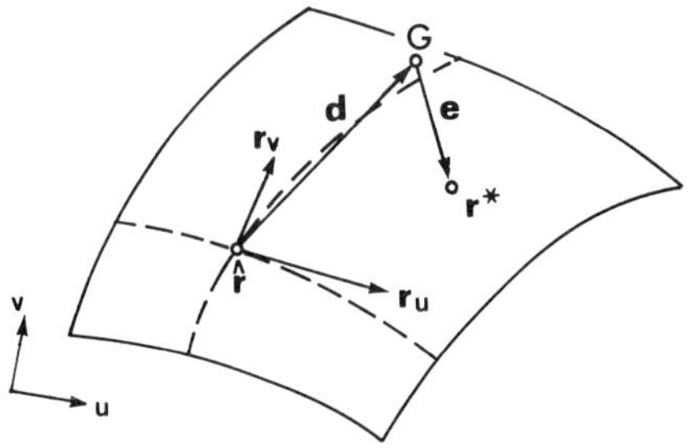

Figure 11.4 Formulation of Jacobian Inversion

Then, we need to minimize the distance $\mathbf{e}$ which is expressed as a linear Taylor expansion:

$$\mathbf{e} = \mathbf{r}(u^*, v^*) - \{\mathbf{r}(\hat{u}, \hat{v}) + \mathbf{d}\}$$
$$\cong (\nabla u \ \mathbf{r}_u + \nabla v \ \mathbf{r}_v) - \mathbf{d}.$$

That is, necessary conditions for minimizing $\mathbf{e} \cdot \mathbf{e}$ are given by

$$\partial(\mathbf{e} \cdot \mathbf{e})/\partial(\nabla u) = 0 \quad and \quad \partial(\mathbf{e} \cdot \mathbf{e})/\partial(\nabla v) = 0.$$

On solving the above linear equations, we get

$$\nabla u = \frac{(\mathbf{r}_u \cdot \mathbf{d})(\mathbf{r}_v \cdot \mathbf{r}_v) - (\mathbf{r}_v \cdot \mathbf{d})(\mathbf{r}_u \cdot \mathbf{r}_v)}{(\mathbf{r}_u \cdot \mathbf{r}_u)(\mathbf{r}_v \cdot \mathbf{r}_v) - (\mathbf{r}_u \cdot \mathbf{r}_v)^2}. \qquad (11.13 - a)$$

$$\nabla v = \frac{(\mathbf{r}_v \cdot \mathbf{d})(\mathbf{r}_u \cdot \mathbf{r}_u) - (\mathbf{r}_u \cdot \mathbf{d})(\mathbf{r}_u \cdot \mathbf{r}_v)}{(\mathbf{r}_u \cdot \mathbf{r}_u)(\mathbf{r}_v \cdot \mathbf{r}_v) - (\mathbf{r}_u \cdot \mathbf{r}_v)^2}. \qquad (11.13 - b)$$

Thus, the Jacobian inversion procedure may be summarized as follows:

Procedure_Jacobian-inversion:

0. Input: a 3D point (G) *and initial domain value* $(\hat{u}, \hat{v})$ *of surface* $\mathbf{r}(u, v)$.

1. Evaluate derivatives $\mathbf{r}_u, \mathbf{r}_v$ *at the current initial point* $(\hat{u}, \hat{v})$.

2. Compute the distance $\mathbf{d} = G - \mathbf{r}(\hat{u}, \hat{v})$.

3. Compute $\nabla u, \nabla v$ *from* (11.13) *and update:* $\hat{u} = \hat{u} + \nabla u$ *and* $\hat{v} = \hat{v} + \nabla v$.

4. If $(|\nabla u| > \varepsilon)$ *or* $(|\nabla v| > \varepsilon)$ *for "small"* ε, *then go to step 1*
else return $u^* = \hat{u}$ *and* $v^* = \hat{v}$.

11.2.4 Tracing Stage

From the *detection stage*, a non-empty *IIP-list (initial intersection point list)* is obtained if the two offset surfaces do intersect with each other. The first step in tracing is to get a domain value tuple from the IIP-list. The selected domain value tuple (r_0, s_0, t_0, u_0) is then used as an initial point of the following *SSI* (*surface/surface intersection*) problem

$$\mathbf{f}^\circ(r,s) - \mathbf{g}^\circ(t,u) = \mathbf{0} \tag{11.14}$$

in the four unknowns r, s, t, u.

In order to solve (11.14), a "guess point" G is generated along the *OSIC (offset surface intersection curve)*. Let δ be the stepping size, that is, the distance between a current SSI point to the next one, then the guess point G may be obtained from the following expression (The $\pm$ sign means that one is to be selected depending on trace direction):

$$G = \mathbf{f}^\circ(r_0, s_0) \pm \delta(\mathbf{m} \times \mathbf{n}) \;/\; |\mathbf{m} \times \mathbf{n}| \tag{11.15}$$

where, $\mathbf{m}, \mathbf{n}$: *unit normals of* **f**,**g** *at the "current" point,*

δ : *stepping distance (fixed input value).*

The next step is to compute domain values for the guess point G by inverting the Jacobians, starting from the domain values (r_0, s_0, t_0, u_0) of the current SSI point. Jacobian inversion may be carried out several times (This is equivalent to finding a surface point which is closest to G). The results are initial domain values $(\hat{r}, \hat{s}, \hat{t}, \hat{u})$ to be used in finding the next SSI point on the OSIC. In order to find domain values satisfying (11.14), a revised guess point $\hat{G}$ is obtained by finding the intersection between the tangent plane to $\mathbf{f}^\circ(r,s)$ at $(\hat{r}, \hat{s})$ and the tangent plane to $\mathbf{g}^\circ(t,u)$ at $(\hat{t}, \hat{u})$. Figure 11.5 shows how to determine the revised guess point $\hat{G}$. Finding a compact algebraic expression for $\hat{G}$ of Fig. 11.5 is a non-trivial exercise which is left to the reader (In fact, a solution is given in §11.3). The next iteration is generated by inverting the Jacobians, and so on. The iterative procedure of solving (11.14) may be summarized as follows:

Procedure_SSI *(surface/surface intersection):*

0. *Input: current domain values* (r_0, s_0, t_0, u_0) *and stepping distance* (δ).
1. *Find a guess point "G" from* (11.15).
2. *Invert the Jacobians at G to obtain its domain values* $(\hat{r}, \hat{s}, \hat{t}, \hat{u})$.
3. *Generate a revised guess point* $\hat{G}$ *as shown in Fig.* 11.5.
4. *Invert the Jacobians at* $\hat{G}$ *to revise the domain values* $(\hat{r}, \hat{s}, \hat{t}, \hat{u})$.
5. *If* $|\mathbf{f}^\circ(\hat{r}, \hat{s}) - \mathbf{g}^\circ(\hat{t}, \hat{u})| > \varepsilon$ *then go to 3, else* return $(\hat{r}, \hat{s}, \hat{t}, \hat{u})$.

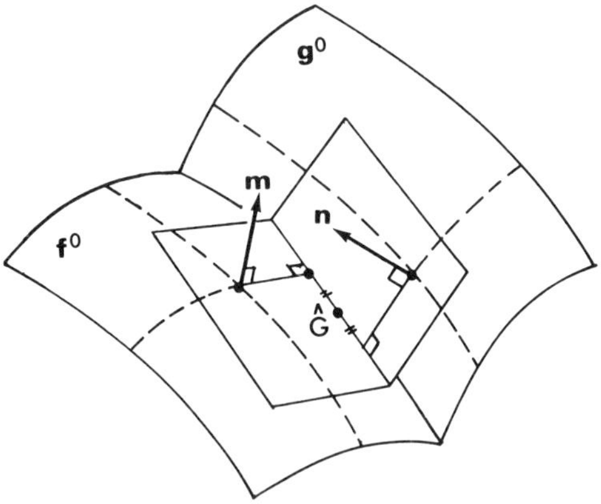

Figure 11.5 Guess Point Determination in SSI

If the above iteration does not converge, the stepping distance δ in (11.5) may have to be reduced temporarily.

The "Procedure-SSI" gives a next SSI point, which completes one stepping along the OSIC. At each stepping, we check to see if the current step (on r, s-domain) crosses any line segment stored in the *IIP-list (initial intersection point list)*. If so, the line segment is removed from the IIP-list. The next stepping is initiated by resetting the "next" SSI point to the "current" SSI point. Tracing is continued (in both directions) until one of the following "stopping conditions" is met:

(a) The two tracings meet together,
(b) Both of the domain value pairs fall outside domain regions, or
(c) The stepping distance δ was "too small".

We will have a closed OSIC *(offset surface intersection curve)* if the tracing is completed with *Case (a)*; otherwise the result is an open OSIC. See Barnhill *et al* (1987) for other aspects of the SSI algorithm.

During the tracing, the domain values of each SSI point along the OSIC are stored in a linked-list called OSIC-list. One complete tracing will results in one OSIC-list. If there are any domain value tuples left in the IIP-list, another OSIC tracing is made until the IIP-list becomes empty. The result is a number of OSIC-lists one for each OSIC.

11.3 CONSTANT RADIUS EDGE BLENDING

This section presents a method of constructing constant radius edge blends between a pair of *base surfaces*, $\mathbf{f}(r, s)$ and $\mathbf{g}(t, u)$. The method of constructing variable radius edge blends will be introduced in the next section. Conceptually, the construction of an edge blend consists of three steps: determination of *offset surface intersection curves (OSIC)*, construction of a conic section curve at each OSIC point, and sweeping of the conic section curves.

11.3.1 Offset Surface Intersection Curves

The first step of constructing an edge blend between the two *base surfaces* $\mathbf{f}(r, s)$ and $\mathbf{g}(t, u)$ is to determine an intersection curves of their offset surfaces $\mathbf{f}^{\circ}(r, s)$ and $\mathbf{g}^{\circ}(t, u)$ with offset distance d. By applying the *offset surface intersection procedure* of the previous section, the intersection curve is obtained in the form of *domain value tuples*

$$\{(r_j, s_j), (t_j, u_j)\}$$

which are stored in an OSIC-list. One OSIC-list is obtained for each offset surface intersection curve (OSIC).

11.3.2 Construction of Conic Section Curves

A tuple of domain values (r_j, s_j, t_j, u_j) stored in an OSIC-list represents a position of the "rolling ball". Since the radius of the ball is equal to the offset distance d in Eqn. (11.1), the center of the ball will be located at the *offset surface intersection point*

$$\begin{aligned} O &= \mathbf{f}(r_j, s_j) + d\mathbf{m}(r_j, s_j) \\ &= \mathbf{g}(t_j, u_j) + d\mathbf{n}(t_j, u_j), \end{aligned}$$

where $\mathbf{m}$, $\mathbf{n}$ are unit surface normals of $\mathbf{f}$ and $\mathbf{g}$, respectively. And the "rolling-ball" will touch the base surfaces at

$$C_0 = \mathbf{f}(r_j, s_j)$$

$$C_2 = \mathbf{g}(t_j, u_j)$$

as depicted in Fig. 11.6.

Let us construct two tangent planes (one for each base surface) at the ball contact points C_0, C_2, and construct a "section plane" π passing through the three points O, C_0, C_2. Then, the intersection point C_1 of the three planes is uniquely determined.

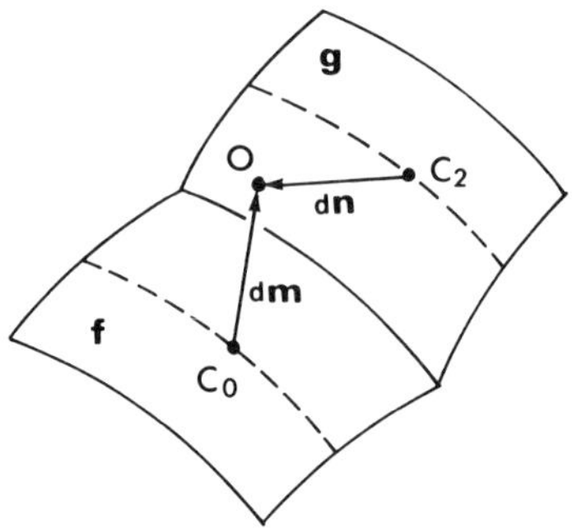

Figure 11.6 Ball-center Point and Ball-contact Points

The relationships among C_0, C_1, C_2, O are depicted in Fig. 11.7, from which one may easily verify that the following holds

$$C_1 = \frac{1}{2}(C_0 + C_2) + \frac{d}{2}\frac{\mathbf{m}\cdot\mathbf{n}-1}{\mathbf{m}\cdot\mathbf{n}+1}(\mathbf{m}+\mathbf{n}), \tag{11.16}$$

where $\mathbf{m}$, $\mathbf{n}$ are unit surface normals and d is the blending radius. By taking C_0, C_1, C_2 as the control vertices of a conic section, we can define a rational parametric equation of the conic section curve (with "section parameter" α):

$$\mathbf{b}(\alpha) = \frac{\omega_0(1-\alpha)^2 C_0 + 2\omega_1(1-\alpha)\alpha C_1 + \omega_2\alpha^2 C_2}{\omega_0(1-\alpha)^2 + 2\omega_1(1-\alpha)\alpha + \omega_2\alpha^2} \;; \quad 0 \le \alpha \le 1, \tag{11.17}$$

where ω_i's are the *weights* of the control points C_i's.

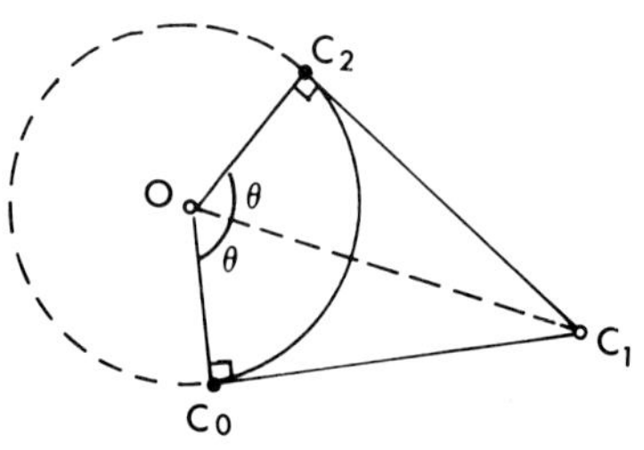

Figure 11.7 Construction of Conic Section Control Points

11.3.3 Sweeping of Conic Section Curves

To obtain a smooth edge blend surface, the ball has to be rolled along while maintanining its contacts with the two base surfaces. In mathematical terms, this action is achieved by sweeping the conic section curve along the ball contact curves. Smooth "ball contact curves" can be obtained by fitting a cubic spline curve (see §4.2.2) through the *domain value tuples* on the domain of each base surface. That is, cubic spline curves are constructed from $\{r_i, s_i\}$ and $\{t_i, u_i\}$ for $i = 0, 1, ..., n$. Let

$$\mathbf{s}(\beta) = (r(\beta), s(\beta)) \quad for\ 0 \le \beta \le 1 \qquad (11.18-a)$$

$$\mathbf{u}(\beta) = (t(\beta), u(\beta)) \quad for\ 0 \le \beta \le 1 \qquad (11.18-b)$$

denote the cubic spline curves with "sweeping parameter" β on (r, s)-domain and (t, u)-domain, respectively (in fact we have $0 \le \beta \le n$, but the range can be normalized). Then the ball contact curves become a function of β. Namely,

$$C_0(\beta) = \mathbf{f}(r(\beta), s(\beta)) \equiv \mathbf{f}(\beta) \qquad (11.19-a)$$

$$C_2(\beta) = \mathbf{g}(t(\beta), u(\beta)) \equiv \mathbf{g}(\beta) \qquad (11.19-b)$$

which would guarantee a position continuity between the *edge blend* and the *base surfaces.*

When the curves in (11.19) are ture "ball contact curves", the middle control point $C_1(\beta)$ can be determined from the relation (11.16). But, the curves in (11.19) are not exactly the ball-contact curves. An exaggerated view of a more realistic situation is shown in Fig. 11.8. In the figure, P_0, P_2 lie on the intersection line of the two tangent planes, and they become the same point (C_1) only when Eqn. (11.19) represents true ball contact curves. If we select C_1 as the mid-point of P_0 and P_2, it can be expressed as

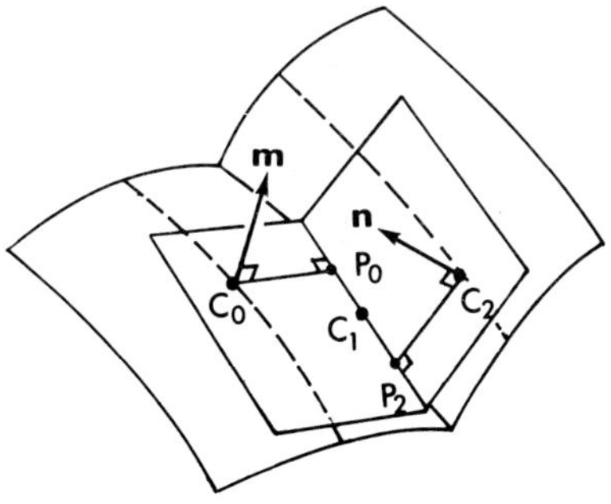

Figure 11.8 Determination of the Middle Control Point

$$C_1(\beta) = \frac{1}{2}[C_0(\beta) + C_2(\beta)] + \frac{1}{2}\frac{1}{(\mathbf{m}\cdot\mathbf{n})^2 - 1}[(\nabla C\cdot\mathbf{n})((\mathbf{m}\cdot\mathbf{n})\mathbf{m} - \mathbf{n}) - (\nabla C\cdot\mathbf{m})((\mathbf{m}\cdot\mathbf{n})\mathbf{n} - \mathbf{m})] \tag{11.20}$$

$$\textit{where,}\quad \nabla C = C_2(\beta) - C_0(\beta),$$

$$\mathbf{m}, \mathbf{n}: \quad \textit{unit normals of } \mathbf{f}(r,s),\ \mathbf{g}(t,u).$$

The above choice for the middle control point $C_1(\beta)$ would guarantee *gradient continuity* between the edge blend and the base surfaces. In practice, however, the use of (11.16) in determining $C_1(\beta)$ would also give a satisfactory result unless the tolerance ε in the CSI and SSI procedures and the stepping distance δ in Eqn. (11.5) are too large.

The weights in the conic section equation (11.17) can be used in adjusting the shape (fullness) of the edge blend surface. The fullness of a conic section is controlled by specifying the "fullness factor". For a given value $(0 < \rho < 1)$ of the fullness factor ρ, the weights are determined from (See §4.4.3)

$$\omega_0 : \omega_1 : \omega_2 = 1 : \rho/(1-\rho) : 1. \tag{11.21}$$

The effect of the "fullness factor" is illustrated in Fig. 11.9. Figure 11.9(a) is a circular blending. Fig.(b) and (c) are non-circular edge blends with $\rho = 0.25$ and $\rho = 0.65$, respectively.

If we want to have a circular arc, which is the most common choice, the *fullness factor* should be set to (See Eqn. (4.51))

$$\rho = \cos\theta/(1+\cos\theta) \tag{11.22}$$

$$\textit{where,}\quad \theta = \cos^{-1}(\mathbf{m}\cdot\mathbf{n})/2,\ (\textit{inverse defined within } (0,\pi)).$$

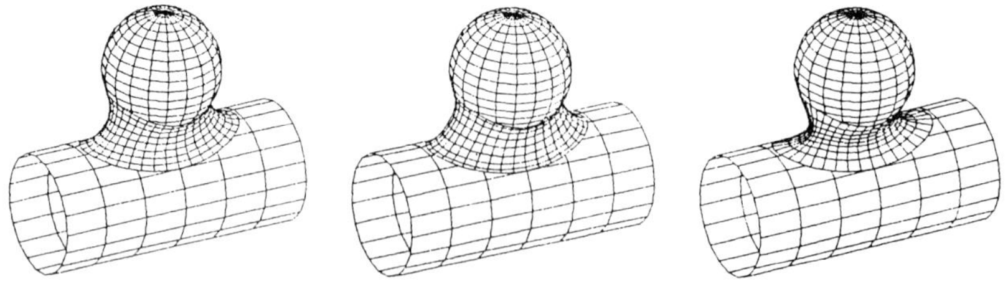

(a) (b) (c)

Figure 11.9 An Example of Constant Radius Edge Blend

From the results of Eqns. (11.18) through (11.21), the *conic section curve* (11.17) now becomes a surface equation for the **edge blend** *(for* $0 \le \alpha,\ \beta \le 1$*)*:

$$\mathbf{b}(\alpha,\beta) = \frac{(1-\alpha)^2 C_0(\beta) + 2\omega(1-\alpha)\alpha C_1(\beta) + \alpha^2 C_2(\beta)}{(1-\alpha)^2 + 2\omega(1-\alpha)\alpha + \alpha^2} \tag{11.23}$$

where, $C_0(\beta),\ C_2(\beta):$ *ball contact curves* (11.19),

$C_1(\beta):$ *middle control point* (11.20),

$\omega = \rho/(1-\rho)$,

$\rho:$ *fullness factor.*

11.4 VARIABLE RADIUS EDGE BLENDING

In general, the *edge blend* equation (11.23) can be treated as a function of *blending radius* d. Thus, the *constant radius edge blend* (11.23) may be denoted as

$$\mathbf{b}(\alpha,\beta;d) \quad for \quad \alpha,\beta \in [0,1],$$

where α is a *section parameter* and β is a *sweeping parameter.* With this notation, a *variable radius edge blend* can be denoted as

$$\mathbf{b}(\alpha,\beta;d(\beta)) \quad for \quad \alpha,\beta \in [0,1],$$

where the *blending radius* is specified as a function of the *sweeping parameter.* In the discussions that follow, we assume that "$d(\beta)$ for $\beta \in [0,1]$" is a non-decreasing function of β.

In practice, however, only the maximum and minimum values of the blending radius function are specified. Let

$d_{min} = d(0)$: *minimum blending radius, and*

$d_{max} = d(1)$: *maximum blending radius.*

The overall procedure of constructing a variable radius edge blend is as follows:

a) An OSIC-list $\{(\bar{r}_i,\bar{s}_i),(\bar{t}_i,\bar{u}_i)\}$ *is obtained by using the SSI algorithm of §11.2 with an "average" offset (ie, blend radius) value* $\bar{d}$.

b) And then, a "true" OSIC-list is constructed by using variable radius values d_i *at each OSIC point.*

c) Finally, a variable radius edge blend surface is defined by sweeping the conic sections along the ball-contact curves.

11.4.1 Construction of Tentative OSIC-List

Let d_{min} and d_{max} denote the minimum and maximum radius values for a variable radius edge blend. A reasonable choice for the *average offset distance* is

$$\bar{d} = (d_{min} + d_{max})/2. \tag{11.24}$$

Using the average offset distance in (11.24), offset surfaces are defined as

$$\mathbf{f}^{\circ}(r, s; \bar{d}) = \mathbf{f}(r, s) + \bar{d}\ \mathbf{m}(r, s) \tag{11.25 – a}$$

$$\mathbf{g}^{\circ}(t, u; \bar{d}) = \mathbf{g}(t, u) + \bar{d}\ \mathbf{n}(t, u) \tag{11.25 – b}$$

where $\mathbf{m}$, $\mathbf{n}$ are unit surface normals of $\mathbf{f}(r, s), \mathbf{g}(t, u)$. Then, by applying the SSI algorithm of §11.2, the following *OSIC-list (offset surface intersection curve list)* is obtained.

$$\{(\bar{r}_i, \bar{s}_i),\ (\bar{t}_i, \bar{u}_i): \quad i = 0, 1, ..., n\}. \tag{11.26}$$

11.4.2 Construction of True OSIC-List

Assuming that the radius of the edge blend varies linearly, a reasonable choice for the radius values d_i at a *tentative* OSIC point $((\bar{r}_i, \bar{s}_i), (\bar{t}_i, \bar{u}_i))$ is given by (note that $d_0 = d_{min}$ and $d_n = d_{max}$)

$$d_i = d_{min} + (d_{max} - d_{min}) * (i/n). \tag{11.27}$$

More elaborate choices for d_i are possible if necessary. Now we construct a plane π passing through the three points

$$\mathbf{f}^{\circ}(\bar{r}_i, \bar{s}_i; \bar{d}) \equiv \mathbf{g}^{\circ}(\bar{t}_i, \bar{u}_i; \bar{d}), \quad \mathbf{f}(\bar{r}_i, \bar{s}_i), \quad and \quad \mathbf{g}(\bar{t}_i, \bar{u}_i)$$

as we did in §11.3.2. The *unit normal* $\mathbf{p}$ of the plane π is given by

$$\mathbf{p} = (\bar{\mathbf{m}} \times \bar{\mathbf{n}})/|\bar{\mathbf{m}} \times \bar{\mathbf{n}}|, \tag{11.28}$$

where $\bar{\mathbf{m}}, \bar{\mathbf{n}}$ are unit surface normals at $\mathbf{f}(\bar{r}_i, \bar{s}_i), \mathbf{g}(\bar{t}_i, \bar{u}_i)$.

Now a circle of radius d_i is constructed on the plane π so that the circle "touches" on both surfaces. This construction allows one to determine the *ball-contact points* on the domains of $\mathbf{f}(u, s)$ and $\mathbf{g}(t, u)$. Determining the *ball-contact point or "true OSIC point"* is equivalent to finding the intersection of three surfaces: the plane π, $\mathbf{f}^{\circ}(r, s; d_i)$, and $\mathbf{g}^{\circ}(t, u; d_i)$.

The intersection problem may be solved by formulating a Newton-Raphson iteration scheme as shown in Fig. 11.10. Let π_f, π_g denote tangent planes of the offset surfaces at the current

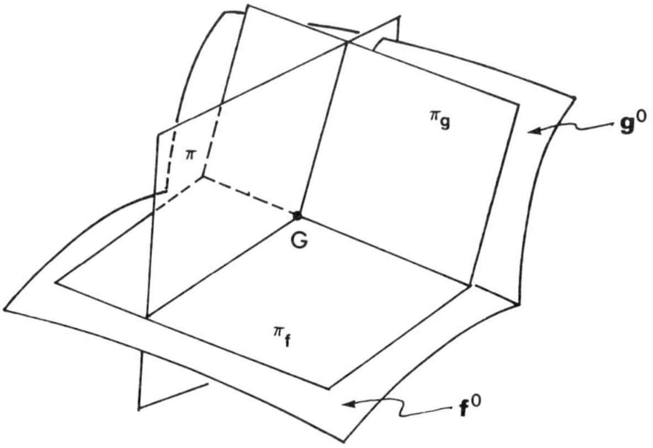

Figure 11.10 Determination of "True" OSIC Point

OSIC point. Then the intersection point G of the three planes π, π_f, π_g is given by (See Eqn. (4.9) of §4.2.2)

$$G = \frac{a(\mathbf{n} \times \mathbf{p}) + b(\mathbf{p} \times \mathbf{m}) + c(\mathbf{m} \times \mathbf{n})}{\mathbf{m} \cdot (\mathbf{n} \times \mathbf{p})} \tag{11.29}$$

where, $a = \mathbf{m} \cdot \mathbf{f}^\circ$,

$b = \mathbf{n} \cdot \mathbf{g}^\circ$,

$c = \mathbf{p} \cdot \bar{\mathbf{f}}$,

$\mathbf{f}^\circ = \mathbf{f}^\circ(r, s; d_i)$,

$\mathbf{g}^\circ = \mathbf{g}^\circ(t, u; d_i)$,

$\bar{\mathbf{f}} = \mathbf{f}(\bar{r}_i, \bar{s}_i)$,

$\mathbf{m} =$ *unit surface normal of* $\mathbf{f}(r,s)$,

$\mathbf{n} =$ *unit surface normal of* $\mathbf{g}(t,u)$,

$\mathbf{p} =$ *unit normal to the plane* π (11.28).

The overall procedure of determining a "true" OSIC point is as follow: Starting from the *"tentative OSIC point"* $((\bar{r}_i, \bar{s}_i), (\bar{t}_i, \bar{u}_i))$ and using the blending radius d_i, a guess point G is obtained from (11.29). And then the OSIC point (ie, domain value tuple) is updated by inverting the Jacobians (See §11.2.4 for Jacobian inversion). Recompute the guess point G

and invert the Jacobians until a "true" ball-contact point is reached. The above process may be summarized as follows:

Procedure-PSSI *(plane/surface/surface intersection):*

0. Input: tentative OSIC-point $(\bar{r}_i, \bar{s}_i, \bar{t}_i, \bar{u}_i)$ *and true offset value* d_i.

1. Initialize: $(r_i, s_i) \Leftarrow (\bar{r}_i, \bar{s}_i)$; $(t_i, u_i) \Leftarrow (\bar{t}_i, \bar{u}_i)$.

2. Find a guess point G using (11.29).

3-1. Using G, invert the Jacobians on $\mathbf{f}(r,s)$ *to revise* (r_i, s_i).

3-2. Using G, invert the Jacobians on $\mathbf{g}(t,u)$ *to revise* (t_i, u_i).

4. If $|\mathbf{f}^{\circ}(r_i, s_i; d_i) - \mathbf{g}^{\circ}(t_i, u_i; d_i)| \leq \varepsilon$ *then return* $((r_i, s_i), (t_i, u_i))$
else go to step 2.

By invoking the above procedure for *each* tuple of domain values in the tentative OSIC-list, a "true" OSIC-list $\{(r_i, s_i), (t_i, u_i);\ i = 0, 1, ..., n\}$ for the variable radius edge blend is obtained.

11.4.3 Construction of Variable Radius Edge Blend

Having determined the "true" OSIC-list (offset surface intersection curve list) for the variable radius edge blend, the procedure of constructing a variable-radius edge blend is the same as that of constructing a constant-radius edge blend discussed in §11.3. That is,

a) Cubic spline curve fitting of domain values: $r(\beta)$, $s(\beta)$, $t(\beta)$, $u(\beta)$.

b) Evaluation of ball-contact curves $C_0(\beta)$, $C_2(\beta)$ *using Eqn.* (11.19).

c) Determination of the middle control point $C_1(\beta)$ *using Eqn.* (11.20).

d) Construction of edge blend surface equation (Eqn. (11.23)).

Examples of variable-radius edge blend are shown in Fig. 11.11.

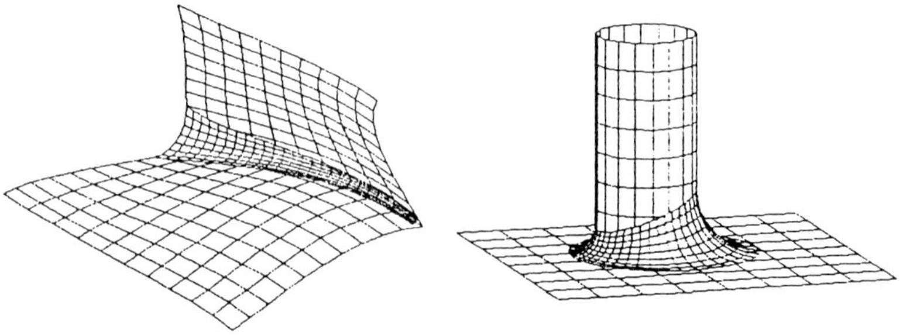

Figure 11.11 Examples of Variable-Radius Edge Blend

11.5 DERIVATIVES OF EDGE BLENDS

It is necessary to be able to evaluate partial derivatives of edge blends if we wish to generate cutter paths for NC machining or to find intersections between edge blends. This section presents methods of evaluating partial derivatives of the edge blend equation (11.23) which is repeated below:

$$\mathbf{b}(\alpha,\beta) = \frac{(1-\alpha)^2 C_0(\beta) + 2\omega(1-\alpha)\alpha C_1(\beta) + \alpha^2 C_2(\beta)}{(1-\alpha)^2 + 2\omega(1-\alpha)\alpha + \alpha^2} \quad for\ 0 \le \alpha, \beta \le 1.$$

The *section direction derivative*, $\partial \mathbf{b}(\alpha,\beta)/\partial\alpha$, is easily evaluated since it is a standard rational (quadratic) Bezier curve.

The *sweeping direction derivatives*, on the other hand, require some efforts. Since the control vertices in $\mathbf{b}(\alpha,\beta)$ are functions of β, we define their derivatives as

$$D_i(\beta) = \partial C_i(\beta)/\partial\beta\ ; \quad i = 0, 1, 2. \tag{11.30}$$

Then, ther sweep-direction derivative $\partial \mathbf{b}(\alpha,\beta)/\partial\beta$ is expressed as

$$\partial \mathbf{b}(\alpha,\beta) = \frac{(1-\alpha)^2 D_0(\beta) + 2\omega(1-\alpha)\alpha D_1(\beta) + \alpha^2 D_2(\beta)}{(1-\alpha)^2 + 2\omega(1-\alpha)\alpha + \alpha^2}. \tag{11.31}$$

For the moment, we assume that the *weight* ω is a constant (ie, not a function of β). When ω is a function of β, the derivative (11.31) will be in a little bit more complicated form involving the $\partial\omega(\beta)/\partial\beta$ term as well. In either case, it is enough to have the derivatives of (11.30) in order to evaluate the sweep-direction derivative (11.31).

Since $C_0(\beta)$ is a curve on the surface $\mathbf{f}(r,s)$, its tangent vector (ie, derivative) is given by (See Eqn. (2.23))

$$D_0(\beta) = \partial C_0(\beta)/\partial\beta = \mathbf{f}_r\dot{r} + \mathbf{f}_s\dot{s} \tag{11.32}$$

$$\begin{aligned} where,\quad \mathbf{f}_r &= \partial\mathbf{f}(r,s)/\partial r, \\ \mathbf{f}_s &= \partial\mathbf{f}(r,s)/\partial s, \\ \dot{r} &= dr(\beta)/d\beta \\ \dot{s} &= ds(\beta)/d\beta \quad (See\ Eqn.\ 11.18). \end{aligned}$$

$D_2(\beta)$ is also expressed similarly.

The derivative of the middle control point $C_1(\beta)$ requires a considerable amount of algebraic manipulation. On rewriting (11.20), we have

$$C_1(\beta) = \frac{1}{2}[C_0(\beta) + C_2(\beta) + E(\beta)]. \tag{11.33}$$

Thus, the derivative of the *middle control point* $C_1(\beta)$ is expressed as

$$\begin{aligned} D_1(\beta) &= \frac{d}{d\beta} C_1(\beta) \\ &= [D_0(\beta) + D_2(\beta) + dE(\beta)/d\beta]/2. \end{aligned} \tag{11.34}$$

From (11.20), the third term of (11.33) is given by

$$\begin{aligned} E(\beta) &= \frac{1}{(\mathbf{m}\cdot\mathbf{n})^2 - 1}[(\nabla C\cdot\mathbf{n})((\mathbf{m}\cdot\mathbf{n})\mathbf{m} - \mathbf{n}) - (\nabla C\cdot\mathbf{m})((\mathbf{m}\cdot\mathbf{n})\mathbf{n} - \mathbf{m})] \\ &= [z(x\mathbf{m} - \mathbf{n}) - y(x\mathbf{n} - \mathbf{m})]/(x^2 - 1) \end{aligned} \tag{11.35 - a}$$

$$\begin{aligned} where,\quad & \nabla C = C_2(\beta) - C_0(\beta), \\ & \mathbf{m}:\ unit\ surface\ normal\ of\ \mathbf{f}(r,s)\ at\ (r(\beta),\ s(\beta)), \\ & \mathbf{n}:\ unit\ surface\ normal\ of\ \mathbf{g}(t,u)\ at\ (t(\beta),\ u(\beta)), \\ & x = \mathbf{m}\cdot\mathbf{n} \\ & y = \nabla C\cdot\mathbf{m} \\ & z = \nabla C\cdot\mathbf{n}. \end{aligned}$$

The derivative of $E(\beta)$ is now evaluated as follows:

$$\begin{aligned} dE(\beta)/d(\beta) = {} & \{\dot z(x\mathbf{m} - \mathbf{n}) + z(\dot x\mathbf{m} + x\dot{\mathbf{m}} - \dot{\mathbf{n}}) - \dot y(x\mathbf{n} - \mathbf{m}) + y(\dot x\mathbf{n} + x\dot{\mathbf{n}} - \dot{\mathbf{m}})\} \\ & /(x^2 - 1) - 2x\dot x\{z(x\mathbf{m} - \mathbf{n}) - y(x\mathbf{n} - \mathbf{m})\}/(x^2 - 1)^2. \end{aligned} \tag{11.35 - b}$$

In the above equation, the derivatives of the scalar variables x, y, z can be evaluated as

$$\dot x = \partial(\mathbf{m}\cdot\mathbf{n})/\partial\beta = \dot{\mathbf{m}}\cdot\mathbf{n} + \mathbf{m}\cdot\dot{\mathbf{n}} \tag{11.36 - a}$$

$$\dot y = \nabla D\cdot\mathbf{m} + \nabla C\cdot\dot{\mathbf{m}} \tag{11.36 - b}$$

$$\dot z = \nabla D\cdot\mathbf{n} + \nabla C\cdot\dot{\mathbf{n}}, \tag{11.36 - c}$$

$$\begin{aligned} where,\quad & \dot{\mathbf{m}} = \partial\mathbf{m}/\partial\beta:\ to\ be\ given\ in\ (11.37), \\ & \dot{\mathbf{n}} = \partial\mathbf{n}/\partial\beta, \\ & \nabla D = D_2(\beta) - D_0(\beta), \\ & \nabla C = C_2(\beta) - C_0(\beta), \\ & C_0(\beta), C_2(\beta):\ as\ defined\ in\ (11.19), \\ & D_0(\beta), D_2(\beta):\ as\ defined\ in\ (11.32). \end{aligned}$$

And, the derivative of the unit normal $\mathbf{m}$ is obtained by applying the chain rule of differentiation (see Fouroki, 1986 for more details).

$$\dot{\mathbf{m}} = \dot{r}\left(\frac{F_r}{|F|} - \frac{(F \cdot F_r)F}{|F|^3}\right) + \dot{s}\left(\frac{F_s}{|F|} - \frac{(F \cdot F_s)F}{|F|^3}\right) \tag{11.37}$$

$$\begin{aligned} \textit{where}, \quad & F = \mathbf{f}_r \times \mathbf{f}_s, \\ & F_r = \mathbf{f}_r \times \mathbf{f}_{rs} + \mathbf{f}_{rr} \times \mathbf{f}_s, \\ & F_s = \mathbf{f}_r \times \mathbf{f}_{ss} + \mathbf{f}_{rs} \times \mathbf{f}_s, \\ & \dot{r} = dr(\beta)/d\beta, \\ & \dot{s} = ds(\beta)/d\beta \quad (\textit{See Eqn.}\ 11.18). \end{aligned}$$

$\dot{\mathbf{n}}$ is also obtained similarly.

From the results in Eqns. (11.35) through (11.37), the *derivative of the middle control point* (11.34) can be determined. In practical applications where sectional shapes of edge blends do not change rapidly, the third term in (11.34) may be neglected in obtaining $D_i(\beta)$. In this case, we have

$$D_1(\beta) = \{D_0(\beta) + D_2(\beta)\}/2.$$

Maintaining circular edge blends is computationally more expensive because the fullness factor ρ is (so is the weight ω in Eqn. (11.23)) continuously changing as the angle θ between the two surface normal vectors $\mathbf{m}$,$\mathbf{n}$ is not fixed (see Eqn. (11.22). This means that the weight ω in (11.23) is also a function of the sweeping parameter β. Let

$$\omega_i = \rho_i/(1 - \rho_i)$$

denote the weight at a sampled ball contact point (r_i, s_i, t_i, u_i), then the functional form of $\omega(\beta)$ is easily obtained by fitting, for example, a cubic spline curve through ω_i.

11.6 BALL-TYPE TRIANGULAR CORNER BLEND

This section presents a procedure for constructing a "corner blend" where three edge blends of identical radius d meet. Let $\mathbf{h}(v, w)$ be a third base surface with unit normal vector $\mathbf{o}(v, w)$ so that its offset surface is given by

$$\mathbf{h}^{\circ}(v, w) = \mathbf{h}(v, w) + d\ \mathbf{o}(v, w).$$

As shown in Fig. 11.12, if *three base surfaces* intersect at a corner point the SSI algorithm will produce three *OSIC (offset surface intersection curve)* lists. In other words, we will have three different lists of domain value tuples as:

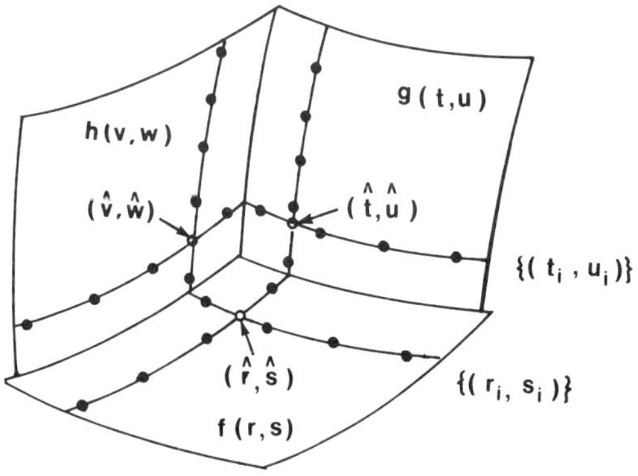

Figure 11.12 Domain Intersection Points at a Corner

$\{(r_i, s_i),\ (t_i, u_i)\}$: *OSIC-list for blending* **f**(r,s) *&* **g**(t,u),
$\{(t_j, u_j),\ (v_j, w_j)\}$: *OSIC-list for blending* **g**(t,u) *&* **h**(v,w),
$\{(v_k, w_k),\ (r_k, s_k)\}$: *OSIC-list for blending* **h**(v,w) *&* **f**(r,s).

In Fig. 11.12, the curves joining the small dots represent ball contact curves, and each pair of *ball-contact curves* forms an *edge blend.* The overall procedure for constructing corner blends consists of three steps:

- *Determination of domain intersection corner points,*
- *construction of boundary curves of corner blend, and*
- *construction of corner blend patch.*

11.6.1 Determination of Domain Intersection Corner Points

On the domain of each base surface, intersection points between a pair of OSICs are identified. Let us call them *domain intersection points.* Namely,

$(\hat{r}, \hat{s})$: *domain intersection point from* $\{(r_i, s_i)\}$ *and* $\{(r_k, s_k)\}$,
$(\hat{t}, \hat{u})$: *domain intersection point from* $\{(t_i, u_i)\}$ *and* $\{(t_j, u_j)\}$, *and*
$(\hat{v}, \hat{w})$: *domain intersection point from* $\{(v_j, w_j)\}$ *and* $\{(v_k, w_k)\}$.

Then the three offset points

$$\mathbf{f}^{\circ}(\hat{r}, \hat{s}), \quad \mathbf{g}^{\circ}(\hat{t}, \hat{u}), \quad \mathbf{h}^{\circ}(\hat{v}, \hat{w})$$

will be very close to each other if they belong to the same corner. This closeness measure is used in identifying the three domain intersection points belonging to the same corner, called *DIC (domain intersection corner) points*. DIC-points are shown in Fig. 11.12 as small circles.

By using the *tentative DIC-points* $(\hat{r}, \hat{s}), (\hat{t}, \hat{u}), (\hat{v}, \hat{w})$ as the initial values, we apply a Newton's iteration method in order to find "true" DIC-points. Let **m**,**n**,**o** denote *unit normal vectors* of the three *base surfaces*, then the *tangent planes* $\pi 1, \pi 2, \pi 3$ of the offset surfaces $\mathbf{f}^{\circ}(r,s), \mathbf{g}^{\circ}(t,u), \mathbf{h}^{\circ}(v,w)$, respectively, are given by (*with* $\mathbf{r} = (x,\ y,\ z)$):

$$\mathbf{m} \cdot \mathbf{r} = \mathbf{m} \cdot \mathbf{f}^{\circ} \ : \text{equation of } \pi 1,$$

$$\mathbf{n} \cdot \mathbf{r} = \mathbf{n} \cdot \mathbf{g}^{\circ} \ : \text{equation of } \pi 2,$$

$$\mathbf{o} \cdot \mathbf{r} = \mathbf{o} \cdot \mathbf{h}^{\circ} \ : \text{equation of } \pi 3.$$

The intersection of the three planes $\pi 1, \pi 2, \pi 3$ at the current *DIC-point* can be expressed as (Compare with Eqn. (11.29))

$$G = \frac{a(\mathbf{n} \times \mathbf{o}) + b(\mathbf{o} \times \mathbf{m}) + c(\mathbf{m} \times \mathbf{n})}{\mathbf{m} \cdot (\mathbf{n} \times \mathbf{o})} \tag{11.38}$$

$$\begin{aligned} \textit{where,} \quad & a = \mathbf{m} \cdot \mathbf{f}^{\circ}(\hat{r}, \hat{s}), \\ & b = \mathbf{n} \cdot \mathbf{g}^{\circ}(\hat{t}, \hat{u}), \\ & c = \mathbf{o} \cdot \mathbf{h}^{\circ}(\hat{v}, \hat{w}), \\ & \mathbf{m}, \mathbf{n}, \mathbf{o} : \ \textit{unit normal vectors at } \ \mathbf{f}(\hat{r}, \hat{s}), \mathbf{g}(\hat{t}, \hat{u}), \mathbf{h}(\hat{v}, \hat{w}). \end{aligned}$$

The Newton's iteration method for finding DIC points may be summarized as:

Procedure_SSSI *(surface/surface/surface intersection);*

0. Input: tentative DIC-points $(\hat{r}, \hat{s}), (\hat{t}, \hat{u}), (\hat{v}, \hat{w})$.

1. Generate a guess point G from (11.38).

2-a. Invert the Jacobians of $\mathbf{f}(r,s)$ *at G to revise the DIC-points* $(\hat{r}, \hat{s})$,

2-b. Invert the Jacobians of $\mathbf{g}(t,u)$ *at G to revise the DIC-points* $(\hat{t}, \hat{u})$,

2-c. Invert the Jacobians of $\mathbf{h}(r,s)$ *at G to revise the DIC-points* $(\hat{v}, \hat{w})$,

3. If the three offset points $\mathbf{f}^{\circ}(\hat{r}, \hat{s})$, $\mathbf{g}^{\circ}(\hat{t}, \hat{u})$, $\mathbf{h}^{\circ}(\hat{v}, \hat{w})$ *are close enough (ie, less than* ε*), then stop, else go to 1.*

11.6.2 Boundary Curves of a Corner Blend Patch

The "true" Dic-points $(\hat{r}, \hat{s}), (\hat{t}, \hat{u}), (\hat{v}, \hat{w})$ obtained from the *SSSI-procedure* corresponds to ball contact points at the corner. That is, *the ball contact points* at the corner are given by

$$\mathbf{v}_1 = \mathbf{h}(\hat{v}, \hat{w}); \quad \mathbf{v}_2 = \mathbf{f}(\hat{r}, \hat{s}); \quad \mathbf{v}_3 = \mathbf{g}(\hat{t}, \hat{u}). \tag{11.39 - a}$$

The above ball contact points will become the *corner points of the triangular corner blend.* Since the edge blends are terminated at the DIC (domain intersection corner) points, they have to be trimmed at the corner. Without a loss of generality, a DIC-point can be made the first domain value tuple of the corresponding OSIC (offset surface intersection curve)-list. Let us define the following:

$\mathbf{b}_1(\alpha, \beta)$: *edge blend between* $\mathbf{f}(r, s)$ *and* $\mathbf{g}(t, u)$;
$\mathbf{b}_2(\alpha, \beta)$: *edge blend between* $\mathbf{g}(t, u)$ *and* $\mathbf{h}(v, w)$;
$\mathbf{b}_3(\alpha, \beta)$: *edge blend between* $\mathbf{h}(v, w)$ *and* $\mathbf{f}(r, s)$;

Then, the $\beta = 0$ *boundary* of an *edge blend* would become a boundary curve of the *corner blend patch.* Namely, we can set

$$\mathbf{e}_i(\alpha) = \mathbf{b}_i(\alpha, 0). \qquad (11.39-b)$$

The *boundary data* (11.39) of a corner blend are shown in Fig. 11.13. For example, $\mathbf{e}_1(\alpha)$ is obtained from (11.23) as

$$\begin{aligned} \mathbf{e}_1(\alpha) &\equiv \mathbf{b}_1(\alpha, 0) \\ &= \frac{(1-\alpha)^2 C_0(0) + 2\omega_1(1-\alpha)\alpha C_1(0) + \alpha^2 C_2(0)}{(1-\alpha)^2 + 2\omega_1(1-\alpha)\alpha + \alpha^2} \quad for\ \alpha \leq\in [0,1] \end{aligned} \qquad (11.40)$$

$$where, \quad C_0(0) = \mathbf{f}(\hat{r}, \hat{s}),$$
$$C_2(0) = \mathbf{g}(\hat{t}, \hat{u}),$$
$$C_1(0) \ as\ given\ by\ (11.16).$$

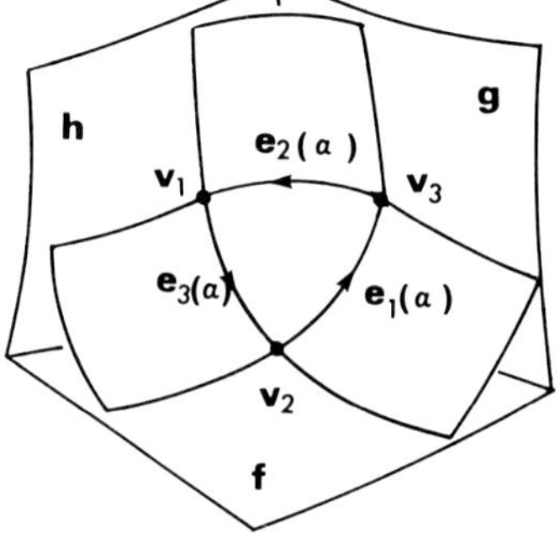

Figure 11.13 Boundary Curves of a Corner Blend

The *corss-boundary tangent* along $\mathbf{e}_1(\alpha)$, for example, is obtained by differentiating (11.23) with respect to β (at $\beta = 0$). The result is the same as (11.40) with the control vertices replaced by their derivatives:

$$\mathbf{d}_1(\alpha) = \frac{(1-\alpha)^2 D_0(0) + 2\omega_1(1-\alpha)\alpha D_1(0) + \alpha^2 D_2(0)}{(1-\alpha)^2 + 2\omega_1(1-\alpha)\alpha + \alpha^2} \tag{11.41}$$

where $D_i(0)$ are as given by (11.32) and (11.34). In both (11.40) and (11.41), the *weight* ω_1 is determined from the *fullness factor* ρ_1 of the edge blend $\mathbf{b}_1(\alpha, \beta)$. That is,

$$\omega_1 = \rho_1/(1-\rho_1).$$

The remaining boundary curves and *corss boundary tangents* $\{\mathbf{e}_i(\alpha), \mathbf{d}_i(\alpha) : i = 2, 3\}$ may be determined in the same way.

11.6.3 Construction of Corner Blend Patch

Having determined the boundary curves $\mathbf{e}_i(\alpha)$ and cross-boundary tangents $\mathbf{d}_i(\alpha)$, the corner blend patch can be expressed as a *convex combination* of linear Taylor interpolants as discussed in §5.3.4. That is, the **corner blend** may be expressed as follows:

$$\mathbf{c}(W) = \textstyle\sum_{i=1}^{3} \gamma_i(W)\{\mathbf{e}_i(\alpha_i) + \lambda_i\ \mu\ \nu_i\ \mathbf{d}_i(\alpha_i)\} \tag{11.42}$$

$$\begin{aligned}
\textit{where,}\quad \alpha_1 &= \lambda_3/(\lambda_3 + \lambda_2),\\
\alpha_2 &= \lambda_1/(\lambda_1 + \lambda_3),\\
\alpha_3 &= \lambda_2/(\lambda_2 + \lambda_1),\\
W &= (\lambda_1, \lambda_2, \lambda_3) : \quad \textit{barycentric coordinates},\\
\gamma_i(W) &= \begin{cases} \frac{(1/\lambda_i)^2}{\Sigma_j (1/\lambda_j)^2}, & \text{if } \lambda_i \neq 0,\\ 1, & \text{if } \lambda_i = 0. \end{cases}
\end{aligned}$$

This is a *triangular convex combination patch* of the type given in (10.22). *Correction factors* ν_i are introduced so that the influence of each cross boundary tangent be equally reflected. A reasonable choice for the **correction factors** is given by

$$\nu_i = |\mathbf{e}_i(0.5) - \mathbf{v}_i| / |\mathbf{d}_i(0.5)|\ ; \quad i = 1, 2, 3. \tag{11.43}$$

Recall that the indices i are ordered in an anti-clockwise sequence and that $\mathbf{v}_i$ is the vertex facing $\mathbf{e}_i(\alpha_i)$ as indicated in Fig. 11.13. The constant μ in (11.42) is called an *expansion factor* (which was not found in (10.22)).

The *expansion factor* μ may be used in controlling the fullness of the corner blend. A reasonable measure of the *fullness* is the *distance between the center point*, $\mathbf{c}(\frac{1}{3},\frac{1}{3},\frac{1}{3})$, *of the corner blend and the center point*, $\hat{\mathbf{f}}^\circ$, *of the "rolling" ball.* Thus, if a *spherical corner blend* is desired, the **expansion factor** is obtained from the following equation:

$$d = |(\bar{\mathbf{e}} + \frac{\mu}{3}\bar{\mathbf{d}}) - \hat{\mathbf{f}}^\circ| \tag{11.44}$$

where, $\bar{\mathbf{e}} = (\mathbf{e}_1(0.5) + \mathbf{e}_2(0.5) + \mathbf{e}_3(0.5))/3,$

$\bar{\mathbf{d}} = (\nu_1\mathbf{d}_1(0.5) + \nu_2\mathbf{d}_2(0.5) + \nu_3\mathbf{d}_3(0.5))/3,$

$d:$ *radius of the rolling ball (blending radius),*

$\hat{\mathbf{f}}^\circ = \mathbf{f}^\circ(\hat{r},\hat{s}))$: *center of the rolling ball at the corner.*

Now we introduce the concept of "fullness" in corner blends. Illustrated in Fig. 11.14 are corner blending surfaces (of the same corner) having different fullness values. Let π be a plane passing through the three ball-contact points $\hat{\mathbf{f}} = \mathbf{f}(\hat{r},\hat{s})$, $\hat{\mathbf{g}} = \mathbf{g}(\hat{t},\hat{u})$ and $\hat{\mathbf{h}} = \mathbf{h}(\hat{v},\hat{w})$, then normal vector $\mathbf{p}$ of the plane π is given by

$$\mathbf{p} = (\hat{\mathbf{g}} - \hat{\mathbf{f}}) \times (\hat{\mathbf{h}} - \hat{\mathbf{f}}). \tag{11.45}$$

Let $\pi1, \pi2, \pi3$ be the tangent planes to the base surfaces $\mathbf{f}(r,s)$, $\mathbf{g}(t,u)$, $\mathbf{h}(v,w)$, respectively, at the ball-contact points. Then the intersection G of $\pi1, \pi2$ and $\pi3$ is given by (See Eqn. (11.38))

$$G = \frac{a(\mathbf{n}\times\mathbf{o}) + b(\mathbf{o}\times\mathbf{m}) + c(\mathbf{m}\times\mathbf{n})}{\mathbf{m}\cdot(\mathbf{n}\times\mathbf{o})} \tag{11.46}$$

where $a = \mathbf{m}\cdot\hat{\mathbf{f}}$, $b = \mathbf{n}\cdot\hat{\mathbf{g}}$, $c = \mathbf{o}\cdot\hat{\mathbf{h}}$, and $\mathbf{m}$,$\mathbf{n}$,$\mathbf{o}$ are unit normal vectors at $\hat{\mathbf{f}}, \hat{\mathbf{g}}, \hat{\mathbf{h}}$. Then, the distance from the center point $\mathbf{c}(\frac{1}{3},\frac{1}{3},\frac{1}{3})$ to the plane π (the maximum being the distance from G to π) can be used as a measure of fullness.

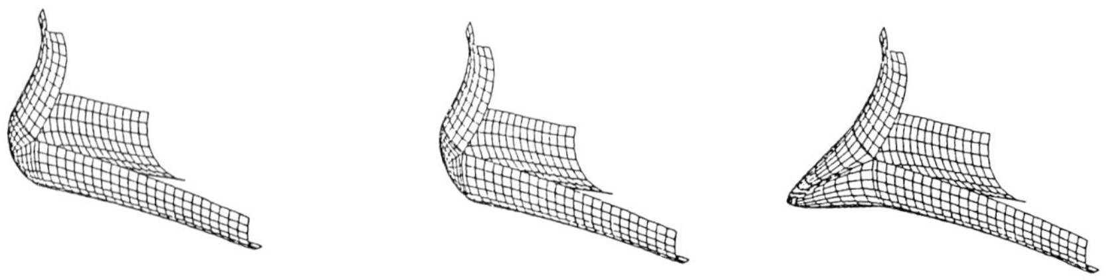

Figure 11.14 Fullness Control in Corner Blending

Thus, we may introduce a "fullness factor" ρ (similar to the one for conic sections) for our corner blend. For a given value of *fullness* ρ, the **expansion factor** μ in (11.42) can be determined by solving the following equation.

$$((\bar{\mathbf{e}} + \mu\bar{\mathbf{d}}/3) - \hat{\mathbf{f}}))\cdot\mathbf{p} = \rho(G - \hat{\mathbf{f}})\cdot\mathbf{p}, \tag{11.47}$$

where $\mathbf{p}$ is the normal vector of π given by (11.45), G is the three planes' intersection point given by (11.46), and ρ is the fullness factor ($0 \le \rho \le 1$). If a spherical corner blend patch is desired, μ is determined from (11.44), otherwise we use (11.47) to obtain a desired fullness (by interactively trying out different values of ρ).

11.7 GENERAL CORNER BLEND SURFACES

This section introduces more general types of corner blends than the one considered in the previous section. One type is the *convex combination*-type corner blend where three or more edge blends having different blend radii may meet. The other is called *edge*-type corner blend which is in fact an edge blend between a base surface and an existing edge blend. A more detailed discussion may be found in Ju and Choi (1990).

11.7.1 Convex Combination-Type Corner Blend

An example of a pentagonal corner blend is shown in Fig. 11.15. The procedure for constructing a general convex combination type corner blend is a straightforward extension of the procedure for ball-type triangular corner blends discussed in the previous section. Assume that n edge blends $\mathbf{b}_i(\alpha,\beta)$ meet at a corner. Let $\{\mathbf{f}_i(u^i,v^i) : i = 1,...,n\}$ denote the *base surfaces*, then their **offset surfaces** are defined as

$$\mathbf{f}_i^{\circ}(u,v;d_i) = \mathbf{f}_i(u,v) + d_i\mathbf{m}_i(u,v) \quad for\ u,v \in [0,1]$$

$$where,\quad \mathbf{m}_i : \ unit\ surface\ normal\ of\,\mathbf{f}_i,$$

$$d_i : \ offset\ distance.$$

In the following discussions, it is assumed that the *edge blends* have already been constructed and a set *DIC-point (domain intersection corner points)* is available. That is, we have

- $(\hat{u}^i,\hat{v}^i)$: *DIC-point on the domain of* $\mathbf{f}_i$ *for i=1,...,n,*
- $\mathbf{b}_i(\alpha,\beta)$: *edge blend between* $\mathbf{f}_i(u,v)$ *and* $\mathbf{f}_{i+1}(u,v)$ *for i=1,...,n.*

The index i is interpreted as "$(i-1)$ *mod n plus* 1".

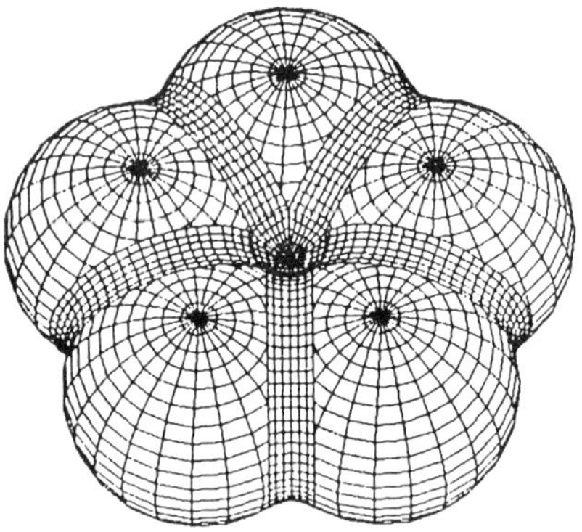

Figure 11.15 Example of a Pentagonal Corner Blend

The first step of constructing a corner blend is to define a smooth curve $\mathbf{b}_i(\alpha(\psi_i), \beta(\psi_i))$ on each edge blend to form a boundary curve $\mathbf{e}_i(\psi_i)$ of the corner blend. That is,

$$\mathbf{e}_i(\psi_i) = \mathbf{b}_i(\alpha(\psi_i), \beta(\psi_i)) : \quad 0 \leq \psi_i \leq 1 \quad for\ i = 1, ..., n \tag{11.48}$$

Compare the boundary curve equation in (11.48) with that in (11.39) of §11.6.2 where the boundary curve was an isoparametric curve at $\beta = 0$ so that $\alpha = \psi$. *Cross boundary tangents along* $\mathbf{e}_i(\psi_i)$ are given by (Compare it with Eqn. (11.41))

$$\mathbf{d}_i(\psi_i) = \frac{(1-\alpha_i)^2 D_0(\beta_i) + 2\omega_i(1-\alpha_i)\alpha_i D_1(\beta_i) + (\alpha_i)^2 D_2(\beta_i)}{(1-\alpha_i)^2 + 2\omega_i(1-\alpha_i)\alpha_i + (\alpha_i)^2}, \tag{11.49}$$

$$for\ \psi_i \in [0,1];\ i = 1, ..., n$$

$$where, \quad \alpha_i \equiv \alpha(\psi_i),$$

$$\beta_i \equiv \beta(\psi_i),$$

$$\omega_i = \rho_i/(1-\rho_i) : \ weights,$$

$$\rho_i : \ fullness\ factor,$$

$$D_j(\beta) : \ derivatives\ of\ C_j(\beta)\ given\ by\ (11.32)\ and\ (11.34).$$

The **corner blend patch** is then constructed by taking a convex combination of the *linear Taylor interpolants* formed by $\mathbf{e}_i(\psi_i)$ and $\mathbf{d}_i(\psi_i)$. That is, we have the following convex combination form:

$$\mathbf{c}(W) = \sum_{i=1}^{n} \gamma_i(W)\{\mathbf{e}_i(\psi_i) + \lambda_i \ \mu \ \nu_i \ \mathbf{d}_i(\psi_i)\} \tag{11.50}$$

where, $\psi_i = \lambda_{i-1}/(\lambda_{i-1} + \lambda_{i+1})$ *for* $i = 1, ..., n,$

μ : *expansion factor,*

ν_i *are correction factors,*

$W = (\lambda_1, \lambda_2, ..., \lambda_n),$

$\gamma_i(W)$: *convex combination weights as given in* (11.42).

The index i is interpreted as *"(i-1) mod n plus 1"*.

As shown in Fig. 11.16, the domain of the corner blend in (11.50) is a *regular polygon.* The coordinate variable λ_i is defined as the distance from a domain point to a side of the regular polygon. When $n = 3$, λ_i's form *barycentric coordinates*, that is,

$$\Sigma\lambda_i = 1.$$

When $n = 4$, the coordinate variables have the following dependency relations:

$$\lambda_1 + \lambda_3 = 1 \quad and$$
$$\lambda_2 + \lambda_4 = 1.$$

The dependency relations for $n \geq 5$ are left to the reader as an exercise.

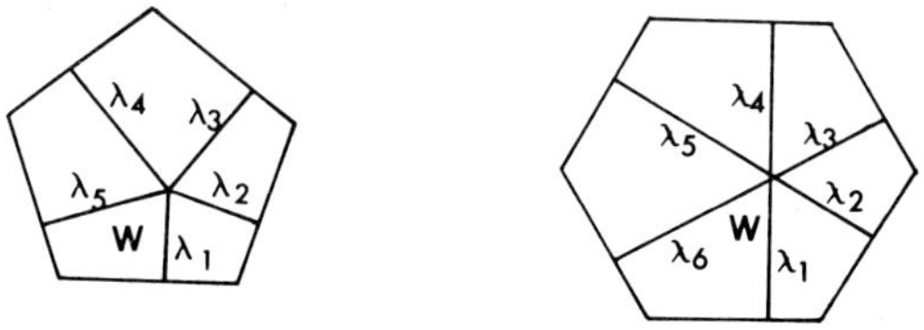

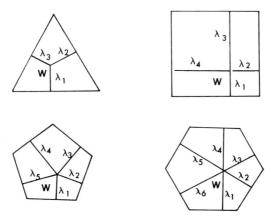

Figure 11.16 Domains of n-sided Corner Blends

In order to evenly reflect the influence of each cross boundary tangent, the magnitudes of the cross boundary tangents $\mathbf{d}_i(\alpha_i)$ should be corrected by the amount ν_i. A reasonable choice for the **correction factors** may be given by

$$\nu_i = |\mathbf{e}_i(0.5) - \bar{\mathbf{e}}|/|\mathbf{d}_i(0.5)| \; ; \quad i = 1, ..., n \tag{11.51}$$

where, $\bar{\mathbf{e}}_i = \Sigma \mathbf{e}_i(0.5)/n$: *centroid of the corner blend.*

The role of the *expansion factor* μ in (11.50) is the same as that for the *ball-type triangular corner blend case* (11.42). In other words, it is related to the "fullness" ρ of the corner blend (See Eqn. (11.47)). Examples of pentagonal corner blends with different fullness factors are shown in Fig. 11.17 (Figure 11.17(a) is the "spherical" corner blend). More details may be found in Ju and Choi (1990). To obtain a "near" spherical corner blend, the *expansion factor* μ may be determined from the following.

$$\bar{d} = |(\bar{\mathbf{e}} + \bar{\lambda}\mu\bar{\mathbf{d}}) - \bar{\mathbf{f}}^\circ| \tag{11.52}$$

where, $\bar{\lambda} = \Sigma\lambda_i/n$,

$\bar{\mathbf{d}} = \Sigma\nu_i\mathbf{d}_i(0.5)/n$,

$\bar{\mathbf{f}}^\circ = \Sigma(\mathbf{f}_i^\circ(\hat{u}^i, \hat{v}^i; \bar{d})/n$: *average offset point, and*

$\bar{d}$: *average edge blend radius.*

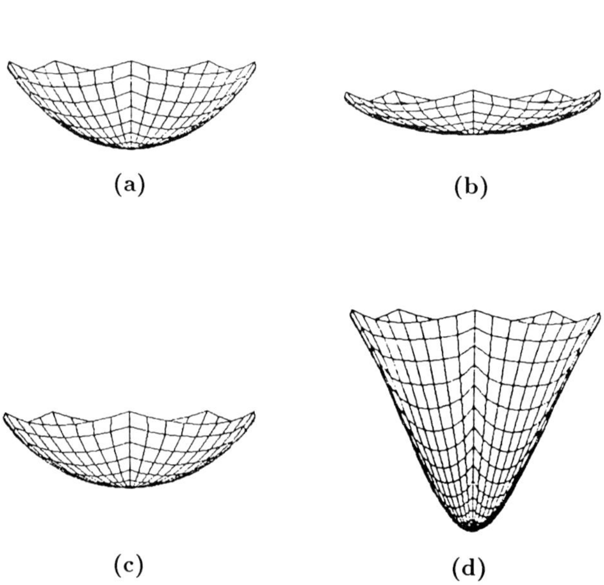

Figure 11.17 Fullness Control in Generalized Corner Blending

Eqn. (11.52) would become (11.44) if $n = 3$ and all the edge blend radii are identical. On solving (11.52), the **expansion factor** for a *spherical corner blend* is given by

$$\mu = \{-(\mathbf{c}\cdot\bar{\mathbf{d}}) + [(\mathbf{c}\cdot\bar{\mathbf{d}})^2 - (\bar{\mathbf{d}}\cdot\bar{\mathbf{d}})(\mathbf{c}\cdot\mathbf{c} - \bar{d}^2)]^{1/2}\}/\bar{\lambda}(\bar{\mathbf{d}}\cdot\bar{\mathbf{d}}) \tag{11.53}$$

where, $\bar{\lambda} = \Sigma\lambda_i/n$,

$\bar{\mathbf{d}} = \Sigma\nu_i\mathbf{d}_i(0.5)/n$,

$\mathbf{c} = \bar{\mathbf{f}}^\circ = \Sigma(\mathbf{f}_i^\circ(\hat{u}^i, \hat{v}^i; \bar{d})/n$: *average offset point*,

$\bar{d}$: *average edge blend radius.*

11.7.2 Edge-Type Corner Blend

Shown in Fig. 11.18 are three *base surfaces* $\mathbf{f}(r,s)$, $\mathbf{g}(t,u)$, and $\mathbf{h}(v,w)$ together with three edge blends $\mathbf{b}_i(\alpha,\beta)$ for $i = 1,2,3$. An *edge-type* corner blend is constructed as an edge blend between the base surface $\mathbf{h}(v,w)$ and the edge blend $\mathbf{b}_1(\alpha,\beta)$. An edge-type corner blend is more "natural" than a convex combination type when a *round* and two *fillets* meet at the corner as in Fig. 11.18 or when there are large differences among the blending radii.

Since the edge-type corner blend is in fact an edge blend, its construction procedure is the same as the one introduced in §11.3 or the one in §11.4. For the corner blend of Fig. 11.18, the edge blend $\mathbf{b}_1(\alpha,\beta)$ acts as the first base surface and the surface $\mathbf{h}(v,w)$ becomes the second base surface. An example of edge-type corner blend is shown in Fig. 11.19.

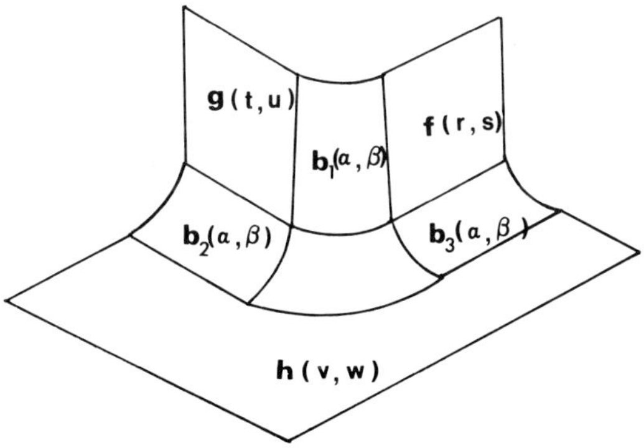

Figure 11.18 Construction of an Edge-type Corner Blend

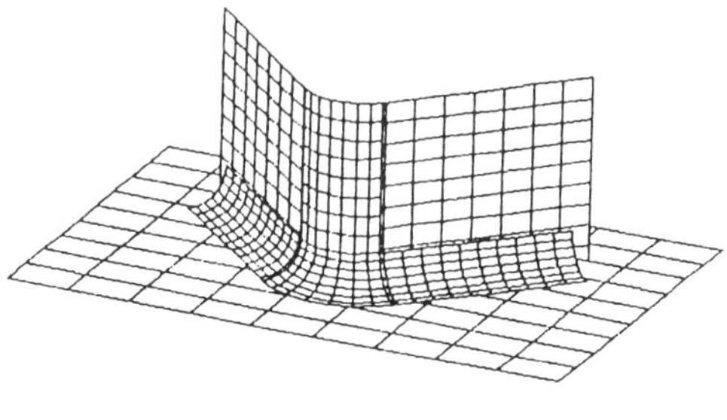

Figure 11.19 An Example of Variable-radius Edge-type Corner Blend

11.8 ROUNDING MODIFICATION OF POLYHEDRON

Presented in this section is a *rounding modification method* proposed by Chiyokura and Kimura (1983). This method has originally been proposed as a *free-form shape design technique*. *Rounding modifications* are carried out in two steps: The *edges* of a polyhedron are modified first to form a *cubic curve-net;* and then the resulting cubic curve-net is interpolated by using the so called *(rectangular) Gregory patch model* introduced in §10.5.

11.8.1 Bezier Approximation of 2D Rounding

A straight line segment joining two *3D points* P_0, P_1 can be represented as a cubic Bezier curve $\mathbf{r}(u)$. Let

$$\mathbf{r}(0) = P_0 \quad \textit{and} \quad \mathbf{r}(1) = P_1,$$

then the *control vertices* of the cubic Bezier curve are given by

$$V_0 = P_0; \quad V_1 = (2P_0 + P_1)/3; \quad V_2 = (P_0 + 2P_1)/3; \quad V_3 = P_1. \tag{11.54}$$

Shown in Fig. 11.20a is a pair of *"base" line segments*, P_0P_1 and P_1P_2, to be *"rounded"*. Let V_0 be the start point of the *round* (of radius d) and V_3 the end point. Then the *half corner angle* ϕ formed by the two line segments can be obtained from

$$\phi = \frac{1}{2} cos^{-1}(\mathbf{a} \cdot \mathbf{b}) \quad 0 \leq \phi \leq 90° \tag{11.55}$$

$$\begin{aligned} where, \quad \mathbf{a} &= (P_0 - P_1)/|P_0 - P_1|, \\ \mathbf{b} &= (P_2 - P_1)/|P_2 - P_1|. \end{aligned}$$

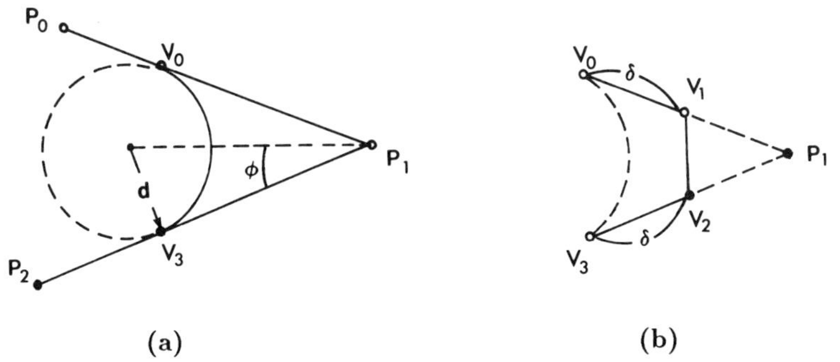

Figure 11.20 Bezier Approximation of 2D Rounding

Further, as shown in the figure, the distance "e" from the *corner point* P_1 to a *"rounding start point"* V_0 or V_3 is given by

$$e = d/tan\phi,$$

where "d" is the rounding radius. Thus, the *rounding start points* are expressed as

$$V_0 = P_1 + e\mathbf{a} \quad and \quad V_3 = P_1 + e\mathbf{b}.$$

From the result (4.53) in §4.4.4, the distance δ between an *end Bezier point* to a *middle Bezier point* is obtained as (See Fig. 11.20b)

$$\delta = \frac{2}{3}c/(1 + sin\phi),$$

where $c = |V_3 - V_0|$. The *middle Bezier points* of the cubic Bezier curve approximating the circular arc are expressed as

$$V_1 = V_0 - \delta\mathbf{a} \quad and \quad V_2 = V_3 - \delta\mathbf{b}.$$

Summarizing the results above, the *control vertices of the cubic Bezier curve* representing the *rounding* between the two line segments, P_0P_1 and P_1P_2, shown in Fig. 11.20 are given by

$$V_0 = P_1 + (d/\,tan\,\phi)\mathbf{a} \qquad (11.56-a)$$

$$V_3 = P_1 + (d/\,tan\,\phi)\mathbf{b} \qquad (11.56-b)$$

$$V_1 = V_0 - \{\frac{2}{3}c/(1 + sin\,\phi)\}\mathbf{a} \qquad (11.56-c)$$

$$V_2 = V_3 - \{\frac{2}{3}c/(1 + sin\,\phi)\}\mathbf{b} \qquad (11.56-d)$$

$$\begin{aligned} where,\quad & d: \ rounding\ radius, \\ & \phi: \ half\ corner\ angle\ given\ by\ (11.55), \\ & \mathbf{a} = (P_0 - P_1)/|P_0 - P_1|, \\ & \mathbf{b} = (P_2 - P_1)/|P_2 - P_1|, \\ & c = |V_3 - V_0|. \end{aligned}$$

It should be noted that it is not always possible to construct a *round* between the two line segments P_0P_1, P_1P_2. A necessary condition for a valid rounding is

$$|P_1 - P_0| \geq (d/tan\phi) \quad and \quad |P_1 - P_2| \geq (d/tan\phi) \tag{11.57}$$

$$\begin{aligned} where,\quad & d: \ rounding\ radius, \\ & \phi: \ half\ corner\ angle\ given\ by\ (11.55). \end{aligned}$$

11.8.2 Construction of Cubic Curve-net

Shown in Fig. 11.21a are two *planar faces* (of a polyhedron) sharing the *common edge* P_1Q_1. We want to construct an *edge blend* of *blending radius "d"* between the two *base surfaces (ie, planes)*. The *"ball contact curves"* are indicated as dashed lines in Fig. 11.21a. By using the results of (11.56), the four Bezier points $V_{00}, V_{01}, V_{02}, V_{03}$ are determined from the two *edges* P_0P_1, P_1P_2. Similarly, the Bezier points $V_{30}, V_{31}, V_{32}, V_{33}$ are defined on the *other edges* Q_0Q_1, Q_1Q_2. See Fig. 11.21b. Finally, the remaining Bezier points are determined by using (11.54).

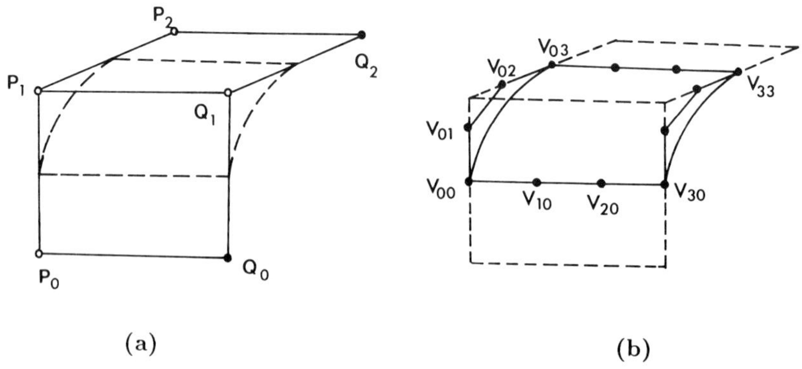

Figure 11.21 Construction of an Edge Blend between two Planes

11.8.3 Construction of Blending Surfaces

The result of a *rounding modification* of *edges* is a *cubic curve-net* as depicted in Fig. 11.21b. In the figure, the *edge blend* is a rectangular patch whose *boundary Bezier points* have already been fixed. Thus, we can construct a "*Gregory patch*" (See Fig. 10.12) by applying the *curve-net interpolation scheme* introduced in §10.5.2.

Another rounding modification example is given in Fig. 11.22. Shown in Fig. 11.22a is a six-sided polyhedron. The three edges marked with "R" are to be rounded with a *constant rounding radius "d"*. The first step is to construct *imaginary ball contact curves* as depicted in Fig. 11.22b (the dashed lines in the figure). And then, *Bezier points* are generated along the *ball contact lines* as depicted in Fig. 11.22c. In the figure, there are three *edge blends* and a *corner blend* which is a triangular patch. Finally, the *cubic curve-net* is converted to a *smooth composite surface* by applying the interpolation schemes introduced in §10.5.

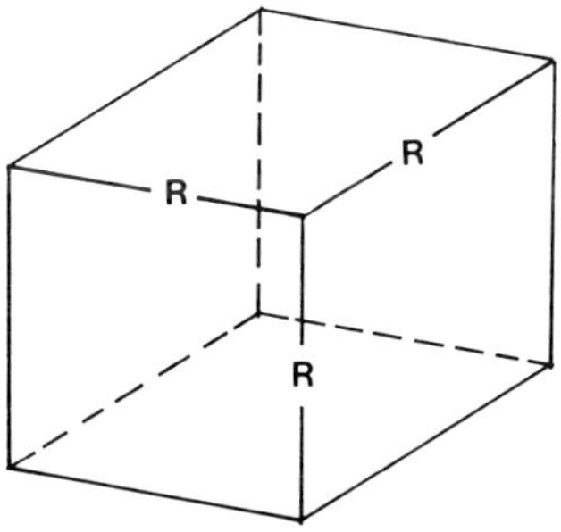

(a)

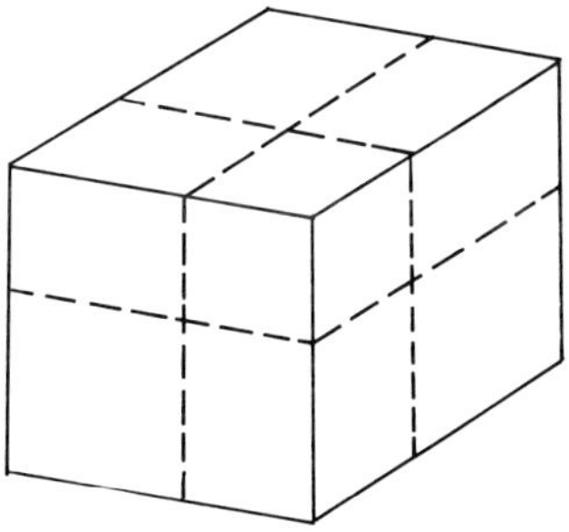

(b)

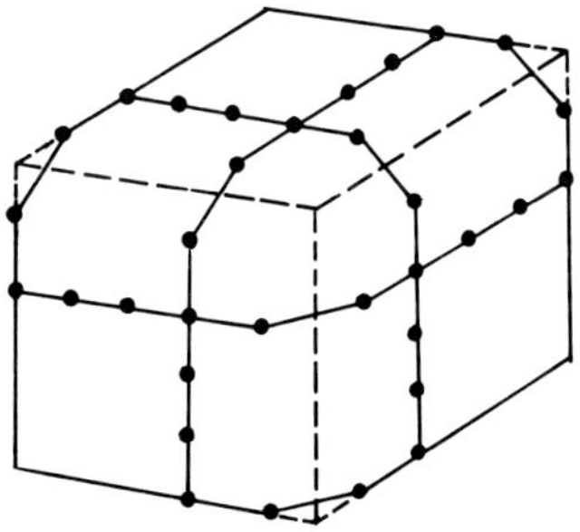

(c)

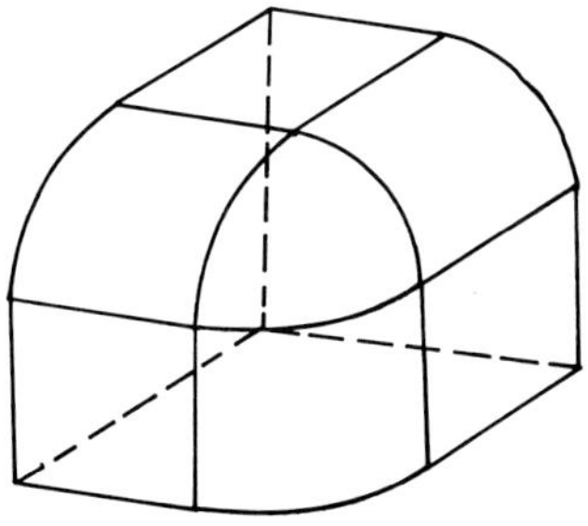

(d)

Figure 11.22 Rounding Modification of a Polyhedron

11.9 DISCUSSIONS

In this chapter, we have introduced schemes for constructing *blending surfaces* of parametric *base surfaces*. The main difficulty with the proposed blending schemes is that the resulting blending surfaces are expressed in non-standard forms. Further research is required to represent (approximately) the blending surfaces in a standard form such as NURB. The subject of implicit surfaces blending has been extensively studied, for example, by Middleditch and Sears (1985), Rockwood and Owen (1987), Hoffmann and Hopcroft (1987), and Holmstrom (1987). The general idea in all the implicit surface blending methods is the use of the *offset surfaces intersection method* in which a blend is defined as an intersection of two offset surfaces of the form $G_1(x, y, z) = s$ and $G_2(x, y, z) = t$, where $G_1 = 0$, $G_2 = 0$ are the base surfaces and s, t are offset parameters. By introducing a functional relation $f(s, t) = 0$ between s and t, a smooth blending surface is obtained.

CHAPTER 12

SURFACE INTERSECTION AND TRIMMED SURFACE

12.1 INTRODUCTION

This chapter presents methods of *SSI (surface/surface intersection)* and the concept of *trimmed surface.* Referring to the "anatomy of surface modeling" (see Fig. 1.9) of Chapter 1, the operations "intersection", "blending", and "trimming" are all related to *SSI.* More specifically, *SSI* is essential to the following application areas, to name a few:

- *construction of blending surfaces (see Chapter 11),*
- *boundary evaluation in solid modeling,*
- *trimmed surface construction in surface (and solid) modeling,*
- *NC cutter path generation,*
- *surface (and solid) evaluation (eg, sectioning and rendering), and*
- *contouring.*

The term **trimming** is used to mean a method of defining sub-areas of a parametric surface. When a parametric surface $\mathbf{r}(u, v)$ is divided into several subareas, each region of the surface can be represented as a *trimmed surface* which is defined as a 2D closed edge list on the u, v-domain of the parametric surface. The *trimmed surface concept* is gaining an increasing popularity, especially, in the "unified solid modeling" community. The term *unified solid model* is used to mean a solid model supporting *sculptured surfaces* (See Chapter 15).

Since a surface may be defined implicitly or parametrically, three cases of SSI arise. They are

a) implicit/implicit,
b) implicit/parametric, and
c) parametric/parametric.

Implicit representation is very convenient in determining whether a given point $(x,\ y,\ z)$ lies on a surface, while *parametric representation* is useful in evaluating a point on the surface corresponding to given values of the independent variables, which is a very important requirement in *CAD/CAM applications*, for example, in *surface rendering* and in NC *machining* of the surface. Since our main interest lies in the parametric surface intersection problem, the implicit/implicit case will not be discussed here. An important case of implicit/parametric SSI is the problem of intersecting a parametric surface with a plane. For the parametric/parametric case, Bezier surface intersectioning will be described in more detail.

12.1.1 Nature of Surface/surface Intersection Problem

The SSI problems to be considered in this chapter are the *implicit/parametric case* and the *parametric/parametric case.* Let

$h(x,y,z)=0$ be an implicit surface, and

$\mathbf{r}(u,v)=(x(u,v),\ y(u,v),\ z(u,v))$ be a parametric surface.

Then, the **implicit/parametric intersection** problem becomes the one solving the following equation:

$$h(x(u,v),\ y(u,v),\ z(u,v)) = 0 \tag{12.1}$$

which is a scalar function of two unknowns u, v. The resulting intersection curve is usually represented as a 2D parametric curve

$$\mathbf{u}(\alpha) = (u(\alpha),\ v(\alpha))$$

in the domain of the parametric surface $\mathbf{r}(u,v)$. Cartesian coordinates of the intersection curve is then given by

$$\mathbf{r}(\alpha) = \mathbf{r}(u(\alpha),\ v(\alpha)),$$

which is a *curve on the parametric surface* $\mathbf{r}(u,v)$. The **parametric/parametric intersection** problem is formulated as

$$\mathbf{f}(r,s) - \mathbf{g}(t,u) = \mathbf{0} \tag{12.2}$$

which in fact represents three scalar equations in the four unknowns.

In both (12.1) and (12.2), an intersection curve is expressed as a *non-linear equation system having one more unknowns than the scalar equations.* Since it is in general not possible to analytically solve an under determined system of nonlinear equations, some form of numerical methods is always required in order to find an intersection curve. Before attempting to applying a numerical method to (12.2), it may be beneficial to try to convert it into the implicit/parametric case (12.1). In theory (Sederberg *et al*, 1984), a parametric polynomial surface can be converted an algebraic (ie, implicit polynomial) surface.

12.1.2 Approaches to Surface/surface Intersection

As discussed in Pratt and Geisow (1986), there are three techniques that can be applied to the SSI problem (12.1) or (12.2). Namely, *a) lattice evaluation method, b) tracing method, and c) subdivision method.*

1) Lattice Evaluation Method:

In **lattice evaluation**, a grid or lattice is defined on the parametric surface to be intersected. That is, the surface is approximated as a polyhedron, as a mesh of line segments, or as a grid of discrete points, in order to obtain a number of sub-problems of lower complexity. The solutions to these sub-problems are then placed together to give a solution to the overall problem.

2) Tracing Method:

Tracing *(or marching)* methods are used when a sequence of intersection points are to be generated by "stepping" along the direction of the intersection curve. The tracing idea is essential in numerically solving a simultaneous equation system. The *tracing direction* is guided either by minimizing some function (*called a penalty function*) or by local differential geometry of the intersecting surfaces.

3) Subdivision Method:

The last technique, **subdivision technique**, is based on the "principle of divide-and-conquer". Namely, the original surface is recursively subdivided until the subdivided patches become simple enough to be intersected. This technique is most appropriate with parametric surfaces which are easily subdivided, such as Bezier and B-spline surfaces.

4) Combined Approach:

In practice, any one or a combination of the three techniques may be employed in an SSI algorithm to suit a specific application. Since SSI algorithms are used in engineering environments, they should be

a) accurate in the usual numerical sense,
b) robust in the sense that they work under any practical circumstances, and
c) fast enough to be used in commercial CAD/CAM systems.

Many SSI methods have been proposed in the literature, but none of them seems to possess the ideal combination of accuracy, robustness, and speed for all types of applications.

In the next section, we consider the problem of intersecting a parametric surface with a plane. The *plane/surface intersection* problem is a special case of the *implicit/parametric surface intersection.* Plane/surface intersection is an essential operation in many CAD/CAM applications such as *NC cutter path generation, surface rendering, and contouring.*

Methods of finding intersections between two parametric surfaces are introduced in §12.3. An SSI algorithm for general *rectangular parametric surfaces* has already been introduced in the previous chapter (See §11.2). Special techniques for intersecting Bezier surfaces and B-spline surfaces will be introduced in §12.3. Various issues in constructing trimmed surfaces are discussed in §12.4.

12.2 SURFACE/PLANE INTERSECTION

The surface/plane intersection problem arises in generating cutter paths, generating sectional views, and contouring. This section presents practical **SPI (surface/plane intersection)** algorithms. The problem is to solve (12.1) with the implicit surface $h(x, y, z)$ given by

$$ax + by + cz - d = 0. \tag{12.3}$$

An SPI algorithm is nothing but a numerical procedure for solving (12.1) with the implicit function $h(x, y, z)$ given by (12.3).

12.2.1 Intersection between a Curve and a Plane

Before presenting SPI algorithms, the problem of finding intersection points between a parametric curve $\mathbf{r}(t)$ and an implicit plane π, as depicted in Fig. 12.1a, is considered first. The *implicit equation of the plane* given by (12.3) may be rewritten as

$$\mathbf{r} \cdot \mathbf{p} = d \quad \textit{with} \quad |\mathbf{p}| = 1 \tag{12.4}$$

$$\textit{where,} \quad \mathbf{r} = (x, y, z) : \textit{ coordinate variables,}$$

$$\mathbf{p} = (a, b, c) : \textit{ unit normal vector of the plane } \pi.$$

Note in (12.4) that we require $a^2 + b^2 + c^2 = 1$. Then from (12.1), the intersection between a plane and a (parametric) curve is given by

$$\begin{aligned} 0 &= \mathbf{r}(t) \cdot \mathbf{p} - d \\ &= a\ x(t) + b\ y(t) + c\ z(t) - d. \end{aligned} \tag{12.5}$$

When $\mathbf{r}(t)$ is a polynomial, its roots are easily obtained. See, for example, Farouki (1987) and Maron (1982, pp 76-85).

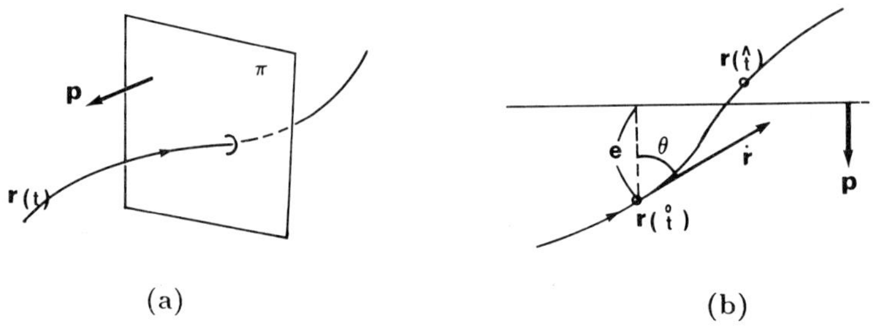

Figure 12.1 Curve/Plane Intersection

For a general parametric curve $\mathbf{r}(t)$, an intersection point if it exists can be obtained by using a Newton-Raphson iteration scheme starting from an *initial guess point* $\mathbf{r}(\overset{\circ}{t})$. As depicted in Fig. 12.1b, one may construct the following linear relation between the *initial guess point* $\overset{\circ}{t}$ and an *estimated intersection point* $\hat{t}$:

$$|\dot{\mathbf{r}}| \cdot (\hat{t} - \overset{\circ}{t}) \cong e / \cos\theta$$

$$where, \quad e = (\mathbf{p} \cdot \mathbf{r}(\overset{\circ}{t}) - d),$$

$$\cos\theta = (\dot{\mathbf{r}} \cdot \mathbf{p}) / |\dot{\mathbf{r}}|.$$

From the above relation, the *estimated intersection point* $\hat{t}$ is obtained as

$$\hat{t} = \overset{\circ}{t} - \{(\mathbf{p} \cdot \mathbf{r}(\overset{\circ}{t}) - d)\} / (\dot{\mathbf{r}} \cdot \mathbf{p}) \tag{12.6}$$

$$where, \quad \dot{\mathbf{r}} = d\mathbf{r}(t)/dt\big|_{t=\overset{\circ}{t}},$$

$$\mathbf{p}, d: \quad coefficients\ of\ plane\ equation\ (12.4).$$

Thus, the iteration scheme for the **curve/plane intersection** can be summarized as follows:

Procedure_CPI *(curve/plane intersection):*

0. *Input:* $\overset{\circ}{t}$ *(initial point),* $\mathbf{p}$ *(unit normal of* π*),* d *(distance of* π*).*
1. *Determine an estimated intersection point* $\hat{t}$ *from* (12.6).
2. *If* $|\mathbf{r}(\hat{t}) \cdot \mathbf{p} - d| \leq \varepsilon$ *then return* $\hat{t}$,

   else $\overset{\circ}{t} = \hat{t}$ *and go to 1.*

One way of obtaining the *initial guess point* $\overset{\circ}{t}$ in the above procedure may be the use of the *lattice evaluation* method. That is, for a sequence of sampled points $\{t_i\}$, if

$$[\mathbf{p} \cdot \mathbf{r}(t_i) - d] \cdot [\mathbf{p} \cdot \mathbf{r}(t_{i+1}) - d] < 0 \tag{12.7}$$

holds, either, t_i or t_{i+1} becomes the *initial guess point.* If we wish to compute all the intersection points of the curve $\mathbf{r}(t)$ with the plane $\mathbf{r} \cdot \mathbf{p} = d$, we have to find all *turning points* t'_j by solving

$$\mathbf{p} \cdot (d\mathbf{r}(t)/dt) = 0 \tag{12.8}$$

and then identify *monotonic segments* of $\mathbf{r}(t)$. For example, let $t \in [t'_1, t'_2]$ be a *monotonic interval,* then there exists an intersection in the interval if

$$[\mathbf{p} \cdot \mathbf{r}(t'_1) - d] \cdot [\mathbf{p} \cdot \mathbf{r}(t'_2) - d] < 0$$

holds. One may employ a *binary search* in the *monotonic interval* to find an *intersection point* $\hat{t}$.

12.2.2 Intersection between a Surface and a Plane

The intersection curve (12.4) between a parametric surface $\mathbf{r}(u, v)$ and the plane π given by (12.3) may be rewritten as

$$\begin{aligned} 0 &= \mathbf{r}(u,v) \cdot \mathbf{p} - d \\ &= a\ x(u,v) + b\ y(u,v) + c\ z(u,v) - d \end{aligned} \tag{12.9}$$

where a, b, c, d are coefficients of the implicit plane equation (12.3).

In solving the *SPI (surface/plane intersection)* problem (12.9), any one of the "intersection techniques" (ie, lattice evaluation, tracing, or subdivision) may be employed. *SPI methods* based on the *tracing technique* will be explained. Two cases may arise in intersecting the *parametric surface* $\mathbf{r}(u, v)$ with the *plane* π. The resulting intersection curve will be either an *open curve* or a *closed curve*. In either case, the overall procedure for solving (12.9) consists of three steps:

a) detection of initial intersection points,
b) marching, and
c) refinement.

1) Open Intersection Curve Case:

To simplify the discussion, we assume that the SPI always results in open curves meaning that the plane must intersect with surface patch boundaries should there be an intersection between the plane and the surface. In this case, *initial intersection points* are obtained by finding intersection points between patch boundary curves and the plane. For this, the *CPI (curve/plane intersection)* procedure of §12.2.1 may be employed.

Having determined initial intersection points, the next step is to define a *marching vector*. As depicted in Fig. 12.2a, the (unit) **marching vector c** is taken to be a vector perpendicular

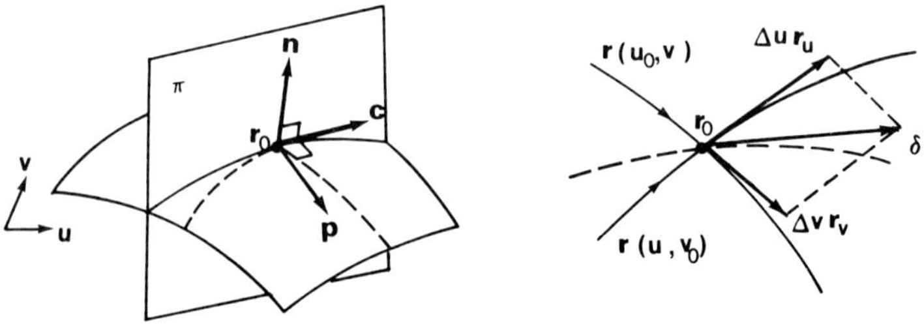

(a) (b)

Figure 12.2 Surface/Plane Intersection (open intersection case) to both the surface normal and plane normal. Namely,

$$\mathbf{c} = \pm(\mathbf{n} \times \mathbf{p})/|\mathbf{n} \times \mathbf{p}| \tag{12.10}$$

where $\mathbf{n} = (\mathbf{r}_u \times \mathbf{r}_v)/|\mathbf{r}_u \times \mathbf{r}_v|$ is the unit surface normal, and $\mathbf{p}$ is the unit normal of the plane π.

The direction of marching (ie, the sign in Eqn. (12.10)) should be carefully selected. Note that the marching vector $\mathbf{c}$ corresponds to the unit tangent vector of the intersection curve.

The next step is to determine a *guess point* for the next intersection along the intersection curve. The increments $\Delta u, \Delta v$ of the domain values are determined by solving the following linear equations (See Fig. 12.2b):

$$\Delta u(\mathbf{p} \cdot \mathbf{r}_u) + \Delta v(\mathbf{p} \cdot \mathbf{r}_v) = 0 \tag{12.11 - a}$$

$$\Delta u(\mathbf{c} \cdot \mathbf{r}_u) + \Delta v(\mathbf{c} \cdot \mathbf{r}_v) = \delta \tag{12.11 - b}$$

where, δ : *3D stepping distance,*

$\mathbf{c}$: *marching vector* (12.10),

$\mathbf{r}_u, \mathbf{r}_v$: *partials of the surface* $\mathbf{r}(u, v)$.

Let (u_0, v_0) denote the current intersection point, then the guess point for the next intersection point is given by

$$\hat{u} = u_0 + \Delta u \qquad \hat{v} = v_0 + \Delta v.$$

Finally the estimated intersection point $(\hat{u}, \hat{v})$ is *refined* to obtain a "true" intersection point (u^*, v^*). For this, one of the isoparametric curves which is more parallel to $\mathbf{p}$ is selected, and then the *Procedure_CPI* of §12.2.1 is applied. The overall procedure for finding an *(open) intersection curve between the plane* π *and the surface* $\mathbf{r}(u, v)$ may be summarized as follows (The initial intersection point is obtained by performing CPI with surface boundary curves):

Procedure_SPI *(surface/plane intersection):*

0. *Input:* $\pi(\mathbf{r} \cdot \mathbf{p} = d)$, $\mathbf{r}(u, v)$, *initial intersection point* (u_0, v_0).
1. *Define marching vector* $\mathbf{c}$ *using* (12.10).
2. *Compute* $\Delta u, \Delta v$ *from* (12.11).
3. *Determine a guess point:* $\hat{u} = u_0 + \Delta u$; $\hat{v} = v_0 + \Delta v$.
4. *If* $[(\mathbf{r}_u \cdot \mathbf{p})/|\mathbf{r}_u| \geq (\mathbf{r}_v \cdot \mathbf{p})/|\mathbf{r}_v|]$ *then*
   perform CPI between $\mathbf{r}(t) \equiv \mathbf{r}(t, \hat{v})$ *and* π *to obtain* $u^* = t^*$, $v^* = \hat{v}$.
   else perform CPI between $\mathbf{r}(t) \equiv \mathbf{r}(\hat{u}, t)$ *and* π *to obtain* $u^* = \hat{u}$, $v^* = t^*$.
5. *If the intersection curve reaches to a surface boundary then stop,*
   else set $u_0 = u^*$, $v_0 = v^*$ *and go to step 1.*

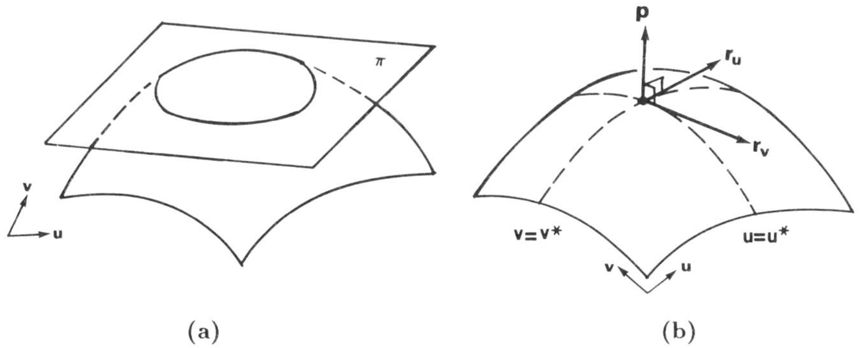

Figure 12.3 Surface/Plane Intersection *(closed intersection case)*

2) Closed Intersection Curve Case:

There may be a case where the plane π intersects the surface $\mathbf{r}(u, v)$ without intersecting surface boundaries as shown in Fig. 12.3a. In this case, it is required to find *extreme points* of $\mathbf{r}(u, v)$ with respect to the plane π. The problem becomes the one of finding a point (u^*, v^*) on the surface where the surface normal is equal to the normal $\mathbf{p}$ of the plane. Thus, we need to solve the following simultaneous equation (See Figure 12.3-b):

$$\mathbf{p} \cdot \mathbf{r}_u = 0 \quad and \quad \mathbf{p} \cdot \mathbf{r}_v = 0 \tag{12.12}$$

where, $\mathbf{p}$: *unit normal of plane*π,

$\mathbf{r}_u = \mathbf{r}_u(u^*, v^*)$: *partial of* $\mathbf{r}(u, v)$ *at an extreme point,*

$\mathbf{r}_v = \mathbf{r}_v(u^*, v^*)$: *partial of* $\mathbf{r}(u, v)$ *at an extreme point.*

Taylor expansions of the partials are given by

$$\mathbf{r}_u(\overset{\circ}{u} + \Delta u, \overset{\circ}{v} + \Delta v) \cong \mathbf{r}_u(\overset{\circ}{u}, \overset{\circ}{v}) + \Delta u \mathbf{r}_{uu}(\overset{\circ}{u}, \overset{\circ}{v}) + \Delta v \mathbf{r}_{uv}(\overset{\circ}{u}, \overset{\circ}{v}), \quad and$$

$$\mathbf{r}_v(\overset{\circ}{u} + \Delta u, \overset{\circ}{v} + \Delta v) \cong \mathbf{r}_v(\overset{\circ}{u}, \overset{\circ}{v}) + \Delta u \mathbf{r}_{vu}(\overset{\circ}{u}, \overset{\circ}{v}) + \Delta v \mathbf{r}_{vv}(\overset{\circ}{u}, \overset{\circ}{v}).$$

On substituting the above relations into (12.12), we obtain the following *Newton's iteration scheme* (with $u^* = \overset{\circ}{u} + \Delta u$ *and* $v^* = \overset{\circ}{v} + \Delta v$):

$$\begin{bmatrix} \mathbf{p} \cdot \mathbf{r}_{uu} & \mathbf{p} \cdot \mathbf{r}_{uv} \\ \mathbf{p} \cdot \mathbf{r}_{uv} & \mathbf{p} \cdot \mathbf{r}_{vv} \end{bmatrix} \begin{bmatrix} \Delta u \\ \Delta v \end{bmatrix} = \begin{bmatrix} -\mathbf{p} \cdot \mathbf{r}_u \\ -\mathbf{p} \cdot \mathbf{r}_v \end{bmatrix}. \tag{12.13}$$

In summary, an *extreme point* (u^*, v^*) may be determined as follows:

0. Input: initial guess point $(\overset{\circ}{u}, \overset{\circ}{v})$ *and a vector* $\mathbf{p}$.

1. Set: $u^* = \overset{\circ}{u}$; $v^* = \overset{\circ}{v}$.

2. Compute increments $\Delta u, \Delta u$ *from (12.13).*

3. Update: $u^* = u^* + \Delta u$; $v^* = v^* + \Delta v$.

4. If the increments are small enough then stop, else go to 2.

Having determined an *extreme point* (u^*, v^*), the next step is to perform *CPI (curve/plane intersection)* using the *isoparametric curves* $\mathbf{r}(u, v^*)$ and $\mathbf{r}(u^*, v)$ in order to find an *initial intersection point* (u_0, v_0). If an *initial intersection point* is obtained the *Procedure_SPI* is invoked in order to find a *closed intersection curve* between the plane π and the surface (In this case, step 5 of the Procedure_SPI needs to be modified so that the iteration stops when it reaches to the initial intersection point).

12.2.3 Intersection between a Surface and a Vertical Plane

We now consider a special case of the SPI problem where the plane π is a vertical plane. It is assumed that the surface $\mathbf{r}(u, v)$ is smooth and that no portion of the surface are vertical. In other words, the entire surface is visible from above. An implication of the assumption is that all the intersection curves, if they exist, are "open". A *vertical plane* may be expressed as (it becomes a line equation)

$$ax + by - d = 0 \ ; \quad a^2 + b^2 = 1. \tag{12.14}$$

Thus, we may use the *SPI (surface/plane intersection)* procedure with $\mathbf{p} = (a, b, 0)$. But, it is beneficial to make use of the special structure of the problem.

Before describing the intersection problem, we introduce a method for finding an intersection between a vertical line $\mathbf{x}^* = (x^*, y^*)$ and a surface $\mathbf{r}(u, v) = (x(u, v),\ y(u, v),\ z(u, v))$. This "vertical line/surface" intersection problem can be formulated as a Newton's iteration scheme. It is called a **2D Jacobian inversion** which may be summarized as follows:

Procedure_2D-Jacobian $(\overset{\circ}{\mathbf{u}}, \mathbf{x}^* \Rightarrow \mathbf{u}^*)$:

0. Input: Guess point $\overset{\circ}{\mathbf{u}} = (\overset{\circ}{u}, \overset{\circ}{v})$ *and a vertical line* $\mathbf{x}^* = (x^*, y^*)$.

1. Determine $\Delta u, \Delta v$ *by solving the following:*

$$\Delta u \ x_u(\overset{\circ}{u}, \overset{\circ}{v}) + \Delta v \ x_v(\overset{\circ}{u}, \overset{\circ}{v}) = x^* - x(\overset{\circ}{u}, \overset{\circ}{v});$$

$$\Delta u \ y_u(\overset{\circ}{u}, \overset{\circ}{v}) + \Delta v \ y_v(\overset{\circ}{u}, \overset{\circ}{v}) = y^* - y(\overset{\circ}{u}, \overset{\circ}{v});$$

2. Update the guess point:

$$\overset{\circ}{u} = \overset{\circ}{u} + \Delta u \ ; \quad \overset{\circ}{v} = \overset{\circ}{v} + \Delta v.$$

3. If $(x(\overset{\circ}{u}, \overset{\circ}{v}) - x^*)^2 + (y(\overset{\circ}{u}, \overset{\circ}{v}) - y^*)^2 > \varepsilon$ *then go to step 1.*

else return $u^* = \overset{\circ}{u}$ *and* $v^* = \overset{\circ}{v}$.

The overall strategy for intersecting a surface with a vertical plane is the same as that for intersecting with a general plane discussed in §12.2.2. That is, an *initial intersection point* $\mathbf{r}(u_0, v_0)$ is *detected* by intersecting the surface boundary curves with the *vertical* plane, and then a next intersection point $\mathbf{r}(u^*, v^*)$ is obtained by *marching* along the intersection curve by the amount of *stepping distance* δ.

In order to compute a *next intersection point*, we need to compute the stepping distance δ' projected on the x, y-plane. It may easily be verified that the projected *stepping distance* is approximately given by

$$\delta' \cong \delta|\mathbf{t} \times \mathbf{n}| \tag{12.15}$$

where, $\delta = |\mathbf{r}(u_0, v_0) - \mathbf{r}(u^*, v^*)|$: *3D stepping distance,*

$\mathbf{r}(u_0, v_0)$: *initial (current) intersection point,*

$\mathbf{r}(u^*, v^*)$: *next intersection point,*

$\mathbf{t} = (b, -a, 0)$

$\mathbf{n}$ = *unit normal vector of the surface* $\mathbf{r}(u, v)$.

Then, the *next intersection point* $\mathbf{x}^* = (x^*, y^*)$ *on the x,y-plane* can be computed from the following relation:

$$\mathbf{x}^* = \mathbf{x}_0 + \delta'\mathbf{t} \tag{12.16}$$

where, $\mathbf{x}_0 = (x(u_0, v_0),\ y(u_0, v_0))$: *initial intersection point on x,y-plane,*

$\mathbf{t} = (b, -a)$: *unit tangent of the intersecting line,*

δ' : *2D stepping distance (projected on xy-plane).*

Finally, the domain intersection point $\mathbf{u}^* = (u^*, v^*)$ corresponding to $\mathbf{x}^*$ is computed by using the *Procedure_2D-Jacobian.*

On summarizing the discussions so far, the overall procedure for finding an intersection curve between a vertical plane (12.14) and a parametric surface $\mathbf{r}(u, v)$ may formally be described as follows:

Procedure_SVPI *(surface/vertical-plane intersection):*

0. *Input: initial intersection point* (u_0, v_0), *stepping distance* (δ).
1. *Compute the projected stepping distance* δ' *from* (12.15).
2. *Obtain the next (2D) intersection point* $\mathbf{x}^*$ *from* (12.16).
3. *Obtain the domain intersection point* $\mathbf{u}^*$ *from the Procedure_2D-Jacobian.*
4. *If* $\mathbf{u}^*$ *is on a boundary curve of the surface then stop,*
   else set $u_0 = u^*$, $v_0 = v^*$ *and go to step 1.*

12.2.4 Intersection between a Surface and a Horizontal Plane

Intersecting a surface $\mathbf{r}(u,v) = (x(u,v),\ y(u,v),\ z(u,v))$ with a horizontal plane is better known as **contouring**. Traditionally, *grid method* which is a form of "lattice evaluation" has been used in contouring. In this method, a rectangular grid is defined on the parameter domain or on the x, y-plane. Let (u_i, v_j) denote a grid point, then a *cell* is defined by four grid points (u_i, v_j), (u_{i+1}, v_j), (u_i, v_{j+1}), $(u_{i+1},\ v_{j+1})$ as depicted in Fig. 12.4. A test is made to check if the horizontal plane $z = h$ intersect an *edge* of the cell. For example, if

$$[z(u_i, v_j) - h][z(u_{i+1}, v_j) - h] < 0 \tag{12.17}$$

then the edge $(u_i, v_j) - (u_{i+1}, v_j)$ intersects with the horizontal plane. The "true" crossing point is found by an inverse interpolation. Since the contour entering the cell must leave by crossing one of the other edges, the second crossing point is similarly located. The process then steps to the neighboring cell across the second edge. When the grid is defined on the x, y-plane, the domain grid point (u_i, v_j) for the x, y grid point (x_i, y_j) is first obtained by using the 2D_Jacobian inversion method of §12.2.3. And then (12.17) is applied.

Since the implicit equation of a *horizontal plane* can simply be expressed as "$z = d$", the intersection between the horizontal plane and the surface $\mathbf{r}(u,v)$ is given by

$$z(u,v) = d \tag{12.18}$$

which may be solved by using the *Procedure_SPI of* §12.2.2 *with* p=(0,0,1).

The *subdivision strategy* has also been applied to *contouring*. A method of contouring based on recursive subdivision is given in Petersen (1984).

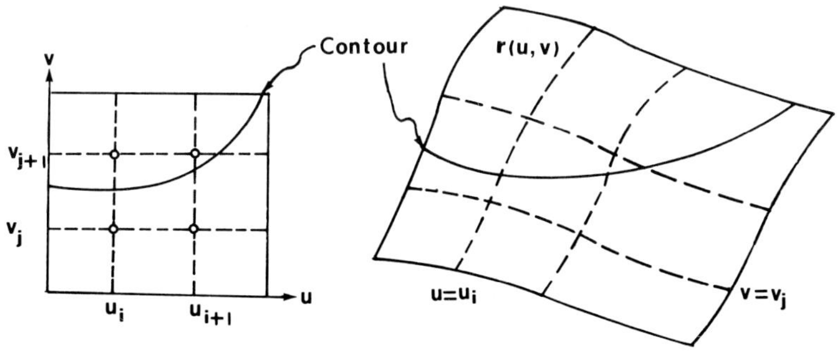

Figure 12.4 Lattice Evaluation Method

12.3 SURFACE/SURFACE INTERSECTION

An *SSI (surface/surface intersection)* algorithm for finding intersection curves between a pair of (rectangular) parametric surfaces has already been described in detail (See §11.2). The SSI algorithm due to Barnhill *et al* (1987) does not require any special structure, such as a surface patch being polynomial. The only requirement for the SSI algorithm to work is that participating patches be *rectangular* C^1 surfaces. The basic strategy of the SSI algorithm is as follows:

a) Detect initial intersection points from an "latice evaluation".
b) Trace out intersection curves using a "tracing method".

In this section, we present another SSI algorithm (Lasser, 1986) which is based on *"subdivision" techniques.* The SSI algorithm to be discussed is used in intersecting Bezier (and B-spline) surfaces. Before describing the SSI algorithm, basic operations needed in the SSI algorithm are introduced first.

12.3.1 Subdivision of Bezier Patch at its Centroid

A Bezier patch is easily subdivided into four *sub-patches* as discussed in §6.3. The subdivision procedure will be explained by using a cubic case as an example. The same holds true for other than cubic case as well. Let us first consider the problem of *subdividing a cubic Bezier curve* $\mathbf{r}(t)$ at $t = 1/2$. Let $\mathbf{r}^a(u)$ and $\mathbf{r}^b(u)$ for $u \in [0,1]$ denote the two *sub-curves* corresponding to the parameter ranges $t \in [0,\ 1/2]$ and $t \in [1/2,\ 1]$, respectively. The *control vertices* are denoted as

V_0, V_1, V_2, V_3: *control vertices of the original curve* $\mathbf{r}(t),\ t \in [0,1]$,
$V_0^a, V_1^a, V_2^a, V_3^a$: *control vertices of the first curve* $\mathbf{r}^a(u),\ u \in [0,1]$,
$V_0^b, V_1^b, V_2^b, V_3^b$: *control vertices of the second curve* $\mathbf{r}^b(u),\ u \in [0,1]$.

Then, from the *de Casteljau algorithm* (6.13), the control vertices of $\mathbf{r}^a(u)$ are given by (See also Fig. 6.2)

$$V_0^a = V_0,$$

$$V_1^a = (V_0 + V_1)/2,$$

$$V_2^a = (V_0 + 2V_1 + V_2)/4, \quad \textit{and}$$

$$V_3^a = (V_0 + 3V_1 + 3V_2 + V_3)/8.$$

The control vertices of the second patch $\mathbf{r}^b(u)$ are also expressed similarly.

An important property of a Bezier curve is that a subdivision results in an affine parameter transformation (Farin, 1988, pp.75-76). What this means is that the *local parameter "u"* and the *global parameter "t"* are linearly related. In general, when a Bezier curve $\mathbf{r}(t)$ for $t \in [0,1]$ is subdivided at $t = c$ *with* $0 < c < 1$ into $\mathbf{r}^a(u_a), \mathbf{r}^b(u_b)$ *for* $u_a, u_b \in [0,1]$, the following hold:

$$\mathbf{r}^a(u_a) = \mathbf{r}(cu_a) \equiv \mathbf{r}(t), \quad and$$

$$\mathbf{r}^b(u_b) = \mathbf{r}(c + (1-c)u_b) \equiv \mathbf{r}(t).$$

The above relations lead to the following:

$$t = cu_a \qquad (12.19-a)$$

$$t = c + (1-c)u_b \qquad (12.19-b)$$

where, $t \in [0,1]$: *global parameter of* $\mathbf{r}(t)$,

c : *subdivision point* $(0 < c < 1)$,

$u_a \in [0,1]$: *local parameter for the interval* $0 \leq t \leq c$,

$u_b \in [0,1]$: *local parameter for the interval* $c \leq t \leq 1$.

When a Bezier patch $\mathbf{r}(u,v)$ for $u,v \in [0,1]$ is subdivided at its centroid, that is subdivided at $u = v = 1/2$, the domain of $\mathbf{r}(u,v)$ is subdivided into four *square regions* as depicted in Fig. 12.5. Each sub-patch is identified by a *pair of binary values* (ψ, φ) called a son-index. For example, the *son-index* (0,0) denotes the *"bottom left"* sub-patch of $\mathbf{r}(u,v)$ corresponding to $0 \leq u,\ v \leq 1/2$. A Bezier patch having control vertices $\{V_{ij}\}$ is subdivided as follows:

a) Subdivided each row "i" of $\{V_{ij}\}$ *as if it is a Bezier curve, and*

b) Subdivided each column "j" of the "subdivided" control vertices.

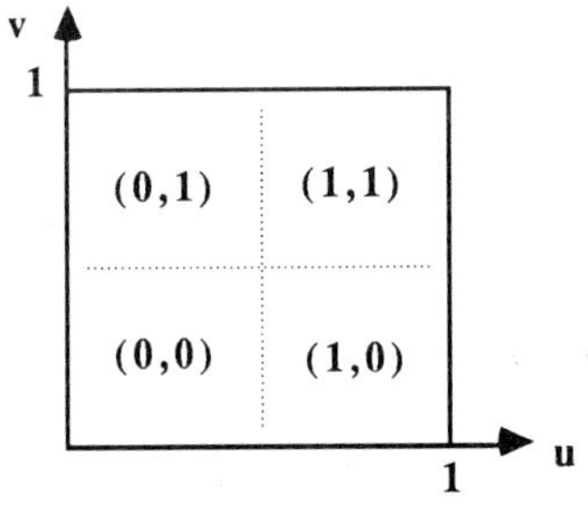

Figure 12.5 Domain Subdivision and Son Index

When the subdivision process is *recursively continued "m" times*, the resulting sub-patches are concisely stored in *quadtree data structure.* A sub-patch in *level "m"* of the quadtree is then identified by a *sequence of m son-indices* $\{(\psi_k, \varphi_k) : k = 1, ..., m\}$ which we call a sub-patch index (SPI).

12.3.2 Separability Test for Bezier Patches

Let $\mathbf{r}^a(u,v)$ and $\mathbf{r}^b(u,v)$ be two Bezier patches defined, respectively, by *nets of control vertices* $\{V_{ij}^a\}$ *and* $\{V_{ij}^b\}$. Then, a sufficient condition for the two Bezier patches to be separable is that the *"convex hulls of the Bezier nets* $\{V_{ij}^a\}$ *and* $\{V_{ij}^b\}$ *are disjoint".* This **separability condition** is a direct result of the *convex hull property* (See §6.5) of a Bezier patch. If two Bezier patches are separable, there are no intersections between them.

As it is not easy to find an *intersection of two convex hulls,* each convex hull is approximated by a majorizing box which is confined in the minimum and maximum x, y, z coordinates of the control vertices defining a Bezier patch. Thus, a quick separability test for two Bezier patches $\mathbf{r}^a(u,v)$ and $\mathbf{r}^b(u,v)$ can be made by checking the *intersection volume of their majoring boxes.* That is, if the intersection volume is found to be empty, the two patches are regarded to be separable. Otherwise the two patches are inseparable.

12.3.3 SSI Algorithm for Bezier Patches

The overall procedure for finding intersection curves of two Bezier surface patches is as follows: (1) Both patches are subdivided simultaneously forming four new subpatches on each surface patch; (2) A *"majorizing box"* is built for each subpatch and a *"separability test"* is performed for each pair of subpatches one from each surface; (3) The *subdivision and separability test* process is continued for each *"inseparable"* subpatch until it becomes *"thin"*; (4) Finally, each pair of *thin subpatches* are intersected.

A *"quick and dirty"* method of estimating the *"thickness"* of a bicubic Bezier patch defined by a *control net* $\{V_{ij}\}$ is as follows:

a) Estimate a center point $\mathbf{c}$ *as an average of the corner vertices:*

$$\mathbf{c} = (V_{00} + V_{03} + V_{30} + V_{33})/4$$

b) Estimate a unit surface normal:

$$\mathbf{n} = [(V_{33} - V_{00}) \times (V_{30} - V_{03})]/|(V_{33} - V_{00}) \times (V_{30} - V_{03})|$$

c) Estimate the thickness τ *of the patch:*

$$\tau = max\{|\mathbf{n} \cdot V_{ij} - \mathbf{n} \cdot \mathbf{c}| \ : \ i, j = 1, 2\}. \tag{12.20}$$

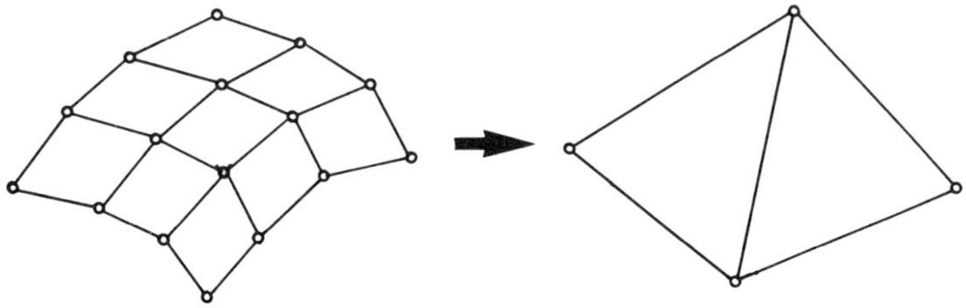

Figure 12.6 A Pair of Triangles Formed by Corner Control Vertices

Actually, the *thickness* τ represents the maximum deviation of the *control net* (not the surface) which should be larger than the thickness of the patch. A patch is called a *"thin"* patch if its thickness τ is less than a specified *tolerance value.*

If a pair of *thin* subpatches are *inseparable*, each of the patches is approximated by a pair of triangles formed by the four corner control vertices as depicted in Fig. 12.6. And then, a *plane/plane* intersection is carried out for a pair of triangles one from each patch. From the plane/plane intersection, *local parameters* u^*, v^* of each intersection point may be obtained. Recall from §12.3.1 that a subpatch at *level "m"* of the *quadtree* is identified by a *SPI (subpatch index)* $\{(\psi_k, \varphi_k) : k = 1, ..., m\}$. Thus, from the results in (12.19), the domain values u, v of the initial Bezier patch corresponding to the intersection point may easily be recovered as

$$\hat{u} = \sum_{k=1}^{m} (1/2)^k \ \psi_k + (1/2)^m u^* \qquad (12.21-a)$$

$$\hat{v} = \sum_{k=1}^{m} (1/2)^k \ \varphi_k + (1/2)^m v^* \qquad (12.21-b)$$

where, $\hat{u}, \hat{v}$: *intersection point (gloval parameters),*

m : *number of subdivisions,*

(ψ_k, φ_k) : *son-index (of Fig. 12.5) at subdivision level k,*

u^*, v^* : *intersection point in local parameters of the "leaf" subpatch.*

More accurate intersection points may be obtained either by increasing *number "m" of subdivisions* or by applying the *Newton's iteration scheme* of §11.2 (ie, by repeatedly applying the *"Procedure_Jacobian-inversion"* of §11.2.3 and *"guess point determination"* of Fig. 11.5).

An essential part of the SSI algorithm is an efficient maintenance of the *quadtrees* storing the subdivided patch information. After the completion of the recursive subdivision process, all the *inseparable thin patches* are stored in *"leaf" nodes* of the quadtrees. In tracing out a sequence of *intersection points* to form an intersection curve, the so called *"neighbor finding technique"* proposed by Samet (1982) may be employed. A key point of the SSI algorithm is the systematic elimination of *separable subpatches* from further consideration, which helps speed up the intersection process. Another merit of the SSI algorithm is in its *robustness* meaning that it *detects* all the intersections of the two Bezier patches.

12.3.4 Intersection of B-spline Surfaces

The same subdivision strategy is directly applicable to intersecting B-spline surfaces as proposed in Peng (1984). However, with the SSI algorithm for Bezier patches on hand, a more practical strategy would be to convert each (composite) B-spline surface into a composite Bezier surface. As discussed in §6.5, the conversion process is easily carried out by using the Boehm's knot insertion algorithm given by Eqn. (6.22).

As discussed in §7.5, a (composite) non-uniform B-spline (NUB) surface is defined in terms of a control net S and u-direction and v-direction knot span vectors $\mathbf{D}_u, \mathbf{D}_v$. For a bicubic $m \times n$ composite NUB surface, the geometric handles are given by

$$\begin{aligned}
\mathrm{S} &= \{V_{ij} : i \in [0, m+2],\ j \in [0, n+2]\} : \textit{control net},\\
\mathbf{D_u} &= \{\Delta_i : i \in [-2, m+1]\} : \textit{u-direction knot span vector, and}\\
\mathbf{D_v} &= \{\nabla_j : j \in [-2, n+1]\} : \textit{v-direction knot span vector.}
\end{aligned}$$

Let us assume that the composite NUB surface has been converted to a composite Bezier surface. Let $\mathbf{r}_n^{ij}(u, v)$ *for* $u, v \in [0, 1]$ be a bicubic NUB patch of the composite NUB surface (See Fig. 6.11). It can be shown that the bicubic Bezier patch $\mathbf{r}_z^{ij}(u, v)$ for the same location in the composite Bezier surface is identical to the NUB patch $\mathbf{r}_n^{ij}(u, v)$. That is, we have

$$\mathbf{r}_n^{ij}(u, v) \equiv \mathbf{r}_z^{ij}(u, v) \quad for \quad u, v \in [0, 1].$$

This is an important (and obvious) result which enables one to use a *domain intersection point* (u, v) of a Bezier patch as that of the corresponding NUB patch. The same should hold true for NURB (non-uniform rational B-spline) surface. Namely, a NURB surface is converted to a (composite) rational Bezier surface, and then intersections (on the u, v domain) of the rational Bezier surfaces are obtained.

12.4 TRIMMED SURFACE

Traditionally, *implicit representations* have been used in solid modeling, while *parametric representations* have mainly been used in surface modeling. In recent years, however, the idea of *unified shape modeling* has emerged. In a unified solid model, *sculptured surfaces* become a generic part of the solid model. As a result, *parametric representation* is becoming *a "must"* in solid modeling as well. In both solid and surface modeling, a parametric surface patch is often intersected with other surfaces, and then only a portion of the surface patch is used in defining a meaningful shape. The portion of the *parametric* surface patch *trimmed-off* by other surfaces is called a *trimmed (parametric) surface.*

12.4.1 Trimmed Surface Definition

An abstract definition of *trimmed surface* is elaborated in Farouki (1987). Let A,B denote solids (ie, regular and compact volume in E^3) so that

$$(A \cup B), \quad (A \cap B) \quad and \quad (A - B).$$

define the volumes generated by the *regularized* Boolean operations of *union, intersection, and difference* of A and B, respectively. For a *solid "X"*, let

I(X) be the interior of X,

E(X) be the exterior of X, and

S(X) be the boundary (surface) of X.

Then, the **exterior trimmed surface** of A with respect to B is defined as

$$S(A > B) \equiv S(A) \cap E(B) \qquad (12.22-a)$$

and the **interior trimmed surface** of A with respect to B is defined as

$$S(A < B) \equiv S(A) \cap I(B) \qquad (12.22-b)$$

The *curve of intersection of A and B* is defined as

$$C(A, B) \equiv S(A) \cap S(B). \qquad (12.23)$$

The boundary of a *combined solid* resulting from a Boolean operation is easily expressed in terms of the *trimmed surfaces* (12.22) and the *intersection curve* (12.23). The boundary of an *union of A and B,* for example, is expressed as

$$S(A \cup B) \equiv S(A > B) \cup S(B > A) \cup C(A, B).$$

In practice, it is convenient to include the intersection curve as part of the trimmed surface. In the case, the *exterior trimmed surface of A* with respect to B is given by

$$S(A \geq B) \equiv S(A > B) \cup C(A, B).$$

However, our main interest is in *trimmed parametric surface*. In the discussions that follow we shall use the term *"trimmed surface"* to mean a *trimmed parametric surface.*

As discussed in §2.4.1, a (rectangular) *parametric surface patch "S"* is a regular mapping $\mathbf{r}(u,v)$ from a *(rectangular) domain "$\mathcal{D}$"* to a 3D space E^3. Namely, a *parametric surface* $\mathcal{S}$ may be denoted as

$$\mathcal{S} : \mathcal{D} \stackrel{\mathbf{r}(u,v)}{\longrightarrow} E^3. \qquad (12.24-a)$$

Let $\mathcal{D}_T$ denote a portion of the domain $\mathcal{D}$ which we call a *trimmed domain,* then a **trimmed surface** $\mathcal{S}_T$ is defined as

$$\mathcal{S}_T : \mathcal{D}_T \stackrel{\mathbf{r}(u,v)}{\longrightarrow} E^3. \qquad (12.24-b)$$

12.4.2 Representation of Trimmed Surfaces

When an *SSI (surface/surface intersection)* is carried out by using a *numerical method,* an intersection curve is represented by a sequence of intersection points $\{(u_i, v_i) : i = 0, 1, ..., n\}$ on the domain $\mathcal{D}$ of each surface patch. An *intersection point sequence* $\{(u_i, v_i)\}$ is called a *branch "$\mathcal{B}$".* The first and last points $(u_0, v_0), (u_n, v_n)$ in a *branch* are called **border points** "$\mathcal{P}_B$" and the rest of the intersection points are called **interior points** "$\mathcal{P}_I$". Thus, a **branch** "$\mathcal{B}$" is defined as *"an $\mathcal{P}_B$ followed by zero or more $\mathcal{P}_I$s and another $\mathcal{P}_B$".* That is, we have

$$\mathcal{B} ::= (\mathcal{P}_B, \{\mathcal{P}_I\}, \mathcal{P}_B). \qquad (12.25)$$

A pair of intersecting patches and a *branch* are depicted in Fig. 12.7.

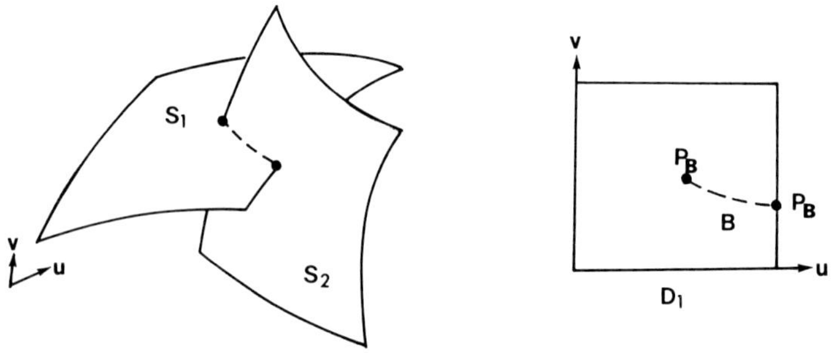

Figure 12.7 A Branch "$\mathcal{B}$" with Border Points "$\mathcal{P}_B$"

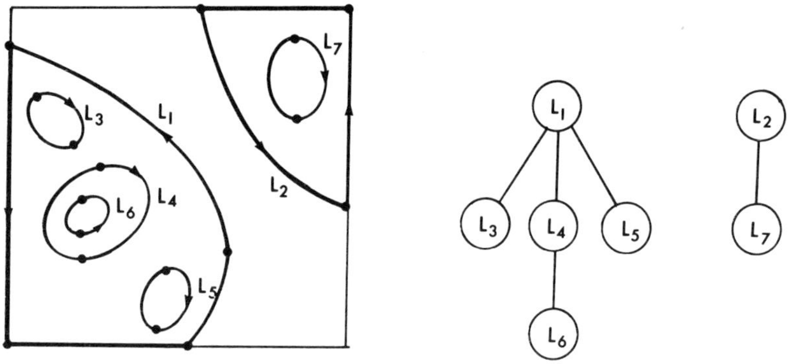

(a) (b)

Figure 12.8 Loops and Loop Structure

By connecting a sequence of *branches*, we may form a *closed loop* on the domain of a parametric surface "$\mathcal{S}$". A **loop** denoted by "$\mathcal{L}$" may be defined as

$$\mathcal{L} ::= \textit{closed sequence of branches} \quad \{\mathcal{B}_i\}. \tag{12.26}$$

A *loop* may become a *trimming boundary* of a *trimmed surface* $\mathcal{S}_T$. In general, there may be more than one *loops* in a domain $\mathcal{D}$. As shown in Fig. 12.8a, *loops* may be disjoint or nested. As a convention, *primary (or periphery) loops* have an anti-clockwise direction and *hole loops* a clockwise direction.

The *loops* on a domain $\mathcal{D}$ may be structured as a *tree* as depicted in Fig. 12.8b. A *tree of loops* is called a **loop structure** "$\mathcal{LS}$". Conceptually, a trimmed surface $\mathcal{S}_T$ is a tuple consisting of the mapping function $\mathcal{S}$ (12.24-a) and a *loop structure* $\mathcal{LS}$. Namely,

$$\mathcal{S}_T ::= (\mathcal{S}, \mathcal{LS}). \tag{12.27}$$

12.4.3 Trimmed Surface and B-Rep Solid

In a *boundary representation (B-Rep) scheme,* there are three types of *topological entities*. They are *vertex "$\mathcal{V}$"*, *edge "$\mathcal{E}$"*, and *face "$\mathcal{F}$"*. A **vertex** $\mathcal{V}$ formed by three *surfaces* $\mathcal{S}_1, \mathcal{S}_2, \mathcal{S}_3$ may be expressed as

$$\mathcal{V} ::= [(\mathcal{S}_1, \mathcal{P}_B), (\mathcal{S}_2, \mathcal{P}_B), (\mathcal{S}_3, \mathcal{P}_B)], \tag{12.28}$$

where "$\mathcal{P}_B$" denotes a *border point* (see Fig. 12.7). An **edge** $\mathcal{E}$ formed by two *surfaces* $\mathcal{S}_1, \mathcal{S}_2$ can similarly be denoted as

$$\mathcal{E} ::= [(\mathcal{S}_1, \mathcal{B}_i), (\mathcal{S}_2, \mathcal{B}_j)], \qquad (12.29-a)$$

where $\mathcal{B}_i, \mathcal{B}_j$ represent *branches* as defined by (12.25). Alternatively, an *edge* may be defined in terms of two vertices. That is,

$$\mathcal{E} ::= (\mathcal{V}_1, \mathcal{V}_2). \qquad (12.29-b)$$

A **face** $\mathcal{F}$ on surface $\mathcal{S}$ may be defined directly as an *edge loop* or indirectly as a *loop structure* $\mathcal{LS}$ on the domain of $\mathcal{S}$. That is,

$$\mathcal{F} ::= \{\mathcal{E}_i\}\ , \quad or \qquad (12.30-a)$$

$$\mathcal{F} ::= (\mathcal{S}, \mathcal{LS}). \qquad (12.30-b)$$

In the above definition, a restriction may be imposed on the *loop structure* in order to have a connected *face*.

An important part of solid modeling is *boundary evaluation*. After a *(solid) Boolean operation* or a *(surface) blending operation*, new boundaries of the resulting solid must be constructed. In general, **boundary evaluation** is carried out in three steps:

a) SSI (surface/surface intersection),
b) SMC (set membership classification), and
c) boundary file construction.

A *boundary evaluation* requires an elaborate manipulation of topological structure (as well as geometry) of solids.

In order to perform a *boundary evaluation* under a *trimmed surface* environment, *branches* "$\mathcal{B}$" on a *domain* "$\mathcal{D}$" may need to be subdivided into **monotonic branches** as discussed in Farouki (1987). Let us assume that the domain $\mathcal{D}$ is a unit square, $(u, v) \in [0,1] \times [0,1]$. Then, an *extreme point* of a branch $\mathcal{B}$ with respect to "u" or "v" is called a *turning point* "$\mathcal{P}_T$". If a *branch* $\mathcal{B}$ has *turning points* $\mathcal{P}_T$, it is subdivided at each turning point to form a sequence of *monotonic branches* (now a turning point $\mathcal{P}_T$ becomes a border point $\mathcal{P}_B$).

It is also necessary to keep a *pointer* to the *counterpart branch*, the branch on the other intersection surface. Let $\mathcal{B}_1$ be the *current branch* on the domain of $\mathcal{S}_1$ that was obtained by intersecting $\mathcal{S}_1$ with another surface $\mathcal{S}_2$. Let $\mathcal{B}_2$ be the branch on the domain of $\mathcal{S}_2$ corresponding to $\mathcal{B}_1$. Then, the branch $\mathcal{B}_1$ may be denoted as

$$\mathcal{B}_1 ::= [\mathcal{P}_B, \{\mathcal{P}_I\}, \mathcal{P}_B, PTR(\mathcal{S}_2, \mathcal{B}_2)],$$

where $\mathcal{P}_B$ and $\mathcal{P}_I$ are *border and interior points* and PTR is a *pointer* to the *counterpart branch*. See Farouki (1987) and Casale (1987) for some detail.

12.5 DISCUSSIONS

SSI (surface/surface intersection) is one of the key operations in surface modeling and in solid modeling. *SSI* is a time consuming operation, and more importantly, it is susceptible to *numerical errors* making a geometric modeler *unreliable.* The two *SSI algorithms,* one by Barnhill et al (1987) and the other by Lasser (1986), found to be very efficient (compared to others tested by the author). However a major drawback in both SSI algorithms is the inability to compute accurately and reliably tangential surface intersections. A solution for this problem is proposed in Markot and Magedson (1989). Another comment concerning the SSI algorithms is that they are not exactly an *"algorithm"* in its exact definition (instead, they are *"strategies"* for finding intersection curves). What this means is that details of the algorithms have yet to be worked out in an actual implementation.

The trimmed surface formulation introduced in this chapter is relatively new. Only recently, *trimmed (parametric) surface* was included in the *IGES* as a *standard geometric entity (entity #144).* The **trimmed surface entity** in the *IGES* (NBS, 1986) has the following structure (In the table, a *"loop"* is a *simple closed curve* which is a *"curve on parametric surface" entity (#142)*):

Index	*Name*	*Type*	*Description*
1	*PTS*	*Pointer*	*Pointer to the surface entity*
2	N_1	*Integer*	*= 0, if periphery loop is the trim boundary* *= 1, otherwise (there are hole loops)*
3	N_2	*Integer*	*number of hole loops in the trimmed surface*
$3+N_1$	*PTPL*	*Pointer*	*pointer to the periphery loop curve*
$4+N_1$	$PTHL_1$	*Pointer*	*pointer to the first hole loop*
...	...		
$3+N_1+N_2$	$PTHL_{n2}$	*Pointer*	*pointer to the last hole loop*

CHAPTER 13

NONPARAMETRIC SURFACE MODELING

13.1 INTRODUCTION

As discussed in §2.4.1, a *(parametric) surface* is the image of a regular mapping $\mathbf{r}(u,v)$ of a set of points in a 2D *domain* $\mathcal{D}$ into E^3. The mapping function

$$\mathbf{r}(u,v) = (x(u,v),\ y(u,v),\ z(u,v))$$

is called the equation of the surface. When the domain $\mathcal{D}$ is defined on the $xy-plane$ of a Cartesian coordinate system, the above surface equation becomes

$$\mathbf{r}(u,v) = (u,\ v,\ z(u,v)) \qquad (13.1-a)$$

which may simply be expressed as

$$z = f(x,y). \qquad (13.1-b)$$

The surface equation given by (13.1-b) is called a **nonparametric representation** of a *surface* $\mathcal{S}$.

Presented in this chapter are practical methods of representing a surface $\mathcal{S}$ in a *nonparametric form* (Choi *et al*, 1988a). Namely, methods for finding a *z-value* and a *surface normal* $\mathbf{n}$ at a given *x,y point* will be presented. If we have a number *nonparametric surfaces* $\mathcal{S}_1$, $\mathcal{S}_2$,...,$\mathcal{S}_n$, we can construct a **compound surface** $\mathcal{CS}$ by applying Boolean operations to the individual surfaces $\mathcal{S}_i$. An example of a *compound surface* is shown in Fig. 13.1.

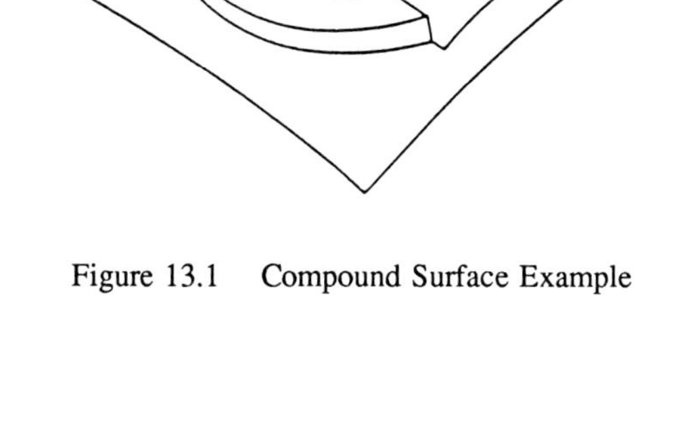

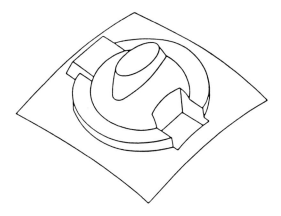

Figure 13.1 Compound Surface Example

13.2 NONPARAMETRIC MODELING OF POLYHEDRON PRIMITIVES

Shown in Fig. 13.2a are simple polyhedron primitives each of which is defined by its vertex points $\mathbf{p}_i$'s. The topological information of a *tetrahedron primitive* is depicted in Fig. 13.2b. Here, we want to determine a *z-value* and *unit normal vector* $\mathbf{n}$ on the boundary (ie, face) of a *convex* **polyhedron primitive** for a given (x, y) value. For this, each convex polyhedron is assumed to be stored in a **B-rep** *(boundary representation) data structure.*

The *geometry* of a polyhedron is initially given by the coordinates of *vertices. The topology* of an *edge* $\mathcal{E}$ is bounded by a counter-clockwise (closed) sequence of *edges* $\mathcal{E}_1, ..., \mathcal{E}_n$. For the tetrahedron shown in Fig. 13.2b, the coordinates of the four *vertices* $\mathcal{V}_1, \mathcal{V}_2, \mathcal{V}_3, \mathcal{V}_4$ are stored in an array *VERTEX*. The edges of the tetrahedron are defined as

$$\mathcal{E}_1 = (\mathcal{V}_1, \mathcal{V}_2); \quad \mathcal{E}_2 = (\mathcal{V}_1, \mathcal{V}_4); \quad \mathcal{E}_3 = (\mathcal{V}_2, \mathcal{V}_3); \quad \mathcal{E}_4 = (\mathcal{V}_2, \mathcal{V}_4); \quad etc.$$

The faces of the tetrahedron are defined as

$$\mathcal{F}_1 = (-\mathcal{E}_1, \mathcal{E}_5, -\mathcal{E}_3); \quad \mathcal{F}_2 = (\mathcal{E}_1, \mathcal{E}_4, -\mathcal{E}_2); \quad etc.$$

For a nonparametric modeling of a convex polyhedron, it is very convenient to store the geometry information of the edges and faces (in the form of line equation and plane equation). In the following, methods of constructing a 2D edge equation of an edge E and a plane equation of a *face* F will be described first.

The projected image of an edge can be represented by a *2D edge equation* of the form

$$ax + by + c = 0. \tag{13.2}$$

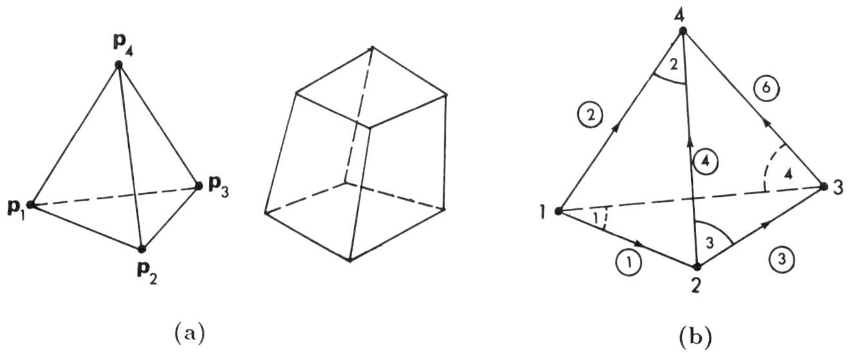

Figure 13.2 Polyhedron Primitives and Topology of Tetrahedron

The 2D edge equation (13.2) may be obtained as follows: Let $\mathbf{p}_1, \mathbf{p}_2$ be the end points of the #1 edge in Fig. 13.2b, and let $\mathbf{x}_i$ denote a projected image of $\mathbf{p}_i = (x_i,\ y_i,\ z_i)$ such that

$$\mathbf{x}_1 = (x_1,\ y_1,\ 0) \quad and \quad \mathbf{x}_2 = (x_2,\ y_2,\ 0).$$

Define a unit vector $\mathbf{m}$ as follows:

$$\mathbf{m} = [\mathbf{k} \times (\mathbf{x}_2 - \mathbf{x}_1)]/|\mathbf{k} \times (\mathbf{x}_2 - \mathbf{x}_1)| \equiv (m_x, m_y, m_z)$$

$$where, \quad \mathbf{k} = (0,\ 0,\ 1).$$

Then the coefficients of (13.2) are given by

$$a = m_x,$$

$$b = m_y, \quad and$$

$$c = -\mathbf{m} \cdot \mathbf{x}_1.$$

It should be noted that the 2D edge equation (13.2) is defined such that, for a 2D point (x, y) located on the "$left$" side of the projected image of the edge, the inequality "$(ax + by + c) > 0$" holds.

The geometry of a face of a polyhedron can be specified by a *plane equation* of the form

$$dx + ey + fz + g = 0 \quad with \quad d^2 + e^2 + f^2 = 1. \tag{13.3}$$

The *face normal* $\mathbf{n} = (d, e, f)$ is defined so that its direction points outward from the solid (tetrahedron). For example, "*#1 face*" of the tetrahedron is defined by the three vertices $\mathbf{p}_1, \mathbf{p}_2, \mathbf{p}_3$. The unit normal $\mathbf{n}$ of the face is then obtained as

$$\mathbf{n} = [(\mathbf{p}_3 - \mathbf{p}_1) \times (\mathbf{p}_2 - \mathbf{p}_1)]/|(\mathbf{p}_3 - \mathbf{p}_1) \times \mathbf{p}_2 - \mathbf{p}_1)|.$$

The above definition of a face equation ensures that "$(dx + ey + fz + g) < 0$" holds if a point (x, y, z) is inside the *convex polyhedron.*

The information (topology and geometry) of the edges and *faces* are stored in arrays *EDGE* and *FACE,* respectively. Shown in Fig. 13.3 are *boundary file data structures* of *EDGE and FACE* for the tetrahedron in Fig. 13.2b. Each pair of *vertex indices* are stored in the first two "*columns*" of *EDGE,* and the coefficients of the *2D edge equation* are stored in the last three *columns.* Stored in the first three *columns* of *FACE* are the indices of edges defining the face (the *minus sign* indicates a reversed edge direction). The coefficients of the face equation (13.3) are stored in the last four *columns.* Also stored are (not shown) the *limiting values of x,y* to define a "*majorizing area*" for the projected image of each convex polyhedron.

	vertex index		2D edge equation		
1	1	2	a_1	b_1	c_1
2	1	4	a_2	b_2	c_2
3	2	3	a_3	b_3	c_3
4	2	4	a_4	b_4	c_4
5	1	3	a_5	b_5	c_5
6	3	4	a_6	b_6	c_6

(a) *EDGE*

	edge index			face equation			
1	-1	5	-3	d_1	e_1	f_1	g_1
2	1	4	-2	d_2	e_2	f_2	g_2
3	3	6	-4	d_3	e_3	f_3	g_3
4	2	-6	-5	d_4	e_4	f_4	g_4

(b) *FACE*

Figure 13.3 Data Structure for the Tetrahedron in Figure 13.2-b

With these boundary data on hand, the procedure for computing the z-value and face normal $\mathbf{n}$ of a convex polyhedron can be described easily. The first step is to check if the *input point* (x^*, y^*) falls inside the *majorizing area.* If it falls outside the majorizing area, no attempt is made to compute the z-value. When the majorizing test is *positive,* the next step is to find an *intersection point* (between the polyhedron and the vertical line $x = x^*$, $y = y^*$). If the "*upper*" surface is to be evaluated, only the faces "*visible*" from above are considered; otherwise, only the *invisible* ones are tested for a possible intersection with the vertical line. The procedure for finding a *z-value and face normal* of a convex polyhedron may be summarized as follows:

Procedure_nonparametric-polyhedron $(x^*,\ y^*,\ sign,\ z^*,\ \mathbf{n},\ j,\ Found)$;

//*Input:* x^*, y^*; *sign*$(= +1$ *for upper surface,* -1 *for lower surface)*//
Found := *False;*
if (x^*, y^*) *falls outside of majorizing area,* **then return**;
for *each FACE "j" of the polyhedron* **do begin**
  if $(f_j * sign) > 0$ **then begin**   //$f_j > 0$ *for a visible face*//
    for *each EDGE "i" of FACE "j"* **do**
      if $f_j * (a_i x^* + b_i y^* + c_i) < 0$ **then goto** *Skip;*
    Found := *True;*   // (x^*, y^*) is inside of *FACE* j //
    $z^* := -(d_j x^* + e_j y^* + g_j)/f_j$;
    $\mathbf{n} := (d_j, e_j, f_j)$;
    return
  end; //*if*//
Skip ://*a jump location*//
end; //*for*//

13.3 NONPARAMETRIC MODELING OF QUADRIC PRIMITIVES

Shown in Fig. 13.4 are **quadric primitives** that may be useful in constructing a *compound surface*. The quadric primitives in the figure are *regular cone, regular cylinder, elliptic cone, elliptic cylinder, sphere, ellipsoid, and paraboloid.* Some of them are defined by cutting off the quadrics with planes perpendicular to the axis.

Let "*h*" denote the *height* of the truncated primitives, then the quadric primitives at their "zero position" may be represented by the following standard equations (See §5.5.1):

- *Cone equation:* (13.4)

  a) side : $(x/a)^2 + (y/b)^2 = (1 - z/h)^2 \quad with \quad 0 \le z \le h$

  b) bottom: $-z = 0 \quad with \; (x/a)^2 + (y/b)^2 \le 1$
- *Cylinder equation:* (13.5)

  a) side : $(x/a)^2 + (y/b)^2 = 1 \quad with \quad 0 \le z \le h$

  b) top : $z - h = 0 \quad with \quad (x/a)^2 + (y/b)^2 \le 1$

  c) bottom: $-z = 0 \quad with \quad (x/a)^2 + (y/b)^2 \le 1$
- *Ellipsoid equation:* (13.6)

  $(x/a)^2 + (y/b)^2 + (z/c)^2 = 1$
- *Paraboloid of revolution* ($a^2 = 4f$ *where f is the focal length):* (13.7)

  a) side : $(x/a)^2 + (y/a)^2 = z \quad with \quad 0 \le z \le h$

  b) bottom: $z - h = 0 \quad with \quad (x/a)^2 + (y/a)^2 \le h$

As depicted in Fig. 13.4, a *quadric primitive* is completely specified by a *center point*, up to three *parameters ("a,b,c" or "a,b,h")*, and up to three vectors. In order to find an

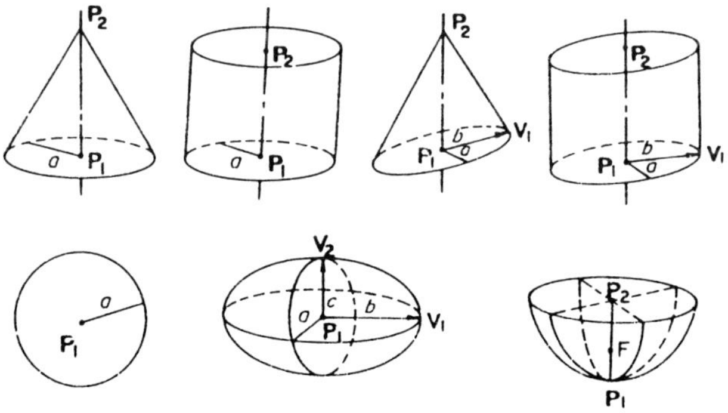

Figure 13.4 Quadric Primitives

intersection between a quadric surface with a vertical line, it is necessary to have an equation of the quadric in 3D space. A method for obtaining an implicit equation of a quadric in a general position will be explained using the ellipsoid as an example.

Let $\mathbf{n}$, $\mathbf{o}$, and $\mathbf{a}$ be the *normal (x-axis), orientation (y-axis), and approach (z-axis) vectors* of a coordinate frame $\mathbf{H}$, and let $\mathbf{p}$ be the *position* of $\mathbf{H}$. Let

$$\mathbf{u} = (u,\ v,\ w,\ 1)^T \quad and \quad \mathbf{r} = (x,\ y,\ z,\ 1)^T,$$

respectively, denote two homogeneous vectors defined with respect to the base coordinate and $\mathbf{H}$, representing the same point in 3D space. Then, as discussed in §2.5.2 (See Eqn. (2.50) and Fig. 2.7), the homogeneous vectors $\mathbf{u}$ and $\mathbf{r}$ are related to each other as follows:

$$\mathbf{r} = \mathbf{H}\ \mathbf{u} \quad or \quad \mathbf{u} = \mathbf{H}^{-1}\mathbf{r} \tag{13.8}$$

which can be rewritten in component forms as (Choi *et al*, 1988a)

$$u = \mathbf{n} \cdot \mathbf{r} - t_x \tag{13.9 – a}$$

$$v = \mathbf{o} \cdot \mathbf{r} - t_y \tag{13.9 – b}$$

$$w = \mathbf{a} \cdot \mathbf{r} - t_z \tag{13.9 – c}$$

$$where, \quad \mathbf{t} = (t_x,\ t_y,\ t_z) = (-\mathbf{n} \cdot \mathbf{p},\ -\mathbf{o} \cdot \mathbf{p},\ -\mathbf{a} \cdot \mathbf{p}).$$

With the transformations given by (13.9), the standard ellipsoid equation (13.6) becomes

$$(\mathbf{n} \cdot \mathbf{r} - t_x)^2/a^2 + (\mathbf{o} \cdot \mathbf{r} - t_y)^2/b^2 + (\mathbf{a} \cdot \mathbf{r} - t_z)^2/c^2 = 1. \tag{13.10}$$

For the *ellipsoid primitive* in Fig. 13.4, which is defined by three scalars a, b, c and three vectors $\mathbf{p}_1, \mathbf{v}_1, \mathbf{v}_2$, the *vectors* defining $\mathbf{H}$ are computed as

$$\mathbf{a} = \mathbf{v}_2/|\mathbf{v}_2|,$$

$$\mathbf{o} = \mathbf{v}_1/|\mathbf{v}_1|,$$

$$\mathbf{n} = \mathbf{o} \times \mathbf{a}, \quad and$$

$$\mathbf{p} = \mathbf{p}_1.$$

To find the z-values on the ellipsoid for a given (x, y) point, (13.10) is rearranged as a quadratic equation of z and then its roots are computed. The larger z-value is taken for *"upper" surface* and the smaller value for *"lower" surface*. Having selected a z-value, the surface normal vector at that point is easily obtained from partial derivatives of the implicit equation (See §2.4.2, Eqn. (2.25)).

13.4 NONPARAMETRIC MODELING OF SWEEP SURFACES

A simple *nonparametric sweep surface* was introduced earlier in §5.4.3. The nonparametric equation given by (5.38) is repeated below as

$$z(x,y) = s(x) + g(y) - g(y_0) \quad for \quad x \in [x_0, x_1],\ y \in [y_0, y_1] \tag{13.11}$$

$$where,\quad z = s(x)\ for\ x \in [x_0, x_1]: \quad section\ curve\ at\ \ y = y_0,$$

$$z = g(y)\ for\ y \in [y_0, y_1]: \quad guide\ curve.$$

In this section, the above nonparametric modeling method is extended so that it can be applied to the construction of a *composite nonparametric surface.* Also to be discussed is a nonparametric modeling of *surface of revolution.*

13.4.1 Hermite-Blended Nonparametric Sweep Surface

Shown in Fig. 13.5 is a sweep surface patch defined by a "guide" curve and two "section" curves. The curve data are

$z = g(y)$ *for* $y \in [y_0, y_1]$: *guide curve located on the yz-plane,*

$z = s_0(x)$ *for* $x \in [x_0, x_1]$: *section curve on the plane* $y = y_0$, *and*

$z = s_1(x)$ *for* $x \in [x_0, x_1]$: *section curve on the plane* $y = y_1$.

Then, from (13.11), two nonparametric sweep surfaces are defined as

$$z_0(x,y) = s_0(x) + g(y) - h_0\ for\ x \in [x_0, x_1],\ y \in [y_0, y_1] \tag{13.12-a}$$

$$z_1(x,y) = s_1(x) + g(y) - h_1\ for\ x \in [x_0, x_1],\ y \in [y_0, y_1], \tag{13.12-b}$$

where $h_i = g(y_i)$ for $i = 0, 1$.

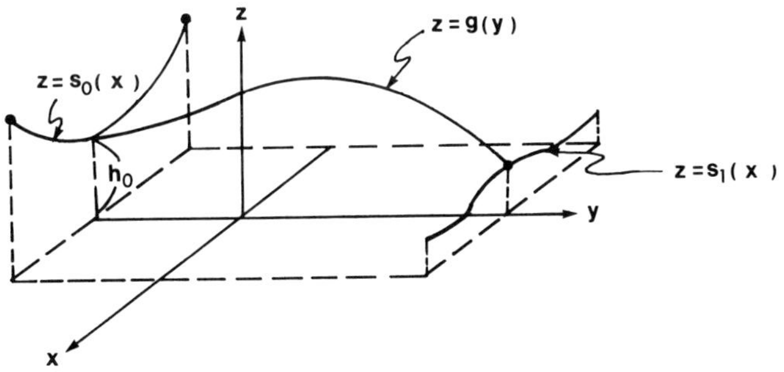

Figure 13.5 Parallel Sweep Surface with Two Section Curves

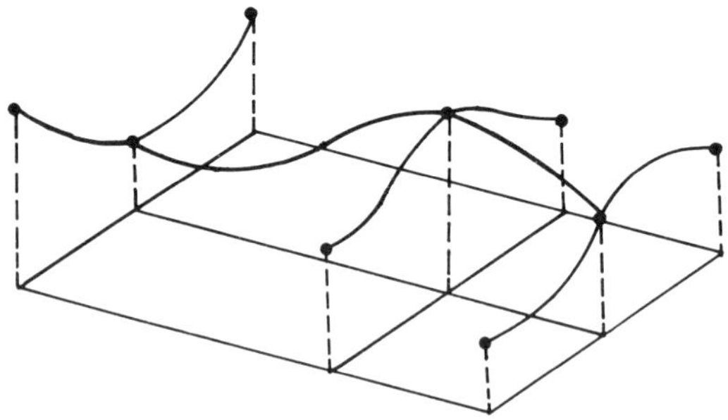

Figure 13.6 Parallel Sweep Surface with a Number of Section Curves

Thus, the parallel sweep surface of Fig. 13.5 can be expressed as a *Hermite blending* of the two nonparametric sweep surfaces in (13.12). That is, *the* **Hermite-blended nonparametric sweep surface** is expressed as

$$\begin{aligned} z(x,y) &= \alpha(v)z_0(x,y) + \beta(v)z_1(x,y) \\ &= g(y) + \alpha(v)[s_0(x) - h_0] + \beta(v)[s_1(x) - h_1] \end{aligned} \qquad (13.13)$$

where, $v = (y - y_0)/(y_1 - y_0) = (y - y_0)/\triangle y,$

$\alpha(v) = 1 - 3v^2 + 2v^3 :$ *Hermite blending function,*

$\beta(v) = 3v^2 - 2v^3 :$ *Hermite blending function,*

$g(y) :$ *guide curve for* $y \in [y_0, y_1],$

$s_i(x) :$ *section curves* $x \in [x_0, x_1]$ *and* $i = 0, 1,$

$h_i = g(y_i)$ *for* $i = 0, 1.$

The domain of the above nonparametric surface is a rectangular region on the xy-plane $(x, y) \in [x_0, x_1] \times [y_0, y_1]$.

The use of the *Hermite blending functions* $\alpha(v), \beta(v)$ ensures a *visual continuity* between neighboring sweep surface patches when a *compound sweep surface* is constructed from a number of section curves as shown in Fig. 13.6. The *surface normal* $\mathbf{n}$ of (13.13) is given by

$$\mathbf{n} = (-\partial z/\partial x,\ -\partial z/\partial y,\ 1) \qquad (13.14)$$

13.4.2 Nonparametric Modeling of Composite Sweep Surface

Now we consider the problem of constructing a *nonparametric composite surface* for the curve-net data of Fig. 13.7 where all the section curves are defined on vertical planes. The restrictions on the curve-net data are

a) The section planes are parallel with either the xz-plane or the yz-plane.

b) The boundary curves on the xy-plane are parallel with x-axis or y-axis except the corner boundary curve which is monotone in x,y directions.

As depicted in the figure, there are three types of patches. They are

(1) *a patch bounded by four vertical section curves,*

(2) *a patch bounded by two section curves and a corner boundary curve, and*

(3) *a patch bounded by three section curves and a boundary line.*

Our goal here is to construct a *visually smooth composite surface* for the curve-net data of Fig. 13.7. The following constructions will guarantee the necessary *visual continuity.*

1) Sweep Surface with Four Vertical Section Curves:

The first type of sweep surface patch is depicted in Fig. 13.8. The four *vertical section curves* are denoted as

$z = s_0(x)$ *for* $x \in [x_0, x_1]$: *"left" section curve on the plane* $y = y_0$,

$z = s_1(x)$ *for* $x \in [x_0, x_1]$: *"right" section curve on the plane* $y = y_1$,

$z = g_0(y)$ *for* $y \in [y_0, y_1]$: *"back" section curve on the plane* $x = x_0$, *and*

$z = g_1(y)$ *for* $y \in [y_0, y_1]$: *"front" section curve on the plane* $x = x_1$.

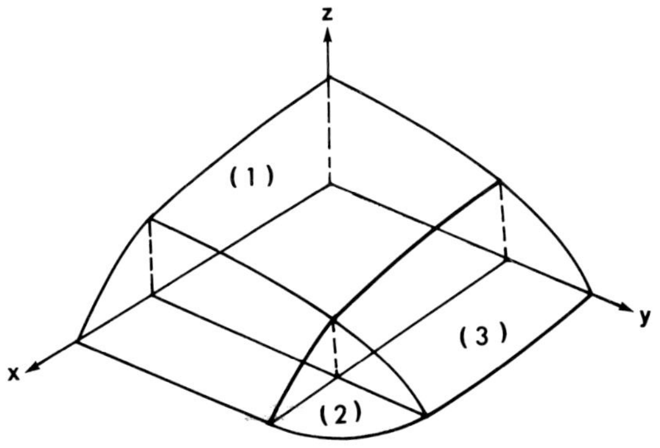

Figure 13.7 Curve-net Data of a Composite Sweep Surface

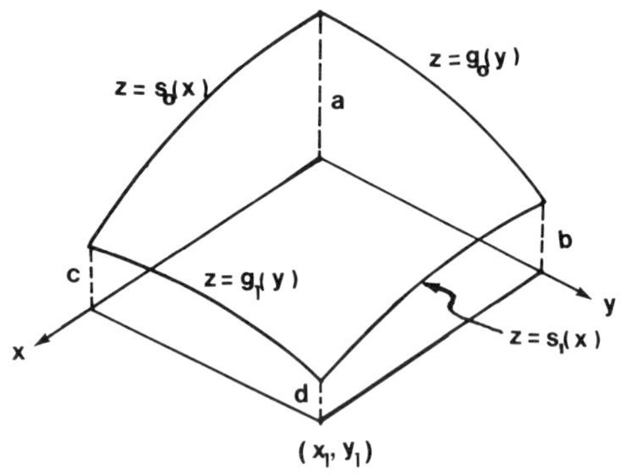

Figure 13.8 A Sweep Surface Patch with Four Vertical Section Curves

Now, we construct two *Hermite blended nonparametric sweep surfaces*, one using $g_0(y)$ and the other using $g_1(y)$. Namely, from (13.13), we get

$$z_0(x,y) = g_0(y) + \alpha(v)[s_0(x) - a] + \beta(v)[s_1(x) - b] \qquad (13.15-a)$$

$$z_1(x,y) = g_1(y) + \alpha(v)[s_0(x) - c] + \beta(v)[s_1(x) - d] \qquad (13.15-b)$$

where a, b, c, d are *corner heights* as depicted in Fig. 13.8. Finally, the two sweep surfaces in (13.15) are blended again in x-direction to obtain

$$\begin{aligned} z(x,y) &= \alpha(u)z_0(x,y) + \beta(u)z_1(x,y) \quad \textit{for } x \in [x_0, x_1],\ y \in [y_0, y_1] \\ &= \alpha(u)\{g_0(y) + \alpha(v)[s_0(x) - a] + \beta(v)[s_1(x) - b]\} \\ &+ \beta(u)\{g_1(y) + \alpha(v)[s_0(x) - c] + \beta(v)[s_1(x) - d]\} \\ &= \alpha(u)g_0(y) + \beta(u)g_1(y) + \alpha(v)s_0(x) + \beta(v)s_1(x) \\ &- [a\ \alpha(u)\alpha(v) + b\ \alpha(u)\beta(v) + c\ \beta(u)\alpha(v) + d\ \beta(u)\beta(v)], \end{aligned} \qquad (13.16)$$

$$\begin{aligned} \textit{where,}\quad & u = (x - x_0)/(x_1 - x_0), \\ & v = (y - y_0)/(y_1 - y_0), \\ & \alpha(t), \beta(t) : \textit{Hermite blending functions as in } (13.13), \\ & a = g_0(y_0); \quad b = g_0(y_1), \\ & c = g_1(y_0); \quad d = g_1(y_1). \end{aligned}$$

2) Sweep surface with Two Vertical Section Curves and A Corner Boundary:

Depicted in Fig. 13.9 is a *"second type"* surface patch (among the three types in Fig. 13.7). The two *vertical section curves* are denoted as

$$z = s(x) \quad for \quad x \in [0, a] \quad and$$

$$z = g(y) \quad for \quad y \in [0, b].$$

The *corner boundary curve* is defined as

$$y = f(x) \quad for \quad x \in [0, a] \quad or$$

$$x = h(y) \quad for \quad y \in [0, b].$$

The domain of the *corner sweep surface patch* defined by the two *vertical section curves* and the *corner boundary curve* is then given by

$$x \geq 0 \quad and \quad 0 \leq y \leq f(x).$$

Then, a *"proportional" sweeping of $g(y)$* in x-direction would generate the following *nonparametric surface*:

$$z_0(x, y) = \frac{s(x)}{c} g\left(\frac{b}{f(x)} y\right) \qquad (13.17-a)$$

$$where, \quad c = g(0) \equiv s(0).$$

An interpretation of the above equation is that it represents an *intermediate cross section* obtained by scaling down $g(y)$: The *scaling factor* in z-direction is *"$s(x)/c$"*, and that in y-direction is *"$f(x)/b$"*. Similarly, a *"proportional" sweeping of $s(x)$* in y-direction would give the following surface equation:

$$z_1(x, y) = \frac{g(y)}{c} s\left(\frac{a}{h(y)} x\right). \qquad (13.17-b)$$

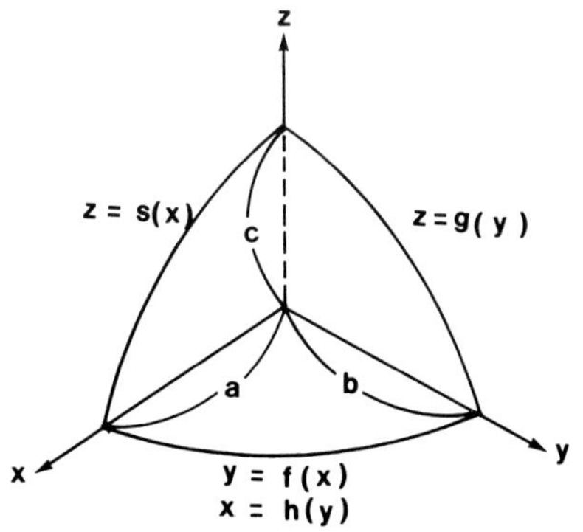

Figure 13.9 A Corner Sweep Surface Patch

Finally, the two surfaces in (13.17) are blended by using the **Brown's blending functions** $\psi(u,v)$, $\varphi(u,v)$ (Barngill, 1977) to obtain a *nonparametric surface equation* for the *corner sweep surface* patch of Fig. 13.9:

$$z(x,y) = \psi(u,v)\frac{s(x)}{c}g(\frac{b}{f(x)}y) + \varphi(u,v)\frac{g(y)}{c}s(\frac{a}{h(y)}x) \qquad (13.18)$$

where, $a, b, c:$ *maximum x,y,z values as depicted in Figure 13.9,*

$s(x):$ *vertical section curve on xz-plane,*

$g(y):$ *vertical section curve on yz-plane,*

$f(x) = h^{-1}(y):$ *corner boundary curve on xy-plane,*

$u = x/a, \quad v = y/b,$

$\psi(u,v) = v^2(1-v^2)/[u^2(1-u^2)+v^2(1-v^2)],$

$\varphi(u,v) = u^2(1-u^2)/[u^2(1-u^2)+v^2(1-v^2)].$

3) Sweep surface with Three Section Curves and A Boundary Line:

Shown in Fig. 13.10 is a *"third type"* sweep surface patch having three vertical section curves and a boundary line. The three *vertical section curves* are denoted as

$z = s(x)$ *for* $x \in [0,a]$: *"left" section curvew on the xz-plane,*

$z = g_0(y)$ *for* $y \in [0,b]$: *"back" section curve on the yz-plane, and*

$z = g_1(y)$ *for* $y \in [0,b]$: *"front" section curve on the plane x=a.*

The domain of the *sweep surface patch* defined by the three vertical section curves and a boundary line is

$$0 \le x \le a \quad and \quad 0 \le y \le b.$$

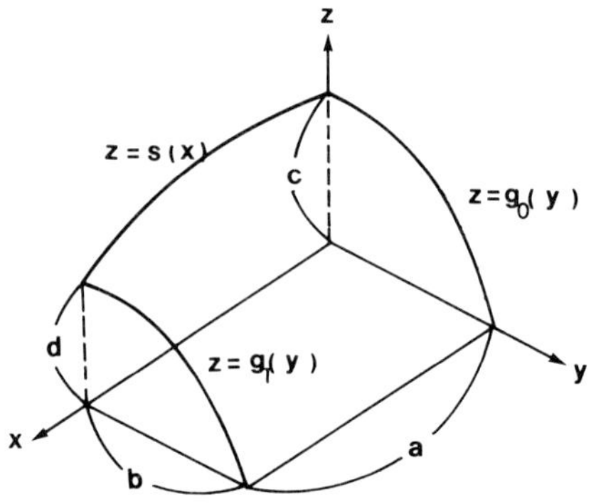

Figure 13.10 A Sweep Surface Patch Having Three Vertical Section Curves

As can be seen in Fig. 13.7, this surface patch (Fig. 13.10) is required to be joined smoothly to both the *rectangular patch* of Fig. 13.8 and the *corner patch* of Fig. 13.9. In other words, it should maintain *visual smoothness* with the *corner patch* (13.18) in x-direction and with the *rectangular patch* (13.16) in y-direction.

Thus, the sweep patch of Fig. 13.10 needs to be modeled as a *"proportional sweep surface"* in x-direction and a *"Hermite blended sweep surface"* in y-direction. From the result of (3.17), a *"proportional" sweeping* of $g_0(y)$ in x-direction may be expressed as

$$z_{x0}(x,y) = \frac{s(x)}{c} g_0(y). \qquad (13.19-a)$$

Similarly, a proportional sweeping of $g_1(y)$ would produce

$$z_{x1}(x,y) = \frac{s(x)}{d} g_1(y). \qquad (13.19-b)$$

The x-direction *nonparametric sweep surface* is then obtained by blending the two sweep surfaces in (13.19). That is, we have *(for $x \in [0,a]$, $y \in [0,b]$)*

$$z_x(x,y) = \alpha(u)\frac{s(x)}{c} g_0(y) + \beta(u)\frac{s(x)}{d} g_1(y) \qquad (13.20-a)$$

$$\begin{aligned} \textit{where,}\quad & u = x/a, \\ & \alpha(u), \beta(u): \textit{ Hermite blending functions as in } (13.13), \\ & c = s(0), \\ & d = s(a). \end{aligned}$$

The y-direction sweep surface is directly obtained from (13.16) by setting "$s_0(x) \equiv s(x)$, $s_1(x) \equiv 0$, $x_0 = 0$, $x_1 = a$, *etc.*" (Compare Fig. 13.8 with Fig. 13.10):

$$\begin{aligned} z_y(x,y) = {} & \alpha(u)g_0(y) + \beta(u)g_1(y) + \alpha(v)s(x) \\ & - [c \cdot \alpha(u)\alpha(v) + d \cdot \beta(u)\alpha(v)] \quad \textit{for } x \in [0,a],\ y \in [0,b] \end{aligned} \qquad (13.20-b)$$

$$\textit{where,}\quad u = x/a, \quad v = y/b.$$

Finally, the two surfaces in (13.20) are blended by using the *Brown's blending functions* $\psi(u,v)$, $\varphi(u,v)$ to obtain

$$z(x,y) = \psi(u,v)\ z_x(x,y) + \varphi(u,v)\ z_y(x,y) \quad \textit{for } x \in [0,a],\ y \in [0,b] \qquad (13.21)$$

$$\begin{aligned} \textit{where,}\quad & u = x/a; \quad v = y/b, \\ & z_x(x,y),\ z_y(x,y): \textit{ x- and y-direction sweep patches in } (13.20), \\ & \psi(u,v),\ \varphi(u,v): \textit{ Brown's blending functions as in } (13.18). \end{aligned}$$

13.4.3 Nonparametric Modeling of Surface of Revolution

Shown in Fig. 13.11a is a 2D *section curve* on the xz-plane. Let us consider a *surface of revolution* obtained by rotating the section curve around the z-axis. As depicted in the figure, let

$$z = s(x) \quad for \quad x \in [a, b] \quad and \quad 0 < a < b$$

denote an *implicit equation* of the section curve. Then, the z-value of the *surface of revolution* rotated around z-axis for a given (x, y)-point is expressed as

$$z = s(\sqrt{x^2 + y^2}). \tag{13.22}$$

The valid domain of the above *nonparametric surface* is

$$a^2 \leq x^2 + y^2 \leq b^2.$$

The *nonparametric surface* (13.22) is undefined outside the above region.

As shown in Fig. 13.11b, the *rotation angle* θ of the *current point* (x, y, z) is given by

$$\theta = atan2(y, x). \tag{13.23}$$

Let $\mathbf{t_1}$ denote the *tangent vector* of the section curve on the xz-plane, then the *sectional tangent* $\mathbf{t_s}$ is obtained by rotating $\mathbf{t_1}$ around the z-axis by θ *degrees*. That is, we have

$$\mathbf{t_s} = \mathbf{t_1}\ ROT(z, \theta) \tag{13.24 - a}$$

where, $\mathbf{t_1} = (1,\ 0,\ \dot{s}(x))$: *tangent vector of the section curve,*

$ROT(z, \theta)$: 3×3 *rotation transformation matrix (See Eqn.* 9.2).

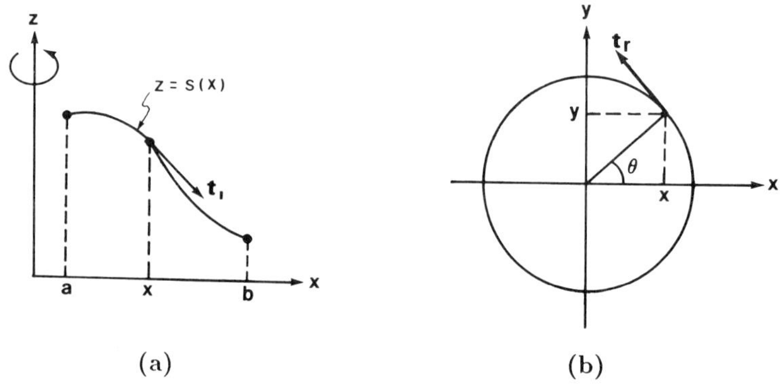

Figure 13.11 Surface of Revolution (around z-axis)

The *rotation vector* $\mathbf{t_r}$ of the *current point* can be expressed as (See Fig. 13.11b)

$$\mathbf{t_r} = (-y,\ x,\ 0). \qquad (13.24-b)$$

Finally, *the surface normal* of the surface of revolution (13.22) is obtained by taking a *vector product* of $\mathbf{t_s}$ and $\mathbf{t_r}$. Namely,

$$\mathbf{n} = \mathbf{t_s} \times \mathbf{t_r} \qquad (13.25)$$

$$where, \quad \mathbf{t_s} = [1\ 0\ \dot{s}(x)]\ ROT(z,\theta) : \ sectional\ tangent,$$

$$\theta = atan2(y,x),$$

$$\mathbf{t_r} = [-y\ x\ 0] : \ rotation\ vector,$$

$$ROT(z,\theta) = \begin{bmatrix} \cos\theta & \sin\theta & 0 \\ -\sin\theta & \cos\theta & 0 \\ 0 & 0 & 1 \end{bmatrix}.$$

When the same section curve "$z = s(x)\ for\ x \in [a,b]$" in Fig. 13.11a is rotated around x-axis, another *surface of revolution* is generated. For a given (x,y)-*point*, z-values on the surface of revolution (rotated around x-axis) can be obtained from

$$z = \pm\sqrt{[s(x)]^2 - y^2}. \qquad (13.26)$$

There are always two z-values for a point (x,y) in the valid domain. The valid domain of the above nonparametric surface is given by

$$x \in [a,b] \quad and \quad [s(x)]^2 \geq y^2.$$

It can be shown that the *surface normal* $\mathbf{n}$ of the *surface of revolution* (13.26) is expressed as

$$\mathbf{n} = \mathbf{t_s} \times \mathbf{t_r} \qquad (13.27)$$

$$where, \quad \mathbf{t_s} = [1\ 0\ \dot{s}(x)]\ ROT(x,\theta) : \quad sectional\ tangent,$$

$$\mathbf{t_r} = [0\ \ -\sin\theta\ \ \cos\theta] : \ rotation\ vector,$$

$$\theta = atan2(z,y),$$

$$ROT(x,\theta) = \begin{bmatrix} 1 & 0 & 0 \\ 0 & \cos\theta & \sin\theta \\ 0 & -\sin\theta & \cos\theta \end{bmatrix}.$$

A *nonparametric* surface of revolution around the y-axis can be constructed the same way.

13.5 NONPARAMETRIC BEZIER PATCHES

In this section, a *nonparametric* version of the *rectangular Bezier patch* (§5.2.3) is introduced. The treatment of the subject in this section is based on that of Farin (1988). Before introducing a nonparametric Bezier patch, a useful identity relation of *Bernstein polynomial* $B_i^n(t)$ (see Eqn. (6.2)) is given first. One may easily verify that the following holds:

$$\sum_{i=0}^{n} \frac{i}{n} B_i^n(t) = t. \tag{13.28}$$

A *nonparametric 2D curve* is expressed as

$$\mathbf{x}(t) = [t \;\; y(t)]. \tag{13.29}$$

If the function y(t) is in a *Bezier form*, it is expressed as

$$f(t) = \sum_{i=0}^{n} b_i B_i^n(t)$$

for some scalar coefficients $\{b_i : i = 0, ..., n\}$. Thus, using the identity (13.28), a **nonparametric Bezier curve** can be written as

$$\mathbf{x}(x) = \sum_{i=0}^{n} \mathbf{b}_i B_i^n(x) \tag{13.30}$$

$$where, \quad \mathbf{b}_i = [\frac{i}{n} \;\; b_i].$$

The "*x-value*" of a 2D *Bezier point* $\mathbf{b}_i$ is called an *abscissa* and its "*y-value*" is called an *ordinate*. The domain of the above nonparametric curve is a *unit interval*, $x \in [0, 1]$. But, as

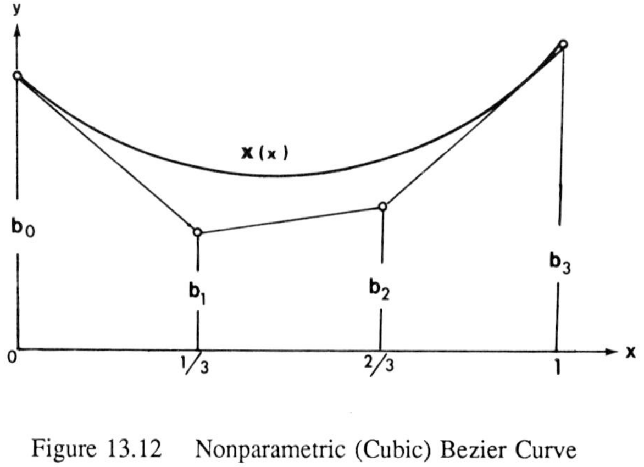

Figure 13.12 Nonparametric (Cubic) Bezier Curve

pointed out in Farin (1988), the domain can take a a general interval $x \in [a, b]$ *for* $a < b$. In this case, the *abscissae values* become

$$a + i(b - a)/n \quad for \quad i = 0, 1, ..., n. \tag{13.31}$$

Now, the *nonparametric Bezier curve* (13.30) is expressed as

$$\mathbf{x}(x) = \sum_{i=0}^{n} \mathbf{b}_i B_i^n((x - a)/(b - a)) \quad for \quad x \in [a, b] \tag{13.32}$$

A *nonparametric (rectangular) Bezier patch* of degrees m, n is defined as a straightforward extension of the *nonparametric curve* (13.30). Namely, we have

$$\mathbf{r}(x, y) = \sum_{i=0}^{m} \sum_{j=0}^{n} \mathbf{r}_{ij} B_i^m(x) B_j^n(y); \quad (x, y) \in [0, 1] \times [0, 1] \tag{13.33}$$

$$where, \quad \mathbf{r}_{ij} = [i/m \;\; j/n \;\; b_{ij}].$$

The 2D points $\mathbf{a}_{ij} = (i/m, \; j/n)$ are called *Bezier abscissae* and b_{ij} are called *Bezier ordinates.* The *unit square domain* of (13.33) can be transformed to a *parallelogram* defined by a point p and two vectors g, h in the x, y-plane as depicted in Fig. 13.13. Then the *Bezier abscissae* of the *nonparametric Bezier patch* are given by

$$\mathbf{a}_{ij} = \mathbf{p} + (i/m)\mathbf{g} + (j/n)\mathbf{h}. \tag{13.34}$$

This is equivalent to the following linear transformation of the domain variables *(from* $\mathbf{u} = (u, \; v) \in [0, 1] \times [0, 1]$ *to* $\mathbf{x} = (x, y)$ *in the parallelogram)*:

$$\mathbf{x} = \mathbf{u}\mathbf{A} + \mathbf{p} \tag{13.35}$$

$$where, \quad \mathbf{A} = \begin{bmatrix} \mathbf{g} \\ \mathbf{h} \end{bmatrix} = \begin{bmatrix} g_x & g_y \\ h_x & h_y \end{bmatrix}.$$

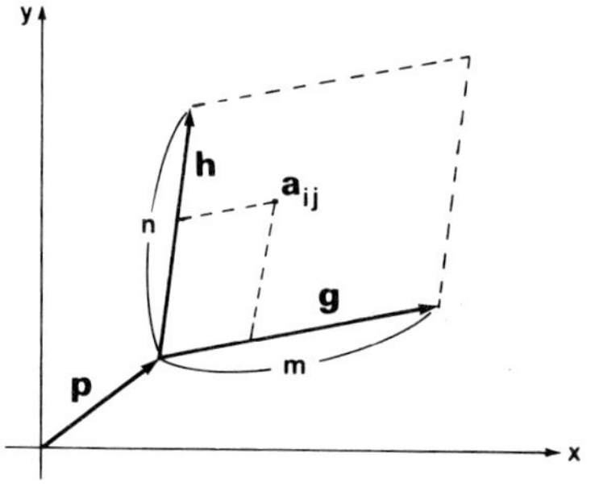

Figure 13.13 Parallelogram Domain

The solution to (13.35) is easily obtained as

$$\mathbf{u} = (u,\ v) = (\mathbf{x} - \mathbf{p})\mathbf{A}^{-1} \tag{13.36}$$

so that the nonparametric Bezier patch (13.33) is evaluated from

$$\mathbf{r}(x, y) = \sum_{i=0}^{m} \sum_{j=0}^{n} \mathbf{r}_{ij} B_i^m(u) B_j^n(v)$$

$$where, \quad \mathbf{r}_{ij} = [\mathbf{a}_{ij} \quad b_{ij}],$$

$$\mathbf{a}_{ij}: \quad Bezier\ abscissae\ (13.34).$$

13.6 NONPARAMETRIC EVALUATION OF PARAMETRIC SURFACE

A *nonparametric evaluation* of a parametric surface patch $\mathbf{r}(u, v)$ involves a *Jacobian inversion.* In finding the z-value for a given point (x, y), one may use the *"Procedure_2D-Jacobian"* in §12.2.3.

13.7 DISCUSSIONS

The topics discussed in this chapter are *nonparametric evaluation of convex polyhedra* and *quadric primitives, nonparametric modeling* of *parallel* and *rotational sweep surfaces, nonparametric Bezier surface,* and *nonparametric evaluation* of parametric surfaces. Since all the nonparametric surfaces are defined on the same domain, that is, on the xy-plane, they can be subjected to Boolean operations to form a *nonparametric compound surface* as detailed in Choi *et al* (1988a). Let z_i denote nonparametric surfaces, then a Boolean expression of the form

$$z_c(x, y) = z_1(x, y) \oplus z_2(x, y) \ominus z_3(x, y) ... \oplus z_n(x, y),$$

for example, can be evaluated as

$$z_c = max(z_n, ..., min(z_3, max(z_2, z_1))...).$$

This formulation is very useful in NC machining of *compound surfaces.* The readers are referred to Choi *et al* (1988a) for details.

CHAPTER 14

DEVELOPMENT AND USE OF CAM SYSTEMS

14.1 INTRODUCTION

Presented in this chapter are issues in developing and utilizing *"surface modeling and machining" software systems* which are usually called **CAM systems**. There are quite a few commercial CAM systems available. Ideally, a CAM system should be able to support all the *"modeling operations"* shown earlier in Fig. 1.9 which is reproduced in Fig. 14.1. The **modeling operations** in Fig. 14.1 have all been covered in this book:

- *Curve fitting methods in Chapter 4,*
- *Surface/surface intersection and trimming in Chapter 12,*
- *Surface fitting methods in Chapters 7 and 8,*
- *Section curve sweeping and blending in Chapter 9,*
- *Curve-net interpolation in Chapter 10,*
- *Blending surface construction methods in Chapter 11, and*
- *Boolean operations and projection in Chapter 13.*

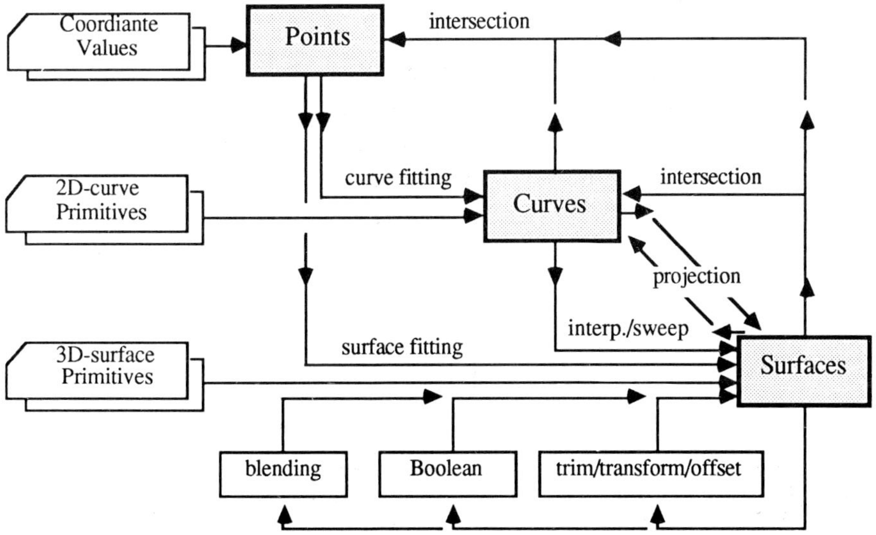

Figure 14.1 Modeling Operations in Surface Modeling

14.2 ARCHITECTURAL ISSUES IN DESIGNING CAM SYSTEMS

The main purpose of a CAM system is to construct a *"computational geometric model"* of a designed shape (not to create a new shape). The designed shape is usually *"described"* in engineering drawings. If the designed shape is already stored in a computer as a complete geometric model, we need a *machining module* in order to generate NC-codes for the machining of the "stored" surface. Until the idealistic *"unified shape modelers"* (see the next chapter) become available to every one, the need for dedicated CAM systems will remain for some time. The situation is analogous to *"machining centers"*. Machining centers have been in existence for decades, but dedicated NC machines are still around and will be.

In this section, we examine some architectural issues of CAM systems. In designing a CAM system, one has to make decisions about the following:

- *Modeling domain: which (or all) of the "modeling operations" in Figure 14.1 should be considered?*
- *Modeling view: surface modeling alone or machining process modeling as well.*
- *Input mode: Interactive versus batch.*
- *Surface geometry: unified versus specialized.*

The issue of **modeling domain** may seem pointless: The more the better. But, in practice, only a subset of the "modeling operations" may be allowed in a CAM system in order to keep the development project in a manageable level.

The issue of **modeling view** and **input mode** has to do with the trade-off between *flexibility* and *efficiency*. An idealistic approach to developing a CAM system is to *"interactively construct"* a complete surface model and then *"automatically convert"* it into machining instructions (ie, NC-codes). This approach may be termed an *"integrated approach"*. In this architecture, a CAM system consists of a *"surface modeler"* and a "CAPP system" as depicted in Fig. 14.2. An integrated CAM system in which surfaces are represented in the so called "z-map" form (Takeuchi et al, 1989) is proposed in §14.5.

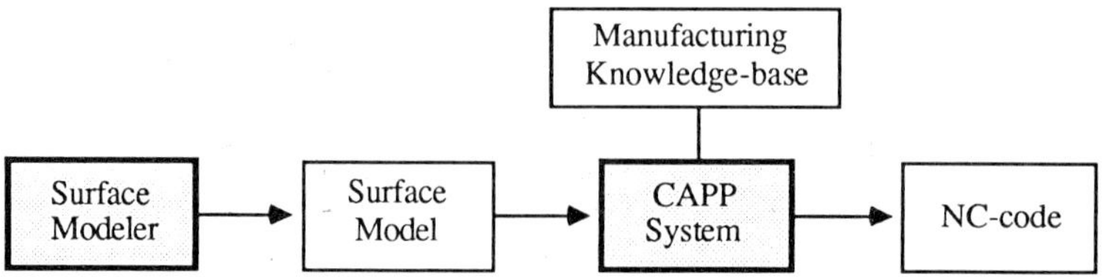

Figure 14.2 Integrated CAM System Architecture

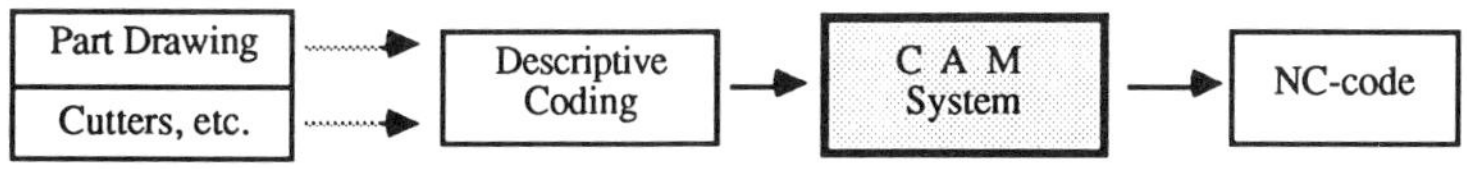

Figure 14.3 Dedicated CAM System Architecture

Another approach (concerning the selection of *modeling view* and *input mode*) is to develop a *"dedicated CAM system"*. This approach is based on the observation that we may not need a *complete model* of a (compound) surface as long as we can produce NC-codes necessary for the machining of the surface. In this approach, we attempt to construct a model of *"surface machining operations"* (instead of a model of the surface to be machined). A *"dedicated CAM system architecture"* is shown in Fig. 14.3. The user of the CAM system has to provide *"cutting plans"* as well as *"surface descriptions"*.

The issue of **surface geometry representation** depends to some extent on the choice of *modeling view*. A *unified representation scheme* is more necessary in the *"integrated architecture"* than in the dedicated case. As with *"unified shape modelers"* (See the next chapter), *NURB (non-uniform rational B-spline)* seems to be the most promising choice. In a dedicated CAM system, *nonparametric surface primitives* are very convenient modeling tools.

A dedicated CAM system is easier to implement (than an integrated one), but it requires the user to have a sound knowledge of NC machining in addition to a basic understanding of surface modeling. As discussed in Chapter 1, an economical machining of sculptured surface requires a good machining process plan so that machining sequences, cutter passes, and cutting conditions are all optimized. Another requirement of a CAM system is concerned with the quality of machining: *Gouging or cutter interference* is one of the most critical problems in sculptured surface machining. When machining cavity regions, the cutter easily invades portions of the surface. Sometimes a considerable amount of gouging comes from cutter deflection. Thus, it is essential to have mechanisms for verifying NC-codes before actual machining is carried out.

In the following (§14.3), we introduce a "dedicated" CAM system named *"SWEEP"* which was developed by the author and his students. "SWEEP" takes a *"batch-type"* input (ie, *part program*) and then produces NC-codes. An "NC verification & editing" system is described in §14.4, and an "integrated" CAM system named "Soft-MASTER" is presented in §14.5.

14.3 A DEDICATED CAM SYSTEM "SWEEP"

Presented in this section is the overall structure of a *"dedicated CAM system"* called "SWEEP" which is being marketed in Korea (by a local company CUBIC TECH Ltd.). At the time of writing this chapter (June 1990), about 30 systems are being used in local die shops. "SWEEP" is by no means an ideal CAM system, but the reader who wants to develop his own system may find the discussion interesting.

Before describing the structure of the CAM system, an explanation of historical backgrounds may be instructional. The development project started in early 1985. The self-imposed constraints on the CAM system were:

(1) *To be used in the machining of injection molding dies and forging dies.*

(2) *Ease of use (prospective users are vocational school graduates).*

(3) *The software should run on IBM/AT PCs under MS-DOS.*

(4) *Fast response time (due to "slow" hardware).*

(5) *Compatible with existing CAM systems already being used in industry.*

On close examination of die surface drawings, we found that most of the sculptured surfaces were described in terms of *cross-section curves*. To meet the three constraints (2), (3), (4), only a minimal set of modeling operations were included in the system. The last requirement was considered important because quite a few *"NC programmers"* had some experience with a Japanese system called DIE-II (Kishi, 1981). A prototype package consisting of about 20,000 lines of TURBO PASCAL codes was completed in early 1987. During the on-site testing period (1987-1988), a considerable amount of rewriting and embellishment was made.

14.3.1 Overall Structure of SWEEP

The overall structure of SWEEP is schematically described in Fig. 14.4. Surface types supported in the CAM system include

- *Parametric sweep surface of §9.2.*
- *Nonparametric sweep surfaces of §13.4.*
- *Hermite blended surface of §9.3.2 and ruled surface (parametric surfaces).*
- *Quadric primitives of §13.3(nonparametric surface).*

Geometric entities supported in the current version of SWEEP are as follows:

- *Basic geometric entities:* P *(points),* L *(lines), and* C *(circles).*
- *Intermediate geometric entities:* CV *(Curves).*
- *Generic curve entities:* SC *(section curve),* BC *(boundary or guide curve).*
- *Base surfaces* (BS) *: nonparametric surfaces.*
- *Sculptured surfaces* (SS) *: parametric surfaces.*
- *CC-Data: a surface point* $\mathbf{p}$ *and unit normal vector* $\mathbf{n}$.

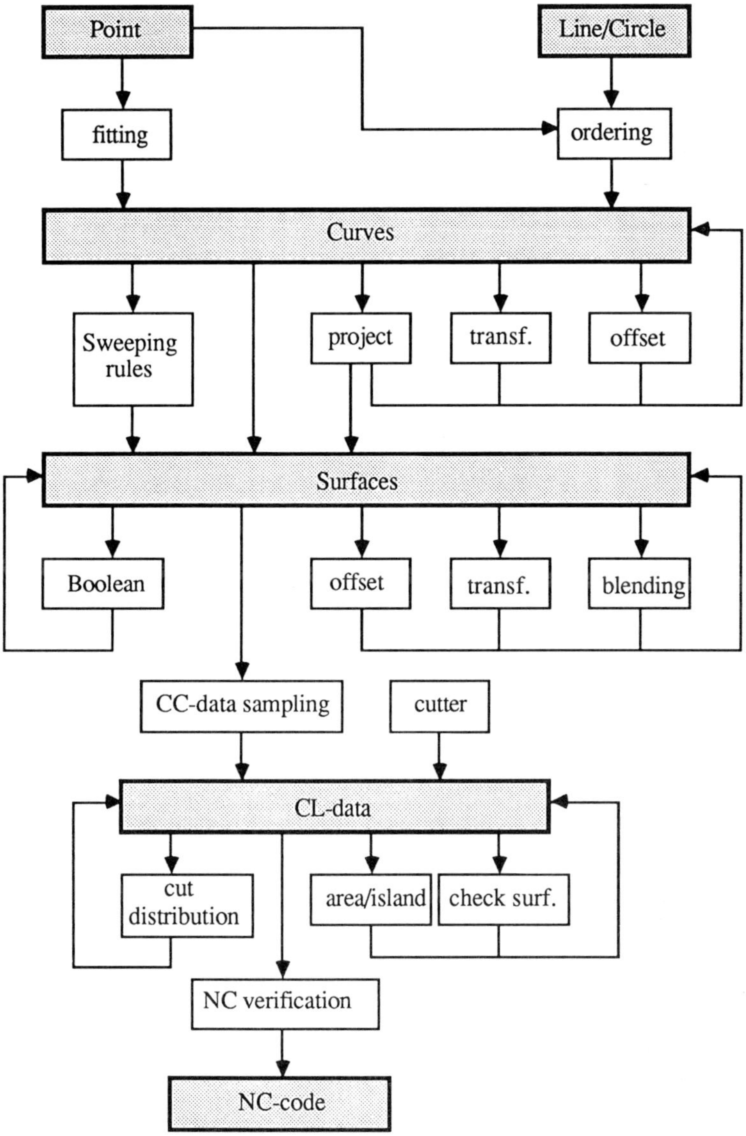

Figure 14.4 Overall Structure of SWEEP

1) Definition of Basic Geometric Entities:

The *basic geometric entities* in SWEEP are *"Points"*, *"Lines"*, and *"Circles"*: There are 13 cases of point definition (from predefined lines and circles), 16 cases of line definition, and 23 cases of circle definition. A Point is a 3D entity where as Lines and Circles are 2D entities. There are two *linked-lists* maintained internally: One for *Points* and the other for *Lines/Circles*. This step is the most time consuming part of *"part programming"* and needs a considerable amount of *arithmetic assignments* and *program control*. *"Arithmetic assignment"* and *"program control"* statements are almost the same as those in BASIC.

2) Definition of Curves:

From the *"basic geometric entities"*, the *intermediate geometric entities "Curves"* are defined: A *Curve* is defined either as an *"ordered sequence"* of *Points/Lines/Circles* (with *rounding and chamfering)* or by *"fitting"* a sequence of *Points*. A *Curve* constructed from the fitting operation is a 3D parametric curve, while the one constructed from the ordering operation is a 2D curve composed of *line/circle segments* (in this case, a sequence of 2D points is converted to a sequence of *"biarcs"* as discussed in §4.4.1). A 2D curve may be subjected to transformation and offset operations.

3) Definition of SCs and BCs:

As discussed in §9.2 and in §13.4, a *sweep surface* is defined in terms of a *"BCs" (guide curve and boundary curves)* and *"SCs" (section curves)*. A "*BC*" or an "*SC*" is defined as a sequence of "*CVs*". Both "*BC*" and "*SC*" can be transformed, but projection operation is allowed only to a "*BC*" to make it a true space curve.

4) Definition of Base Surfaces:

A set of *"nonparametric surface primitives"* (see Chapter 13) is supported in SWEEP so that a 2D curve is *"projected"* onto them to make it a 3D space curve. A (compound) nonparametric surface is called a *"Base Surface (BS)"*. In general, a *"Base Surface"* is defined from a collection of *"nonparametric surfaces" via* suitable Boolean operations.

5) Definition of Sculptured Surfaces:

By applying a suitable sweeping rule to the *"BCs"* and *"SCs"*, a *"Sweep Surface (SS)"* is defined. Also supported in SWEEP are the *"Hermite blended surface"* of § 9.3.2 and *"ruled surfaces"* for which the 3D parametric curves constructed from the fitting operation is used. Both *"SS" (sweep surface)* and *"BS" (base surface)* may be subjected to offset and transform operations. Only limited blending operations are supported (actually *"SCs" or "BCs"* are blended instead of the surfaces).

6) Computation of Cutter Paths:

Through the CC-data sampling operation, a sequence of *"cutter-contact data"* (ie, cutter contact points and unit surface normal vectors) can be generated. A point on the sweep surface for a given parameter value (u, v) is obtained from the expressions (Eqs 9.6 and 9.12) in §9.2. Surface normals can be evaluated by taking partial derivatives of the expressions. More details may be found in Lee (1990). Combined with the cutter information, *CC-data* are converted to *"CL-data"* for the machining of the surface. The resulting CL-data may be processed further to

- *obtain cut distribution for rough-cut machining,*
- *restrict cutting regions (within "areas" or outside "islands") or*
- *protect a portion of the part surface designated as check surface.*

7) NC-Code Generation:

Before producing NC-codes, the CL-data have to be verified for possible *"cutter interference"* and *"programming errors"*. The NC verification function is not an integral part of SWEEP. It is a companion package supporting *"NC machining simulation"* and *"CL-data editing"* functions. Here the row stock and intermediate volumes left by machining are modeled as *nonparametric surfaces* (See §13.3). The basic strategy of machining simulation is similar to the one proposed by Jarard and Drysdale (1988).

8) Part Program Syntax:

The SWEEP *part program syntax* is summarized in Fig. 14.5. As shown in the figure a typical part program may consist of the following six *"blocks"*:

- *Part P/G identification,*
- *2D geometry definition block,*
- *Curve definition block,*
- *Surface definition block,*
- *Cutting condition block, and*
- *Cutting command block.*

The sequence of individual statements is not necessarily in that order.

A target postprocessor is specified in the *"Part P/G id. block"*. Cutting conditions such as cutter geometry, feed rate, spindle speed, shrinkage factor, roughing allowance, etc. are given in the *"Cutting condition block"*. Actual cutting instructions are specified in the *"Cutting command block"*. Any surfaces defined in the *"Surface def. block"* can be cut. As many surfaces as we wish can be defined and *machined* one by one (by *resetting it*).

```
Part P/G Identification

    PART = (part number)
    MACH = (post-processor name)

2D Geometry (basic entities) Definition

    P1 = (point definition)
    L2 = (line definition)
    C3 = (circle definition)
    ...

Curve Definition

    CV1 = ...
    ...

Surface Definition

    SIM1 = (nonparametric simple sweep surfaces)
    SPH1 = (sphere - nonparametric quadrics)
    ...
    SC1 = (section curve definition)
    BC1 = (boundary & guide curve definition)
    ...
    SS = (sweep surface definition)
    ...

Cutting Conditions

    CUTTER =
    FRAT = (feed rate)
    ...

Cutting Commands

    CUT, SS, ...
         PATH = (tool paths and pitch intervals, etc)
         AREA = (area cutting)
         ISLAND = (prohibited area)
         CHECK = (check surface)
         HLIMIT = (cut distribution)
    RESET
    ...

End of Part P/G

    END
```

Figure 14.5 Overview of SWEEP Part Program Syntax

14.3.2 Software Structure of "SWEEP"

The software package consists of seven primary modules operating under the main PASCAL procedure. A primary module consists of a number of individual procedures some of which are grouped into submodules. The seven primary modules and their submodules are as follows:

(1) User interface module:
- *Menu system submodule,*
- *Editor submodule, and*
- *Parser submodule.*

(2) Mathematical library module:
- *Point definition submodule,*
- *Line definition submodule,*
- *Circle definition submodule, and*
- *Curve definition submodule.*

(4) Surface module:
- *Base surface submodule,*
- *Connection (Hermite and ruled) surface submodule, and*
- *Sweep surface submodule.*

(5) Machining condition module:
- *Post processor submodule.*

(6) Graphics module:

(7) Device control module:
- *Plotter drive submodule,*
- *Printer drive submodule, and*
- *Puncher/DNC submodule.*

Each module has its own data structure. All the data structures are simple ones except that for *"Sweep Surfaces"*. A data structure for a sweep surface is shown in Fig. 14.6. Due to the memory restriction, all the *Lines/Circles* are stored in a single linked list which is pointed by *"CVs"*. A sequence of *CVs* forming a *generic curve* (ie, *SC* or *BC)* is stored in a linked list which again is pointed by the *generic curve*.

For example, a *"parallel sweep surface"* $\mathbf{r}(u,v)$ defined by a *"BC"* $\mathbf{b}(v)$ and an *"SC"* $\mathbf{s}(u)$ is evaluated as follows: As discussed in §9.2, the surface equation is expressed as

$$\mathbf{r}(u,v) = [\mathbf{s}(u) - G]\ SWEEP\ (\phi, \theta) + \mathbf{b}(v) \qquad (14.1)$$

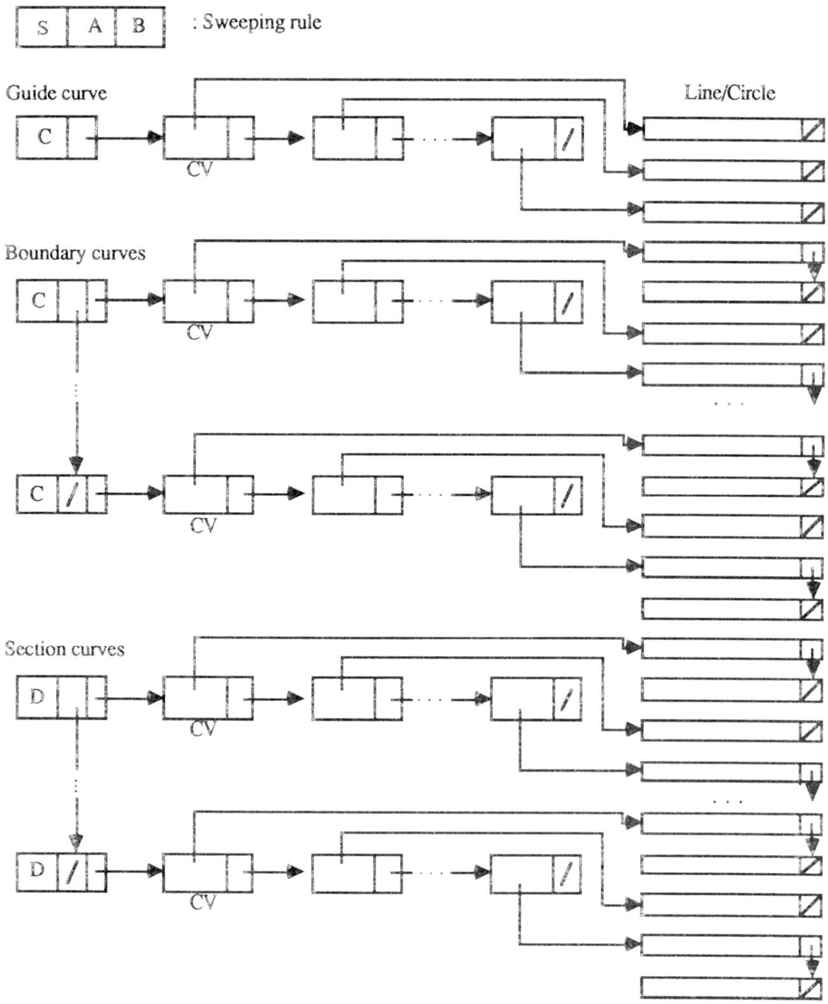

< c.f > A,B : Center of rotation or intersection angle (see Fig. 9.10)
C : Information on transformation and projection, etc
D : Intersection angle and guide point G (see Fig. 9.11)
S : Sweeping rule

Figure 14.6 Data Structure of a Sweep Surface

In order to evaluate (14.1), the following have to be given:

Sweeping rule (PARallel),

$G = (x_g,\ y_g,\ 0)$*: guide point on the section plane (See Fig. 9.11),*

ϕ*: intersection angle on the secton plane (See Fig. 9.11),*

$\mathbf{M}$*:* (3×3) *transformation matrix for the guide curve, and*

θ*: intersection angle on the base plane (See Fig. 9.10).*

Let us assume that the "BC" consists of two *Curves* $CV1$, $CV2$ and the "SC" is defined by the *Curve* $CV3$. These data are stored in the data structure of Fig. 14.6 as follows *(CVi are pointers to the "line/circle" linked list)*

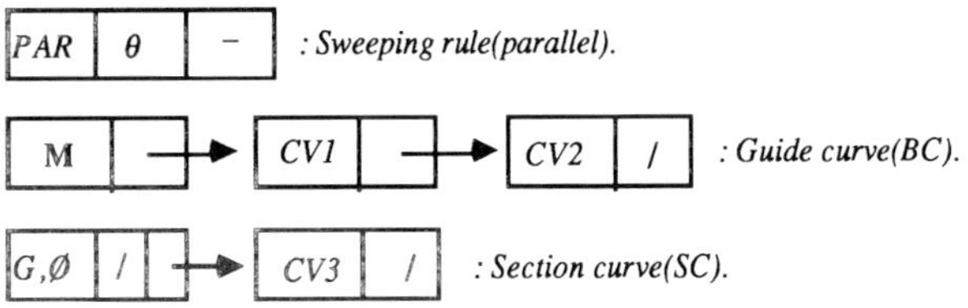

The first step to evaluate (14.1) is to compute the *arc lengths* of the "BC" and "SC" by adding up individual *curve segments (ie, line/circle segments)* belonging to them. The *"arc lengths"* obtained are stored elsewhere. The next step is to find the positions on "SC" and "BC" corresponding to given parameter values $u = u^*$ and $v = v^*$. For example, $\mathbf{s}(u^*)$ is a point on the "SC" satisfying the following:

$$\frac{(accumulated\text{-}arc\text{-}length)}{(total\text{-}arc\text{-}length)} = u^*.$$

Having determined $\mathbf{s}(u^*)$ in this way, the sweep surface is evaluated from (14.1) as follows:

$$\mathbf{r}(u^*, v^*) = [\mathbf{s}(u^*) - G]\ SWEEP(\phi, \theta) + \mathbf{b}(v^*)\ \mathbf{M} \qquad (14.2)$$

where $SWEEP(\phi, \theta)$ is a *"sweep transformation"* matrix given by (9.2) and $\mathbf{M}$ is a transformation matrix for the "BC".

The *gradients* of the "SC" and "BC" at $\mathbf{s}(u^*)$ and $\mathbf{b}(v^*)$, respectively, are easily obtained from the stored information. Let $\dot{\mathbf{s}}(u^*)$ denote the gradient of the "SC", then the *u-direction derivative* of $\mathbf{r}(u, v)$ is given by

$$\partial\mathbf{r}(u^*, v^*)/\partial u = \dot{\mathbf{s}}(u^*)\ SWEEP(\phi, \theta). \qquad (14.3)$$

The *v-direction derivative* is expressed similarly.

14.3.3 NC Machining Example

Examples of *"sweep surfaces"* were given earlier in Chapter 9. The surfaces in Figs. 9.1, 9.2 and 9.3 were produced by the CAM system SWEEP (Cubic, 1989). Presented in this section is a die surface machining example. Shown in Fig. 14.7 is a rough sketch of a part (it is called *"linker side bar"*) which is to be produced by forging. Both the upper and lower dies are to be machined by *EDM (electro-discharge machining).* We now describe an approach to generating NC cutter paths for machining an *EDM electrode* (for the upper die cavity).

The *upper surface* depicted in Fig. 14.7 is a *"compound surface"* composed of a number of *"surface features"*. The surface was described in a blue print in terms of a number of sectional views and feature curves. On neglecting the four holes and sides, the upper surface could be modeled as a *"BS" (base surface)* composed of seven *nonparametric surface primitives.* Thus, it was decided that the electrode is to be machined in three phases:

- *Machining of the nonparametric "base surface",*
- *Machining of the four holes, and*
- *Machining of the sides.*

Plottings of CL-data for the *"base surface", "holes", and "sides"* are shown in Fig. 14.8a, -b, and -c, respectively. The combined cutter paths are shown in Fig 14.8d.

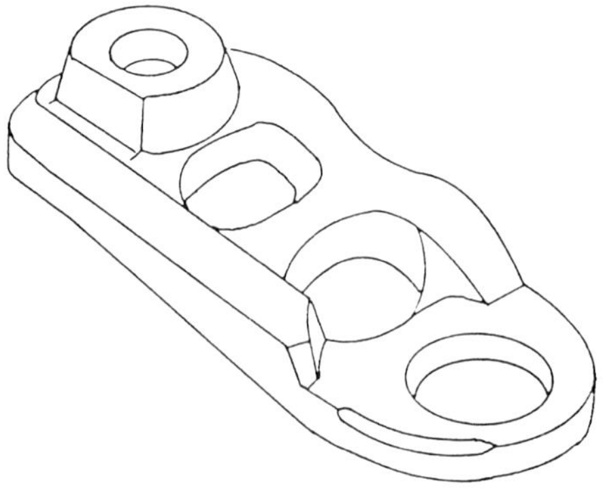

Figure 14.7 An Example Part to be Machined

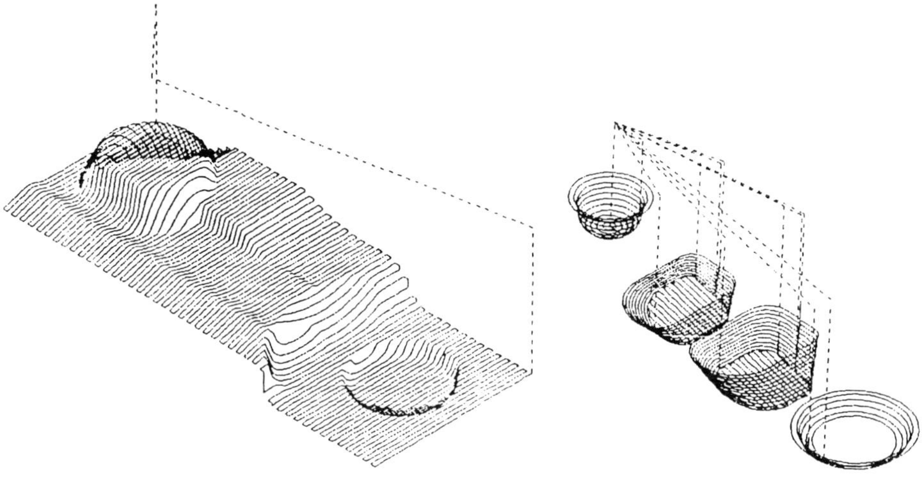

(a) (b)

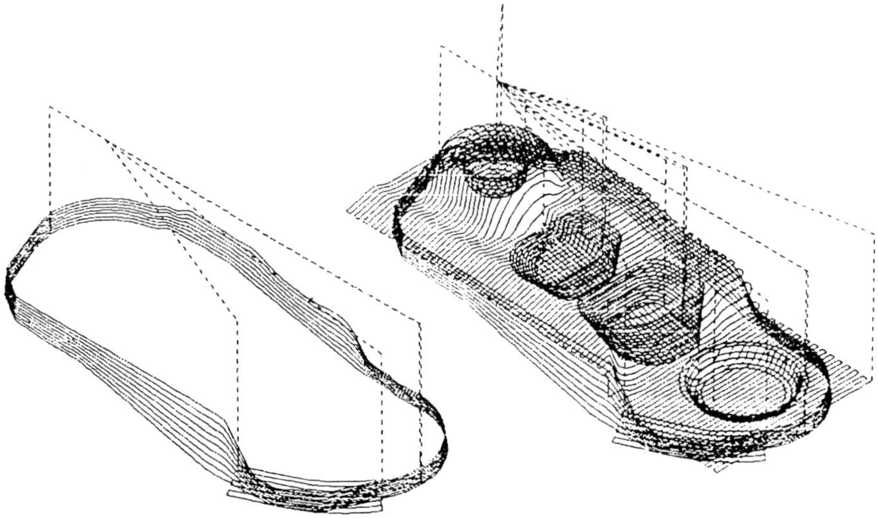

(c) (d)

Figure 14.8 CL-Data for an NC Machining of the Part in Fig. 14.7

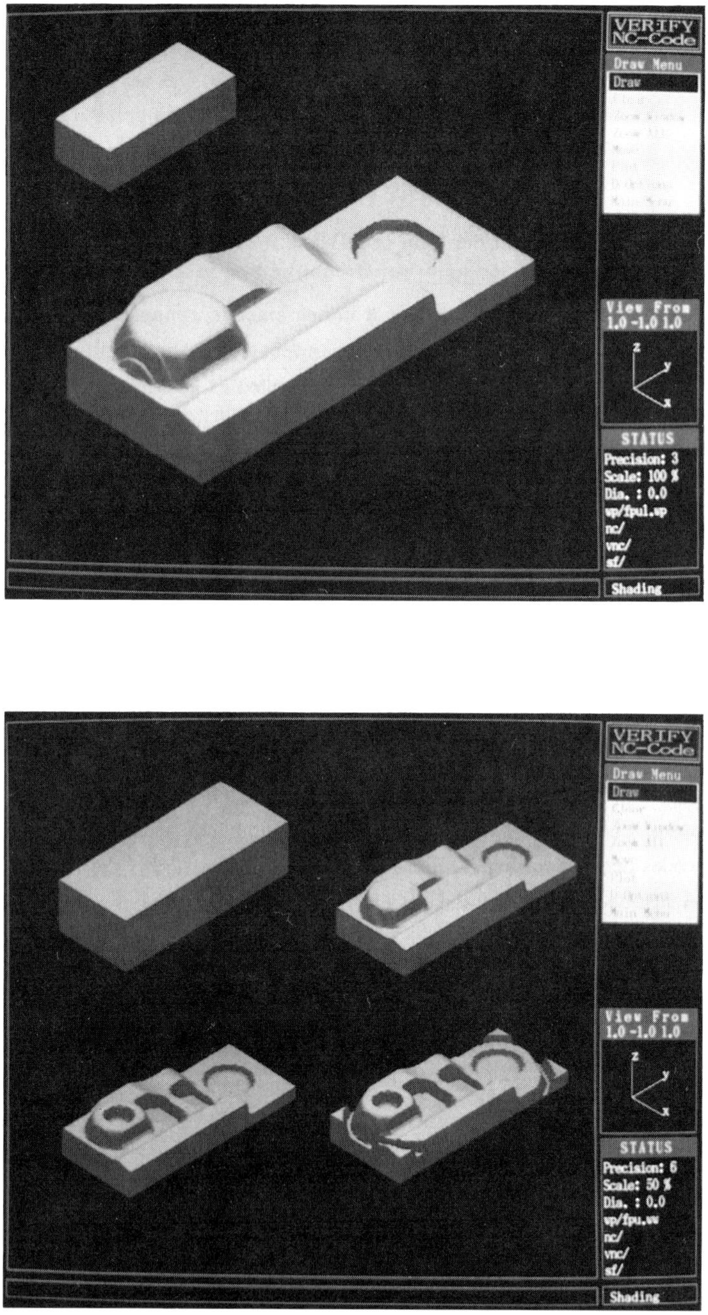

Figure 14.10 Shaded Display of Cutting Process Simulation

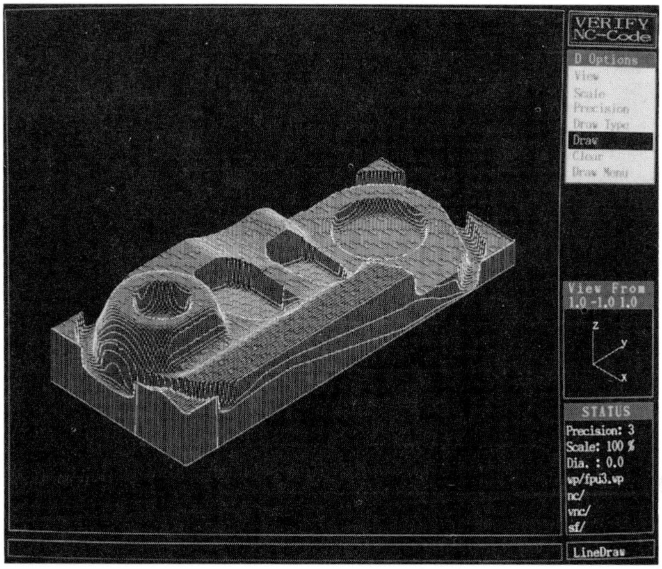

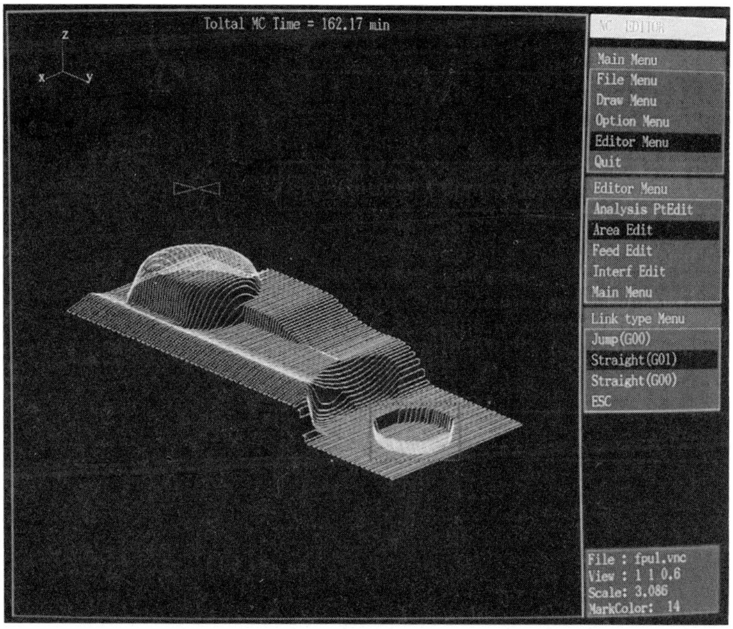

Figure 14.11 Wireframe Display of Machined Shape and Tool Paths

14.4.2 Z-map Representation and Cutting Process Simulation

The Z-map representation of a surface is a special form of "discrete" nonparametric representation in which the z-values of the surface are given only at "grid points" on the xy-plane. Recall from Chapter 13 that a nonparametric surface is expressed as a "mapping" from xy-plane to z-axis:

$$z = f(x, y).$$

The **Z-map representation** of a surface is defined as a set of "tuples"

$$\{x(i),\ y(j),\ z(i,j)\} \quad for \quad i \in [0, I],\ j \in [0, J],$$

where i, j are "grid indices" for x, y and I, J are positive integers denoting "grid limits". When a square grid of size "g" is used, x, y-coordinate values are expressed as

$$x(i) = x(0) + g * i \quad : \quad y(i) = y(0) + g * j. \tag{14.4}$$

An example of Z-map representation is depicted in Fig. 14.12.

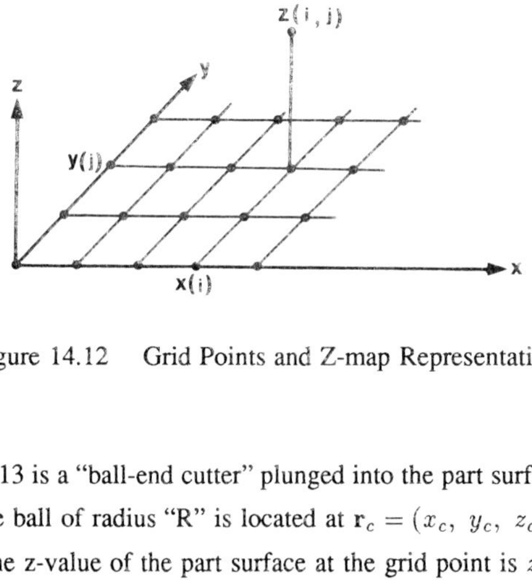

Figure 14.12 Grid Points and Z-map Representation

Shown in Fig. 14.13 is a "ball-end cutter" plunged into the part surface. As depicted in the figure, the center of the ball of radius "R" is located at $\mathbf{r}_c = (x_c,\ y_c,\ z_c)$ and a "grid point" is located at $(x_p,\ y_p)$. The z-value of the part surface at the grid point is z_p. Then, the "revised" part surface z'_p at the grid point $(x_p,\ y_p)$ generated by the ball-end cutter is computed as follows:

If $w = sqrt[(x_p - x_c)^2 + (y_p - y_c)^2]$ *is less than R, then*

$$z'_p = min(z_p, z_b) \ with\ z_b = z_c - R * sin\,\theta, \tag{14.5}$$

where $\theta = cos^{-1}(w/R)$.

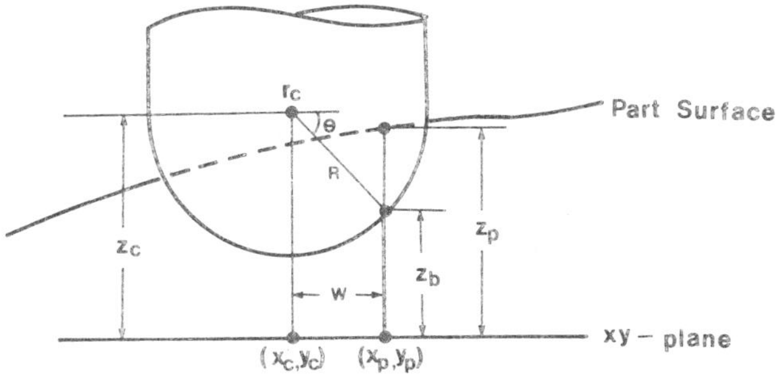

Figure 14.13 Z-map Representation of Ball-end Cutter

The "bottom" surface of the volume generated by moving the ball-end cutter from a "start" point $\mathbf{r}_s$ to an "end" point $\mathbf{r}_e$ can be represented as a "union" of a cylinder and two balls. Figure 14.14 shows the cutter movement in which the following data are given:

$\mathbf{r}_s = (x_s,\ y_s,\ z_s)$: center of the ball at the start of movement,
$\mathbf{r}_e = (x_e,\ y_e,\ z_e)$: center of the ball at the end of movement,
$(x_p,\ y_p)$: a grid point on the xy-plane.

Further, let us define

$\mathcal{L}$: 3D line segment joining $\mathbf{r}_s$ and $\mathbf{r}_e$, and
$\mathcal{L}_\mathcal{P}$: 2D "projected" line of $\mathcal{L}$.

Then, the distance w from the grid point (x_p, y_p) to the 2D line $\mathcal{L}_\mathcal{P}$ is given by

$$w = abs[\Delta x * y_p + \Delta y * x_p + x_s * y_e - x_e * y_s]/d_p \quad (14.6)$$

$$where, \quad \Delta x = (x_e - x_s); \quad \Delta y = (y_e - y_s),$$
$$d_p = sqrt(\Delta x^2 + \Delta y^2).$$

The relative distance h of the grid point (x_p, y_p) from (x_s, y_s) in the direction of the projected cutter path $\mathcal{L}_\mathcal{P}$ is given by

$$h = [\Delta x * (x_p - x_s) + \Delta y * (y_p - y_s)]/d_p. \quad (14.7)$$

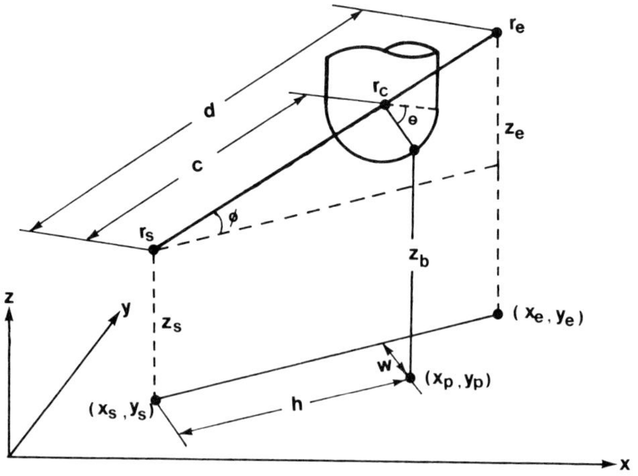

Figure 14.14 Ball-end Cutting Process Simulation

When the distance "w" from the grid point (x_p, y_p) to the "projected" cutter path $\mathcal{L}_{\mathcal{P}}$ is less than the ball radius "R", the z-map value z_b of the cylindrical surface of the ball-swept volume at the grid point can be expressed as (See Fig. 14.14)

$$z_b = z_s + c * sin\,\phi - R * cos\,\phi * sin\,\theta \tag{14.8}$$

$$where, \quad c = (h - R * sin\,\phi * sin\,\theta)/\,cos\,\phi,$$

$$h : as\ given\ by\ (14.7),$$

$$\phi = cos^{-1}(d_p/d) : inclination\ angle\ of\ the\ tool\ path\ \mathcal{L},$$

$$\theta = cos^{-1}(w/R),$$

$$d_p = sqrt(\triangle x^2 + \triangle y^2),$$

$$d = |\mathbf{r}_e - \mathbf{r}_s| = sqrt(\triangle x^2 + \triangle y^2 + \triangle z^2),$$

$$w : as\ given\ by\ (14.6),$$

$$R : radius\ of\ the\ ball.$$

The "revised" z-map value z'_p at the grid point is then obtained as

$$z'_p = min(z_p,\ z_b),$$

where z_p is the z-value of the initial part surface at the grid point. More details may be found in Jerard and Drysdale (1988).

14.5 AN INTEGRATED CAM SYSTEM "SOFT-MASTER"

This section introduces an approach to implementing the "integrated CAM system architecture" of Fig. 14.2. The proposed scheme which is based on the Z-map representation of surfaces (and solids) is depicted in Fig. 14.15. An integrated CAM system, dubbed "SOFT-MASTER", having the structure of Fig. 14.15 is being developed by the author and his students. The name "soft-master" was given because the "Z-map nonparametric model" in Fig. 14.15 can play the role of a (hard) "master model" in car-body development and production processes. The **Z-map representation** has some distinct features:

1. Its data structure is very simple to represent and easy to understand.
2. It is very versatile in that virtually all die and mold surfaces can easily be modeled as Z-maps.
3. It gives robustness in many geometric operations such as offsetting, booleaning, hidden line removal, and local modifications.
4. Its simple data structure allows one to develop efficient algorithms (In this respect, more research is needed).

On the other hand, the Z-map representation has the following limitations:

1. A large amount of memory is needed to store the Z-map data.
2. Only "discrete" data points of a surface are stored.
3. Only the surfaces which are "visible" from above may be represented.

However, the above limitations found to be non-critical in the CAM of free-formed die and mold surfaces.

14.5.1 Structure of SOFT-MASTER

Returning back to the structure of the "SOFT-MASTER" depicted in Fig. 14.15, the *integrated CAM system* consists of three "modules": *Z-map generation module, Z-map manipulation module*, and *NC-code generation module.* A similar system configuration is also proposed in Takeuchi *et al* (1989).

1) Z-Map Generation:

The Z-map model of a surface can be obtained from *NC-codes, solid models*, or *surface models.* Alternatively, the Z-map surface may directly be constructed from "interactive modeling". By performing "cutting simulation" (See §14.4.2), a Z-map model of the "machined surface" is easily obtained. This capability provides a convenient "interface" with commercial (dedicated) CAM systems. A direct link to commercial CAD systems is made possible through the "Z-map converter" in which a ray-casting method is used for converting implicit surfaces (and CSG solids) to Z-maps and the "nonparametric evaluation method" of §13.6 is used for parametric surface conversions.

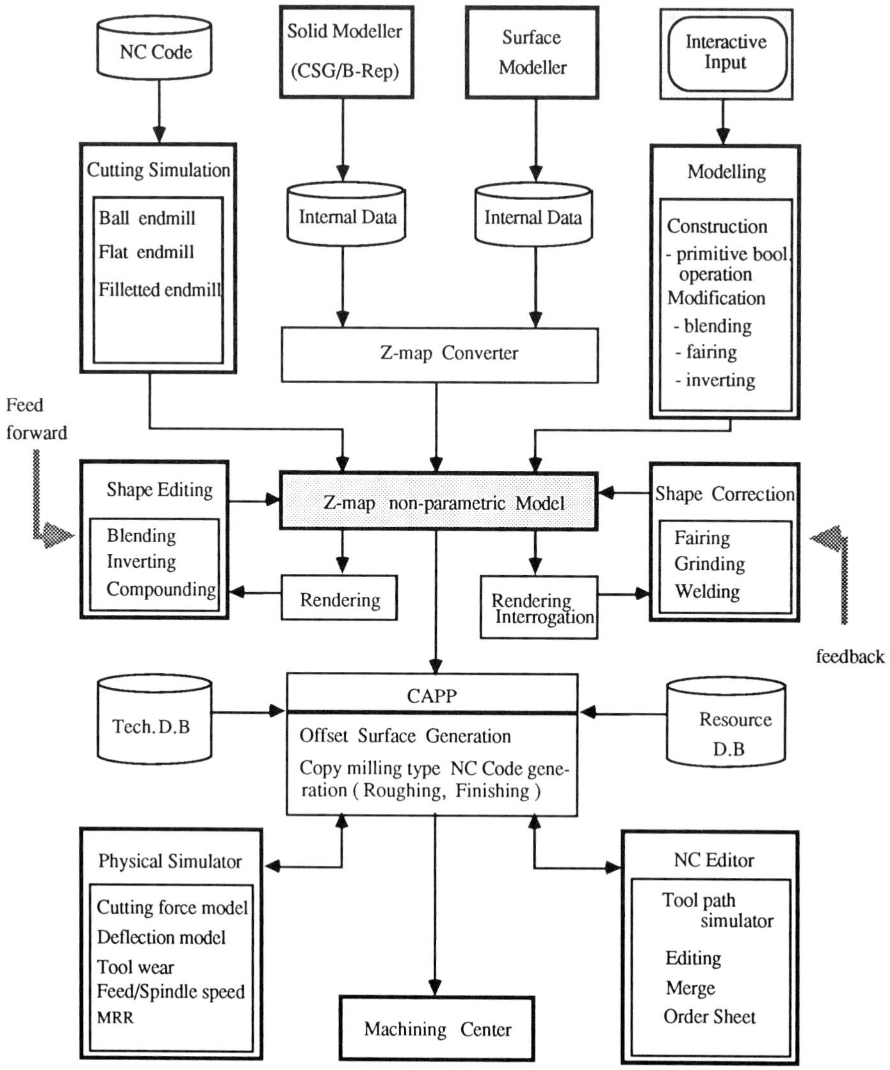

Figure 14.15 Overall Structure of the SOFT-MASTER

2) Z-Map Manipulation:

There are four types of shape operations supported by the Z-map manipulation module: They are

- shape editing,
- shape corrections,
- shape rendering, and
- geometry interrogation.

In "shape editing", blending surfaces (ie, rounding and filleting surfaces) are constructed if needed; a Z-map surface may be "inverted" and offsetted to obtain a mating die surface (eg, a lower mold cavity surface is obtained from the upper core surface); and local modifications to an existing Z-map surface are made via "compounding" (or Boolean) operations.

The "shape correction" routine handles surface fairing, surface "grinding" (ie, removal of small cusps or protrusions), and "welding" (ie, filling in small indents or pit-holes). The "rendering" routine produces shaded color images and hidden-line-removed wireframe displays, and the "interrogation" routine is used in examining or inspecting the Z-map surfaces (eg, cross sectional views, coordinate values on the surface, and curvature plotting).

3) NC-code Generation:

There are four types of functions to be performed by the *NC-code generation module*: They are

- tool path (ie, CL-data) generation for NC milling,
- tool path editing,
- "physical simulation" of end milling processes, and
- CAPP (computer automated process planning).

CL-data for ball-end milling are easily generated by offsetting the part surface by the amount of the ball-radius. This mode of tool path generation is equivalent to "copy milling". CL-data generation for a "flat-end mill" requires an "inverse offset" (See §14.5.2) of a flat "disk", and the generation of CL-data for a "filleted-end mill" may be carried out as a combination of ball-end milling and flat-end milling. The tool path editing function is the same as that of the "NC verification and editing" system introduced earlier.

For "physical simulation" of cutting processes, the MRR (metal removal rate) during the cutting process has to be computed. The Z-map representation is found to be very handy for such a computation. The CAPP for free-formed mold surface machining is becoming an important research issue. Here, we need a "technical knowledge base" about machining, a cutting tool database, a feature extraction mechanism, and a force model of the machining process.

14.5.2 Basic Operations in the SOFT-MASTER

Important geometric operations on Z-map surfaces are *tool path generation*, construction of *blending surfaces, cutting simulation*, generation of a *mating die surface* having uniform clearance, *shape corrections*, etc. Among these operations, tool path generation and surface blending are direct result of "surface offsetting". In the following, a method of offsetting Z-map surfaces is described.

The Z-map offsetting method is sometimes called the **inverse offset method** a 2D view of which is depicted in Fig. 14.16. In this scheme, the offset surface of "offset distance R" is defined as the envelope of the "balls" having radius R centered at the "Z-map points". That is, let

$$\{x(m),\ y(n),\ z(m,n)\}$$

denoted the "current" Z-map point, then the z-value $z^\circ(m,n)$ of the offset surface at the current grid point $(x(m),\ y(n))$ is given by

$$z^\circ(m,n) = \underset{(i,j)\in I(m,n:R)}{maximum} \{z(i,j) + h(i,j,m,n:R)\} \tag{14.9}$$

$$\begin{aligned} where,\quad & I(m,n:R) = \{(i,j)|(\Delta x_{im}^2 + \Delta y_{jn}^2) \le R^2\}, \\ & h(i,j,m,n:R) = sqrt[R^2 - (\Delta x_{im}^2 + \Delta y_{jn}^2)], \\ & \Delta x_{im} = [x(m) - x(i)], \\ & \Delta y_{jn} = [y(n) - y(j)], \\ & R:\ offset\ distance. \end{aligned}$$

Let "g" denote the *grid size*, then the above expression needs about $(R/g)^2$ evaluations for a point on the ball. That is, the "time complexity" becomes the square of R/g or $O((R/g)^2)$. When R=40mm and g=0.5mm, for example, we need about 6,400 evaluations! Then, a Z-map surface of size 30cm by 30cm (defined on a 600×600 grid) would need more than two billion evaluations, which may take "forever" on an ordinary minicomputer.

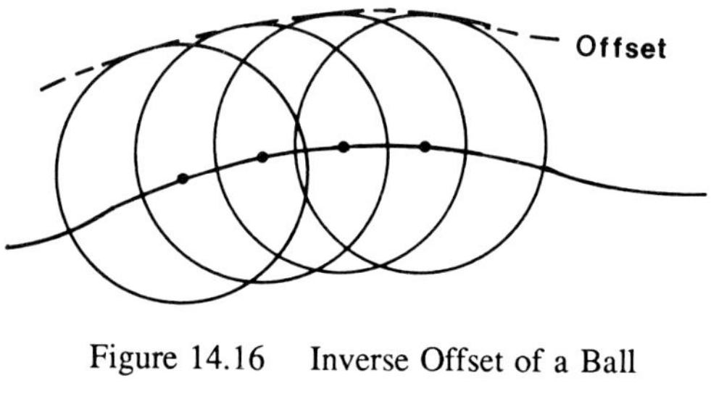

Figure 14.16 Inverse Offset of a Ball

Thus, it is necessary to improve the efficiency of the expression (14.9). First, the square root terms $h(\ )$ in (14.9) are stored in an array H$[\Delta m, \Delta n]$ at the beginning of the offset operation. That is, for fixed values of "R" and "g", we have (noting that $[x(m) - x(i)] = g * (m - i)$)

$$\begin{aligned} h(i,j,m,n:R) &= sqrt[R^2 - (\Delta x_{im}^2 + \Delta y_{jn}^2)] \\ &= sqrt\{R^2 - g^2 * [(m-i)^2 + (n-j)^2]\} \qquad (14.10) \\ &\equiv H[\Delta m, \Delta n] \quad for \quad (\Delta m^2 + \Delta n^2) \le (R/g)^2 \end{aligned}$$

$$where, \quad \Delta m = |m - i|,$$
$$\Delta n = |n - j|.$$

Note in (14.10) that only a quarter of the indices are needed in the array H. By using (14.10), the example offset case (ie, R=40, g=0.5, 600 × 600 grid) took about 20 hours to complete on a 2.3 MIPS engineering work station.

The next step of improvement is to reduce the number evaluations in (14.9). On a smooth surface, the number of evaluations can be reduced to only a few (from 6,400!), with an extreme case of 1 when the surface is locally flat and horizontal. With this improvement, the test example took about 7 minutes.

For ball-end milling, the offset surface becomes the "CL-data surface". Rounding of a Z-map surface is achieved by a "downward" offset followed by an "upward" offset, and filleting is achieved by an upward offset followed by a downward offset as depicted in Fig. 14.17. Inverting a Z-map surface is a trivial operation. That is, an inverted Z-map $z'(i,j)$ is expressed as

$$z'(i,j) = z_0 + [z_0 - z(i,j)],$$

where z_0 is a reference height value. Shown in Fig. 14.18 are the "initial" Z-map surface, tool paths for a ball-end cutter, the Z-map surface with fillets, and an inverted Z-map surface.

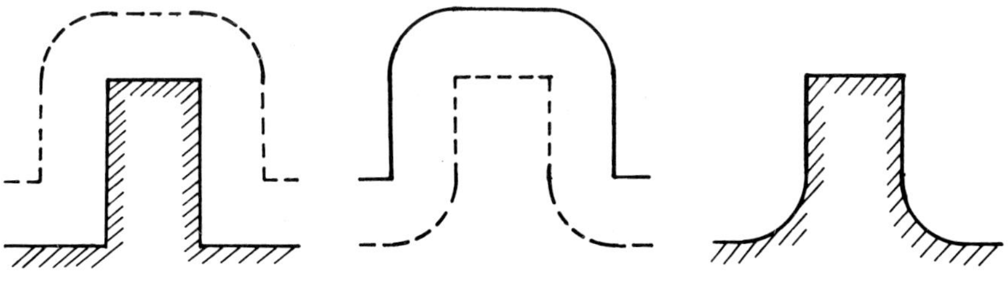

Figure 14.17 Filleting of a Z-map Surface

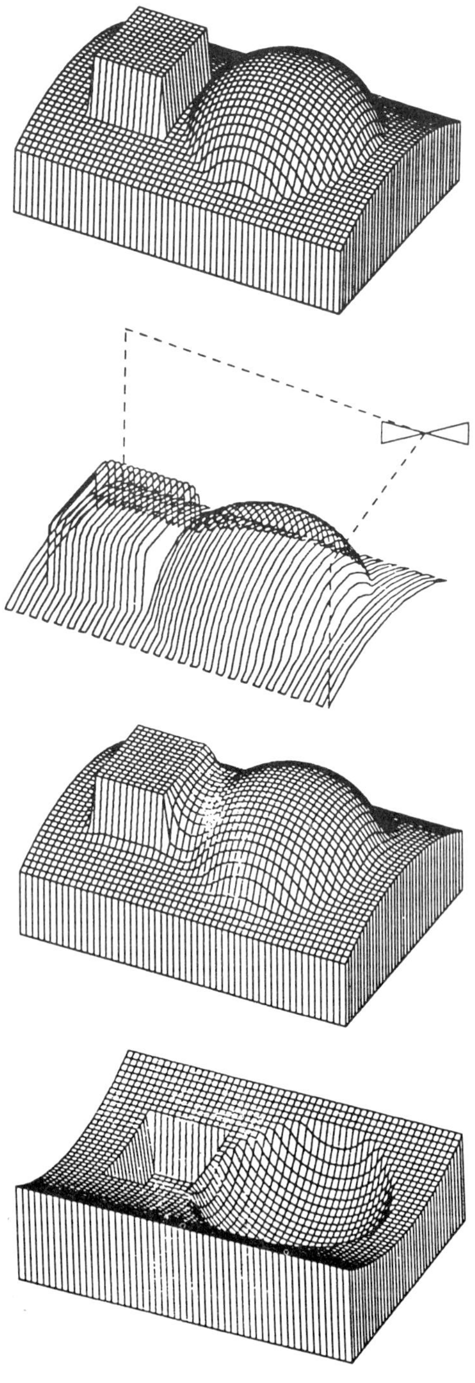

Figure 14.18 Some Geometric Operations on Z-map Surface

14.5.3 SOFT-MASTER and Concurrent Engineering in Die Manufacturing

The author is enthusiastic about the potentials of the "SOFT-MASTER". Its structure is well suited for the **concurrent engineering** of die manufacturing processes. The design/production cycle of car body dies, for example, consists of the following (Okamoto *et al*, 1989):

1. Style design.
2. Car body design including surface fairing/breaking and drafting.
3. **Master model** production (and style evaluation).
4. Die (deep drawing dies) design.
5. Die surface modeling and process planning.
6. Die machining.
7. Die surface finishing (polishing) and die-set assembly.
8. Tryout and modification.

The above eight steps are "serial" in nature, and it is the aim of "concurrent engineering" to make them "parallel" (or overlap) as much as possible while minimizing "disturbances" (ie, chances of "un-do" and "re-do"). The objective here is to reduce the "total lead-time". In order to achieve the objective, the following requirements should be met:

- There should be "feed-forward" mechanisms so that design changes be immediately "broadcast" to succeeding steps.
- The "feed-back" loops should be shortened as much as possible.
- The activities in each stage can be started (and continued) without a complete set of input data from its preceding steps.

The Z-map model in the SOFT-MASTER may play the two critical roles: It can serve the role of the "master model"; it can be used as a central "receiving and storage" station for all the feed-forward and feed-back information. Further, the Z-map model can store intermediate (and incomplete) shape information and may produce absolutely "gouge-free" NC-code (just like in copy milling).

14.6 DISCUSSIONS

The purpose of this chapter was to share the author's experience on developing "CAM systems" with the readers. Three types of CAM software have been introduced: a dedicated CAM system "SWEEP"; an "NC-code verification and editing" software; an integrated CAM system "SOFT-MASTER". The CAM system SWEEP which is running on IBM PCs (and compatibles) was developed for a local S/W firm, and it is successfully being marketed by the company. The NC-code verification software, writen in C and running on an EWS (engineering work station), was developed recently for a manufacturer of injection molding dies to be used as an in-house system. The "integrated" CAM system SOFT-MASTER is in an early stage of development (as of June 1990).

It is debatable whether it is worthy spending so much time in developing "commercial" software systems in academia. The author was very luck to have the opportunity to work with some of the outstanding and highly motivated students here at KAIST (Korea Advanced Institute of Science and Technology). What the author has learned from the development projects is that the "theory" part (of surface modeling) seems to be much easier than the "implementation" part. It would be safe to say that one may never fully appreciate the "theory" without implementing it.

To the author's opinion, the art of surface modeling is based on the premise that "surfaces" should be "describable". The surface description may rarely be given in a complete form, however. Another point to make is that a "described surface" (whether it is described in a drawing, sketched on a sheet of paper, displayed on a CRT screen, or stored in computer) is a "model" of the (conceptual) *surface*. A *model* is by definition an *abstracted* (or simplified) form of the "real" object. In order to "mathematically" describe a "physical" surface (whether it is real or imaginary), some forms of "mathematical" equations are needed. The terms "mathematical equation" and "mathematical model" (or simply "surface model") are used interchangeably. Every type of "surface model" (eg, Bezier, B-spline, etc.) has its own "modeling domain" and "range". The **modeling domain** means the "shape descriptions" that can be accepted by the *surface model*, and the **modeling range** refers to the "resulting shape". In this respect, one can not say that one model type is always better than the other; it all depends on what we want to do with the surface models. For example, NURB surface model is very popular (and indeed very versatile), but it can not represent an ellipsoid exactly (at least the author does not know how).

CHAPTER 15

UNIFIED SHAPE MODELING FOR CIM

15.1 INTRODUCTION

In this final chapter, we depart from our main subject *"surface modeling for CAD/CAM"* in order to contemplate on an "advanced" subject *"unified shape modeling for CIM"*. The purpose of this chapter is to propose a "user's view" of unified shape modeling which could be used as "design specs" in developing a *unified shape modeler*. Also to be discussed are "architectural issues" in unified shape modelers.

The term *"unified shape model"* has been used for some time, but its exact definition is not fully given yet. The same seems to be true with the term *"CIM"*. Nevertheless, we will use the terms without attempting to give exact definitions. Since the term *CIMS (computer integrated manufacturing system)* was coined in mid-seventies, it seems to be becoming the concern of almost every discipline in engineering. Certainly, CIM has many facets, but the common denominator in all CIM applications is the need for *"computer-understandable models"* of the products to be manufactured. In this respect, a *computer-understandable model* supporting diverse CIM applications may be called a **unified shape model**.

15.2 TRADITIONAL CAD/CAM CONCEPT

Traditionally, there have been five types of "computer-aided" software tools being used in the *design and manufacture* of mechanical parts. They are

- *Solid modelers supporting polyhedra and quadric primitives (with blending),*
- *Surface modelers supporting free-formed (or sculptured) surfaces,*
- *CAM systems for the machining of free-formed surfaces,*
- *2D-CAD systems for wireframe drafting, and*
- *Part programming systems (or languages).*

The above software tools are usually called *"CAD/CAM" systems*. These software tools have extensively been used in industry. They are quite versatile tools, but each type of the "CAD/CAM systems" has its specific application domain as will be shown shortly.

From the *"demand side"*, mechanical parts manufacturing may be classified into *"prismatic solid part"* manufacturing and *"curved object"* manufacturing. Examples of the former type are machinery components and those of the latter include sheet metal panels in a car and injection molding dies. In both types of mechanical parts, a typical *"design/machining cycle"* may be broken down into the following six phases:

Design/Mnfg Phases	*Prismatic solid part CAD/CAM*	*Free-formed surface CAD/CAM*
Conceptual design	————	————
Basic shape design	*solid modeler*	————
Detailed design	*2D CAD system*	*surface modeler*
Drafting	*2D CAD system*	*2D CAD system*
Process planning	————	————
NC programming	*Part P/G system*	*CAM software*

Figure 15.1 Primary Application Areas of Traditional CAD/CAM Systems

- *conceptual design,*
- *basic function and shape design,*
- *detailed shape design,*
- *drafting,*
- *machining process planning, and*
- *NC programming.*

As pointed out earlier, each type of *"CAD/CAM systems"* has its own application domain. Primary application areas (ie, phases in design/machining cycle) of the traditional CAD/CAM systems are summarized in Fig. 15.1.

15.3 CAD/CAM CONCEPT IN CIMS

There are quite a few *CIMS (computer integrated manufacturing system) architectures* proposed in the literature. A view of CIMS due to Gunn (1987) is depicted in Fig. 15.2. In this view, a *CIMS* consists of the following three primary subsystems:

- **product/process** *design subsystem,*
- **equipment** *(production processes) subsystem, and*
- *manufacturing* **plan/control** *subsystem*

which should be tightly coupled by an *"information handling and processing"* subsystem.

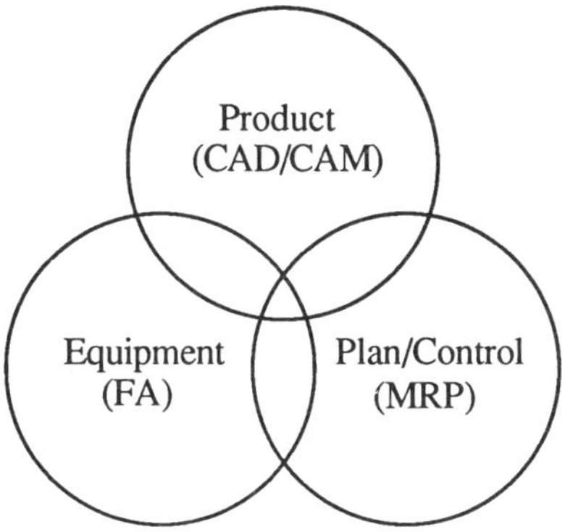

Figure 15.2 Primary Subsystems in a CIMS

Among the three CIM subsystems in Fig. 15.2, the *"Product subsystem"* is most closely related to *geometric modeling*. Here, the key word is *"CAD/CAM"*. In a typical "curved object manufacturing", the activities in the *"Product subsystem"* include

(1) Part (final shape) design,

(2) Preform design,

(3) Die design,

(4) Die surface machining, and

(5) Assembly planning.

The primary goal of CAD/CAM is to boost productivity and improve quality by applying "computer-aided" technologies to the above activities. Further, all the manufacturing activities should be carried out in an "integrated" way. Thus, it is first necessary to have a *"CIM architecture"* for the *"Product subsystem"* so that a new *"CAD/CAM concept"* can be developed based on the CIM architecture. Such a *CIM-oriented CAD/CAM concept* is proposed in Fig. 15.3. Brief explanations of the major activities under the *"CIM-oriented CAD/CAM concept"* are given below:

1) Part Design:

Designing is by nature a creative process which is difficult to automate. A designing process involves manipulation of complex geometric shapes and various types of *"engineering analysis"*. Another essential element of part design is *"aesthetic evaluation"* of the designed part. Thus, a CAD/CAM system should be able to support all the design activities in an integrated manner.

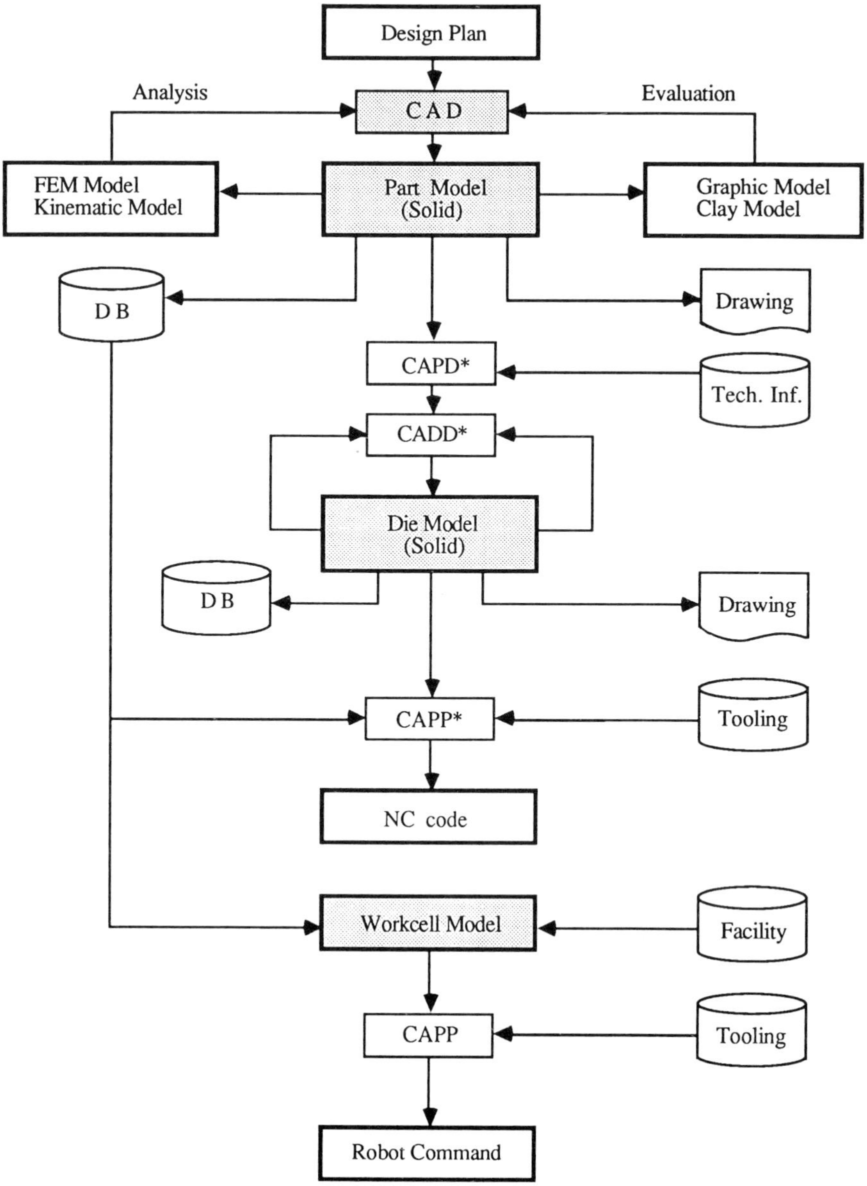

CAPD (Computer automated preform design)
CADD (Computer automated die design)
CAPP (Computer automated process planning)

Figure 15.3 CIM-Oriented CAD/CAM Concept

2) Preform Design:

The design engineer usually designs only the final shape of a part leaving manufacturing details at the discretion of a manufacturing engineer. Unless the part is to be produced by machining directly from a standard row stock, the first decision a manufacturing engineer has to make is how to prepare a *preform* for machining. Even when the final shape is to be produced by metal forming, the preform for forming has to be prepared one way or another. This **preform design** is an essential step in most metal forming, especially in forging and deep drawing. It is usually an iterative process meaning that each forming operation requires its own preform.

The subject of **CAPD (computer automated preform design)** has been studied quite extensively, see Richards (1985) and Tang and Oh (1988) for example, but the research seems to be still in its conceptual stage mainly because of lack of suitable representation schemes for mechanical parts having complex shapes. Here we need a *unified solid modeler* supporting relevant feature extraction functions.

3) Die Design:

The next step in the design/manufacturing cycle for curved object is the **die design** stage. Die design usually requires extensive engineering analyses which again require *"computer understandable models"* of the parts (and the dies). *CADD (computer automated die design)* is not a new concept, but without a unified shape model, it will remain just a concept.

4) Die Surface Machining:

The geometrical shape of the dies are produced by a sequence of machining operations. The plans for the die machining operations may be generated *"computer-automatedly"* if we have *"computer understandable models"* of the dies. *CAPP (computer automated process planning)* for die surface machining is an untapped research area.

5) Assembly Planning:

In order to fully support a CIM, the planning of assembly operations (to be performed by robots) also needs to be *"computer-automated"*. The *CAPP* for robot assembly planning is better known as *"task-level off-line programming"* (Bradly *et al*, 1982). *"CAPP for robots"* is also a well known concept as discussed in Matos (1987), but its real progress would depend pretty much on the availability of suitable representation schemes for workcells as well as the parts to be assembled.

15.4 REQUIRED FUNCTIONS OF A UNIFIED SHAPE MODELER

In this section, we will attempt to derive *required characteristics of a shape modeler* supporting the various CIM functions shown in Fig. 15.3. One may identify the following list of *requirements* by examining the *"CIM-oriented CAD/CAM concept"* of Fig. 15.3.

(1) High modeling power,
(2) Ease of modeling (ie, shape design),
(3) Manufacturing planning support,
(4) Ease of interface with application programs,
(5) Handling of non-geometric attributes,
(6) High performance.

In the following, each of the requirements will be explained briefly.

1) High Modeling Power:

In order to support a wide range of manufacturing applications, a key requirement is the ability to represent solid objects having sculptured surfaces. A shape modeler having this capability is often called a **unified solid modeler**. It should have wide *"modeling domain"* and *"modeling range"* meaning that the shape modeler should be able to take any *"shape description"* and generate a wide *"range of shapes"*.

2) Ease of Shape Design:

Once the range of representable solid shapes is increased, it would become impractical to construct solid models by relying on geometric input mechanisms alone. As a result, it is imperative for the *unified shape modeler* to have mechanisms for *designing with features.* The idea of **"feature-based design"** has long been advocated, but the definition of a feature has yet to be formalized (Gandhi and Myklebust, 1989). In general, however, one may think of three categories of features for *unified shape modeling:*

- *Solid primitive features,*
- *Surface features (ie, handles), and*
- *Manufacturing process-related features.*

Thus, a *unified shape modeler* should be equipped with a *"feature/geometry processor"* which can accept the three types of feature input and convert them into internal shape models.

3) Manufacturing Planning Support:

Another critical requirement of a *unified shape modeler* is that it should support a variety of *manufacturing applications* such as *"preform design", "die design" and "process planning".* In Fig. 15.3, these activities are named as *"CAPD", "CADD", and "CAPP"*, respectively.

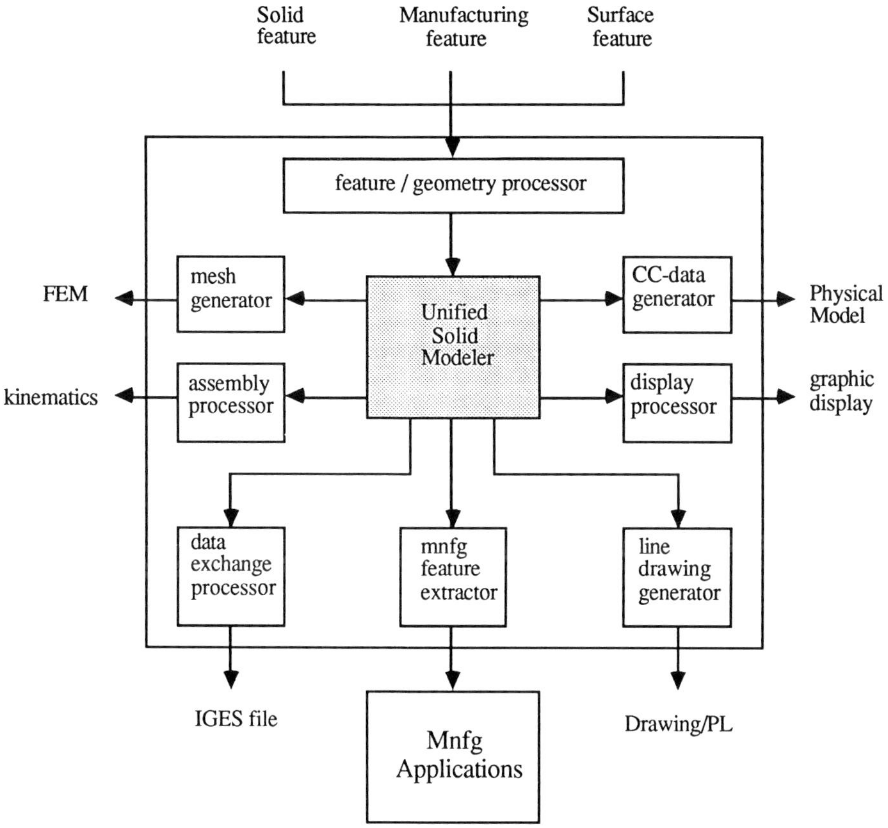

Figure 15.4 Required Functions of Unified Shape Modeler

Each of the *"computer-automated"* activities requires comprehensive **feature extraction schemes.** Thus, the structure of *internal solid models* must be carefully designed so that *feature extraction* is facilitated.

4) Interface with Application Programs:

In order to fully support the *"CIM-oriented CAD/CAM concept"*, the shape modeler should be easily interfaced with a wide variety of *"application programs"*. A partial list of application programs are given below:

- *Mesh generation for FEM analysis.*
- *Assembly modeling for kinematic analysis and assembly evaluation.*
- *CC-Data (cutter contact data) generation for machining physical models.*
- *High performance rendering (graphic display).*
- *Line drawing generator for drafting.*

The *"required functions"* of the *unified shape modeler* discussed so far are schematically shown in Fig. 15.4 (See the previous page).

The shape modeler should be able to represent a *"product"* (not just a shape of the product). Thus, it is necessary to be able to accommodate *tolerancing information* and *non-geometric attributes* such as surface properties, desired machining characteristics, and loads and restraints for FEM (finite element method) analysis (Miller, 1989). In order for a shape modeler to be used in production environments, the performance of the shape modeler becomes a critical factor. Important performance measures are

- *Response time,*
- *Storage efficiency, and*
- *Robustness and accuracy.*

15.5 BASICS OF SOLID REPRESENTATION SCHEMES

Before presenting *architectural issues in unified shape modelers*, a brief review on solid representation schemes is given first. This is by no means a complete discussion on the subject of solid modeling. The interested reader is referred to, for example, Mantyla (1988). Among the various representational forms (Requicha and Voelcker, 1983), the *CSG (constructive solid geometry)* and *B-Rep (boundary representation)* schemes are dominantly being used in commercial CAD/CAM systems because of their generality and the availability of useful algorithms.

In a **CSG scheme,** a solid is defined in terms of Boolean combinations of simple *solid primitives* such as blocks, spheres, cones, and cylinders. Thus, a model for the solid can be conveniently described by a *tree data structure* with its *terminal nodes* representing solid

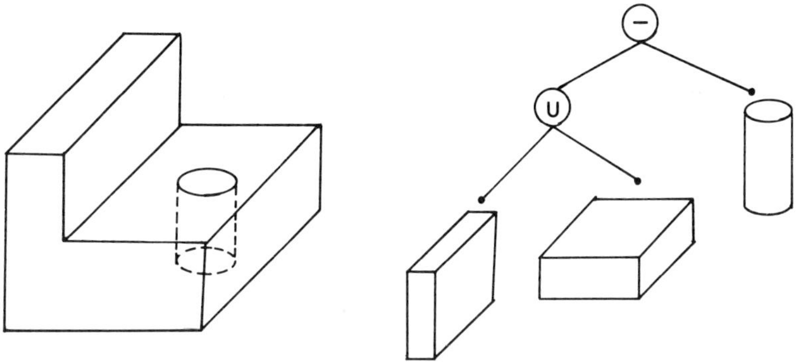

Figure 15.5 CSG Model of an Object

primitives and *nonterminal nodes* denoting Boolean set operations. An example of a CSG model is depicted in Fig. 15.5 where the object is constructed by *"unioning"* the two blocks and then *"differencing"* a cylinder.

A *"B-Rep model"* of the object shown in the above figure is depicted in Fig. 15.6. In a **B-Rep scheme**, a solid object is viewed as consisting of one or more *"shells"*; a **shell** is defined as a closed and connected set of *"faces"*; a **face** is bounded by one or more *"edge loops"*; an **edge-loop** is formed by a closed and connected set of *"edges"*; an **edge** is (topologically) bounded by two *vertices*. The topological information of a B-Rep model is stored in a *graph data structure*. Also to be stored in the B-Rep data structure are the *"adjacency relationships"*

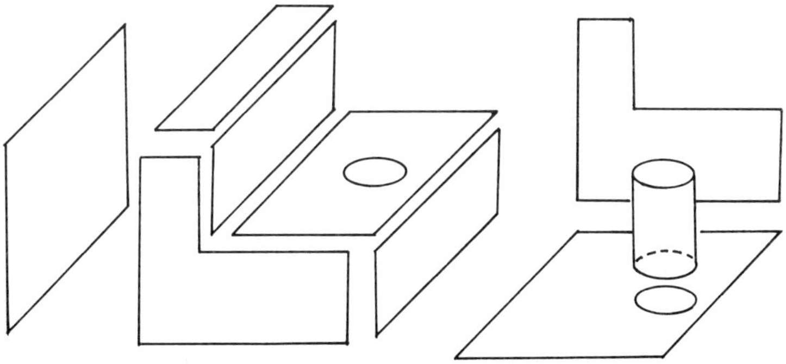

Figure 15.6 B-Rep Model of the Object in Fig. 15.5

describing how the *elements* (ie, faces, edges, and vertices) are connected. A B-Rep data structure for a polyhedron was given earlier in §13.2 (Figs. 13.2 and 13.3).

There are nine types of adjacency relationships (because there are three types of elements), but not all of them are stored explicitly. In classical B-Rep modelers, the adjacency relationships are derived from the well known **winged-edge data structure** in which the following data are stored for each edge (Miller, 1989):

- *two vertices defining the edge,*
- *two faces* $\mathcal{F}_1$ *and* $\mathcal{F}_2$ *that meet at the edge, and*
- *four of its adjacent edges: the "next edge" in a counter-clockwise or clockwise traversal about either* $\mathcal{F}_1$ *or* $\mathcal{F}_2$*.*

The B-Rep data structure is usually constructed and manipulated by using **Euler operators** (Mantyla, 1986 and 1988).

A CSG model is usually evaluated by using a *ray tracing* method. CSG is well suited for *"set partitioning"* operations. Application examples involving set partitioning operations include *collision detection, global feature recognition, computation of volumetric properties, and color-shading.* On the other hand, an *"edge-based"* B-Rep representation is ideally suited for operations that require an orderly traversal of the outer boundary of a solid object. Application examples involving *"boundary traversal operations"* include *CC-data generation for NC machining* and *local feature extraction* (Choi *et al*, 1984).

15.6 ARCHITECTURAL ISSUES IN UNIFIED SHAPE MODELERS

With a basic understanding of solid representation schemes, one may be able to propose architectures for a unified shape modeler so that it can support the diverse CIM requirements depicted in Fig. 15.4. The proposed architectures are an extension of the ones presented in Miller (1989). The requirements of the unified shape modeler may be recapitulated as follows:

- *It should support free-form surfaces (unified).*
- *It should support "feature-based design" and "parametric design".*
- *It should support "manufacturing feature extraction".*
- *It should be able to handle "non-geometric" attribute data.*
- *It should be "efficient".*
- *It should be reliable and "robust".*

The issue of accommodating free-form surface geometry has long been studied and different approaches have been proposed. See, for example, Varady and Pratt (1984), Casale (1987), and Chiyokura (1988). Some commercial CAD/CAM systems are already equipped with free-form geometry. Examples include GEOMOD (Boyse and Gilchrist, 1982) and CIMPLEX (Markot and Magedson, 1989). General edge-based data structures for free-form geometry also are in Weiler (1985). Thus, it is expected that the *"requirement of free-form geometry"* can be easily accommodated in the existing *edge-based data structure* by adopting the **parametric trimmed surface** method discussed in §12.4.

The requirement of parametric design can be met better with CSG (than with B-rep) as pointed out by Miller (1989), but the feature-based design and manufacturing feature extraction require both representations. Since it would be impractical to define all the standard features *a priori*, the shape modeler may need a *"feature library"* in which *"user-defined"* features are to be stored. Non-geometric attribute data are handled better in B-Rep (than in CSG).

From the above observations, one may conclude that we need a **dual architecture** supporting both CSG and B-Rep at the same time. As discussed in §15.5, B-Rep is ideally suited for operations requiring an orderly traversal of the outer boundary of a solid object, while CSG is well suited for such applications as *collision detection, global feature recognition, computation of volumetric properties, and color-shading.* Unfortunately, however, *"B-Rep to CSG conversion"* is not possible with today's technology. Thus, we may have to be satisfied with a *primary CSG architecture* or *a primary B-rep architecture.*

Proposed in Fig.s 15.7 and 15.8, respectively, are *"feature-based primary CSG architecture"* and *"feature-based primary B-rep architecture"* for a unified shape modeler. In both architectures, the shape modeler is provided with a *"feature/geometry processor"*, a *"feature library file"*, and a *"manufacturing feature extractor"*. As explained in Miller (1989), the *primary CSG architecture* allows the user to construct, query, and edit the *CSG trees.* The *B-rep graphs,* which are automatically obtained from the CSG trees, can be queried (as indicated by the arrow to *"designer"*) but are not allowed to be modified in order to maintain consistency between the two models. In the *primary B-rep architecture,* the CSG model is used only as a construction mechanism. Once a significant modification is made on the B-rep model, the CSG model would become invalid.

In order to support the diverse CIM applications, the shape modeler should have an **open architecture** meaning that the shape modeler is structured such that it can be used as *"toolkits"* (not as a *"turnkey system"*). Another architectural issue is whether to allow **nonmanifold objects**. Data structures for *nonmanifold geometric models* are available in the literature (Y. Choi, 1989), but their usefulness has yet to be justified. Other architectural issues include the handling of **unevaluated forms** and the utilization of **approximate models** in order to improve the efficiency (in terms of storage and response time) of the shape modeler.

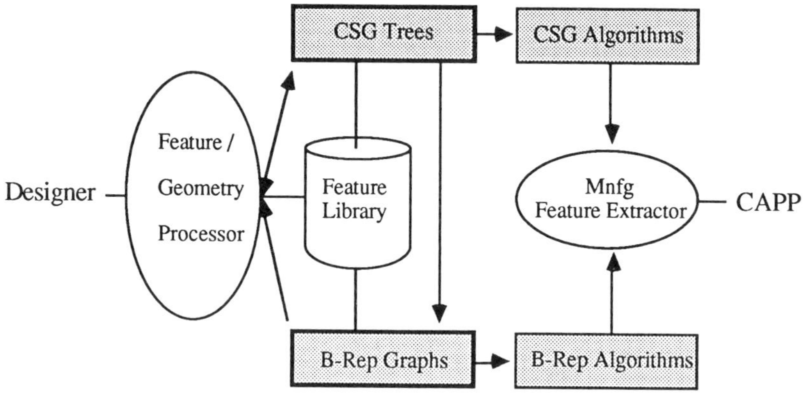

Figure 15.7 Feature-based Primary *CSG* Architecture

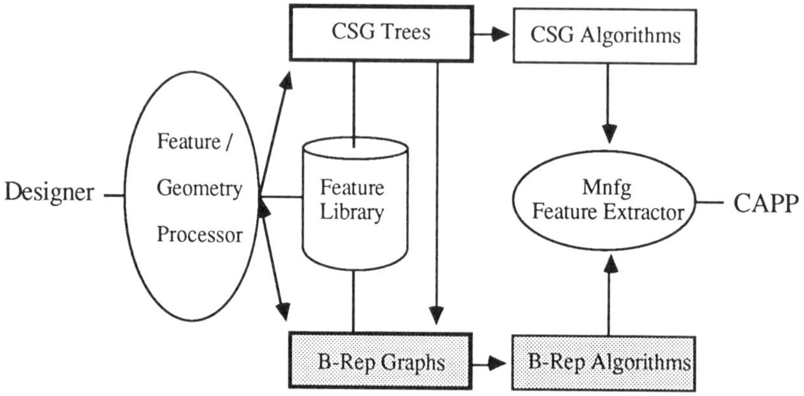

Figure 15.8 Feature-based Primary *B-Rep* Architecture

15.7 NURB AS A UNIFIED GEOMETRY OF A SHAPE MODELER

There are many types of surface forms available for use in free-form shape modeling: Popular ones include *rectangular and triangular Bezier patches, B-spline patches, Coons patches, rectangular and non-rectangular Greogory patches* (all of them have been introduced in this book), *β-spline surfaces, and υ-spline surfaces.* Each surface form has its own strength (and weakness), but supporting different types of surface forms will pose some difficult problems for the architect of a geometric modeling system. Thus, it is desirable to adopt a single internal representation of surface geometry.

The most popular choice for a *unified surface geometry* is *the non-uniform rational B-spline (NURB)* representation. As discussed in §3.4.3, Bezier models are special cases of non-uniform B-spline models. NURBs can exactly represent conic sections and rotational sweep surfaces obtained by rotating conic sections. NURBs can be used for representing most geometry formats used in existing CAD/CAM systems, and is the most flexible transfer and storage format existing for free-form surface geometry (Dokken *et al*, 1989).

A *composite NURB surface* of degrees *"d"* and *"e"* is completely defined by the following *geometric handles* (Choi *et al*, 1990b):

$$\begin{aligned}
\mathbf{S} &= \{V_{ij} : i = 0, ..., m;\ j = 0, ..., n\} : \textit{control net},\\
\mathbf{W} &= \{w_{ij} : i = 0, ..., m;\ j = 0, ..., n\} : \textit{weights},\\
\mathbf{D_u} &= \{\Delta_{-d+1}, ..., \Delta_{-1}, \Delta_0, ..., \Delta_{m-1}\} : \textit{u-direction knot span vector, and}\\
\mathbf{D_v} &= \{\Delta_{-e+1}, ..., \Delta_{-1}, \Delta_0, ..., \Delta_{n-1}\} : \textit{v-direction knot span vector.}
\end{aligned}$$

Let $V_i = (x_i, y_i, z_i)$ be a 3D control vertex, then a *homogeneous control vertex* is defined as

$$V_i^h = (w_i x_i,\ w_i y_i,\ w_i z_i,\ w_i) \quad \textit{with } w_i \neq 0. \tag{15.1}$$

Then, a NURB curve of degree *"d"* is concisely expressed in a matrix form as

$$\mathbf{R}_i^d(u) = \mathbf{U\ N\ H} \equiv (x(u),\ y(u),\ z(u),\ h(u)) \tag{15.2}$$

$$\begin{aligned}
\textit{where,}\quad & \mathbf{U} = [1,\ u,\ u^2, ..., u^d],\\
& \mathbf{N} : (d+1) \times (d+1) \quad \textit{coefficient matrix},\\
& \mathbf{H} = [V_i^h,\ V_{i+1}^h, ..., V_{i+d}^h]^T.
\end{aligned}$$

The NURB coefficient matrix N for quadratic and cubic cases were given earlier in (3.27) and (3.28), respectively. N for higher than cubic cases may be derived from the method presented in Choi *et al* (1990b). Derivatives of (15.2) are easily evaluated as discussed earlier in §3.5.2.

Similarly, a **NURB surface patch** of degree *"d", "e"* is expressed as

$$\mathbf{R}_{i,j}^{d,e}(u, v) = \mathbf{U\ N_u\ H\ (N_v)^T\ V^T} \tag{15.3}$$

$$
\begin{aligned}
where, \quad & \mathbf{U} = [1,\ u,\ u^2,\ ...,u^d], \\
& \mathbf{V} = [1,\ v,\ v^2,\ ...,v^e], \\
& \mathbf{N_u} = (d+1) \times (d+1)\ \textit{coeff. matrix with} \Delta_i, \\
& \mathbf{N_v} = (e+1) \times (e+1)\ \textit{coeff. matrix with} \nabla_j, \\
& d, e : \quad \textit{degrees in u- and v-direction, respectively}, \\
& \mathbf{H} = \{V^h_{r,c} : r = i, ..., i+d;\ c = j, ..., j+e\}, \\
& V^h_{r,c} = (w_{r,c}x_{r,c},\ w_{r,c}y_{r,c},\ w_{r,c}z_{r,c},\ w_{r,c}), \\
& V_{r,c} = (x_{r,c},\ y_{r,c},\ z_{r,c}) : \ \textit{3D control vertex}, \\
& w_{r,c} : \quad \textit{weight}.
\end{aligned}
$$

As discussed in Choi *et al* (1990b), the matrix form of NURB surface is very efficient compared to the direct evaluation of the knot insertion algorithm or the Cox-deBoor form (3.20). For example, when 100 points are sampled from a bicubic NURB surface patch, the evaluation of the matrix form takes about 12% of the execution time for evaluating the knot insertion algorithm.

Even with its promising features, NURBs alone can not accommodate all the needs of a unified shape modeler. For example, offsetting or blending of NURB surfaces would result in non-NURB geometry. Thus, a unified shape modeler should be provided also with a **parametric evaluator** which takes all queries involving *"non-NURB" geometry* and branches to appropriate *representation specific evaluators* so that application software needs no knowledge of the actual representation of the geometry. *NURB geometry* is handled by NURB algorithms.

15.8 DISCUSSIONS

The ultimate goal of geometric modeling is to develop an idealistic *"unified shape modeler"* supporting the diverse CIM functions. There are a list of issues in developing such a shape modeler : Design with features, handling of non-homogeneous geometry, manufacturing feature extraction, and open architecture, to name a few. In this final chapter, we have tried to propose an idealistic model of a *"unified shape modeler for CIM"* and to identify related research issues. As far as surface modeling is concerned, the topics discussed in earlier chapters are also relevent to unified shape modeling.

APPENDIX A

TRIDIAGONAL MATRIX SOLUTION

Presented in this appendix is an approach to solving a linear equation system whose coefficient matrix is *"tridiagonal"*. The solution technique to be introduced is due to Spath (1974). Let's consider a system of "$n+1$" linear equations of the form

$$\mathbf{A}\ \mathbf{x} = \mathbf{d}, \tag{A.1}$$

where

$$\mathbf{A} = \begin{bmatrix} b_0 & c_0 & 0 & & & & & \\ a_1 & b_1 & c_1 & 0 & & & & \\ 0 & a_2 & b_2 & c_2 & 0 & & & \\ & 0 & \cdot & \cdot & \cdot & & & \\ & & & & \cdot & \cdot & \cdot & 0 \\ & & & & 0 & a_{n-1} & b_{n-1} & c_{n-1} \\ & & & & & 0 & a_n & b_n \end{bmatrix}, \tag{A.2}$$

$$\mathbf{x} = \left[x_0\ x_1\ x_2 \ldots x_{n-1}\ x_n\right]^T,$$

$$\mathbf{d} = \left[d_0\ d_1\ d_2 \ldots d_{n-1}\ d_n\right]^T.$$

In order to meet the *"diagonal dominance" condition,* we need to have

$$|b_i| \geq |c_{i-1}| + |a_{i+1}| \tag{A.3}$$

with the inequality holding for at least one "i".

The coefficient matrix $\mathbf{A}$ can be decomposed into a product of two *"bidiagonal"* matrices $\mathbf{L}$ and $\mathbf{U}$ as follows:

$$\mathbf{A} = \mathbf{L}\ \mathbf{U}. \tag{A.4}$$

Then, the solution to (A.1) can be obtained by solving the two linear equation systems $(A.5)$ and $(A.6)$. That is, first solve the following for $\mathbf{y}$

$$\mathbf{L}\ \mathbf{y} = \mathbf{d} \tag{A.5}$$

and then solve the following for $\mathbf{x}$

$$\mathbf{U}\ \mathbf{x} = \mathbf{y} \tag{A.6}$$

Now, let the *bidiagonal* matrices **L**,**U** have the following forms:

$$\mathbf{L} = \begin{bmatrix} \beta_0 & & & & & & \\ \alpha_1 & \beta_1 & & & & & \\ & \alpha_2 & \beta_2 & & & & \\ & & \cdot & \cdot & & & \\ & & & \alpha_{n-1} & \beta_{n-1} & \\ & & & & \alpha_n & \beta_n \end{bmatrix} \qquad (A.7)$$

$$\mathbf{U} = \begin{bmatrix} 1 & \gamma_1 & & & & \\ & 1 & \gamma_2 & & & \\ & & 1 & \gamma_3 & & \\ & & & \cdot & \cdot & \\ & & & & 1 & \gamma_n \\ & & & & & 1 \end{bmatrix} \qquad (A.8)$$

On multiplying **L** and **U** and then equating the result with **A**, the elements of the bidiagonal matrices are obtained as follows:

$$\alpha_i = a_i \quad for \quad i = 1, ..., n \qquad (A.9)$$

$$\gamma_{i+1} = c_i/\beta_i \ ; \quad \beta_{i+1} = b_{i+1} - a_{i+1}\gamma_{i+1} \quad for \quad i = 0, ..., n-1 \qquad (A.10)$$

$$with \quad \beta_0 = b_0.$$

Now we are ready to solve (A.5) and (A.6). The solution to (A.5) is obtained after a *"forward substitution"* pass ($with \quad y_0 = d_0/\beta_0$):

$$y_i = (d_i - \alpha_i y_{i-1})/\beta_i \quad for \quad i = 1, ..., n. \qquad (A.11)$$

By combining (A.11) and (A.10) to eliminate α_i, β_i, we have

$$\gamma_{i+1} = c_i/(b_i - a_i\gamma_i); \quad y_i = (d_i - a_i y_{i-1})/(b_i - a_i\gamma_i) \quad for\ i = 1, ..., n \qquad (A.12)$$

$$with \quad \gamma_1 = c_0/b_0 \quad and \quad y_0 = d_0/b_0.$$

(A.6) is similarly solved by a *"backward substitution"* pass ($with\ x_n = y_n$):

$$x_i = (y_i - \gamma_{i+1}x_{i+1}) \quad for \quad i = n-1, ..., 0. \qquad (A.13)$$

Thus, the solution to (A.1) is obtained from (A.12) and (A.13).

REFERENCES

1 Avriel, M., 1976. *Nonlinear Programming: Analysis and Methods.* Prentice-Hall, U.S.A.

2 Barnhill, R. E., 1977. Representation and approximation of surfaces. In *Mathematical Software* III, J. R. Rice (ed.), Academic Press, U.S.A.

3 Barnhill, R. E., Farin, G., Jordan, M., and Piper, B. R, 1987. Surface/surface intersection. *Computer Aided Geometric Design*, 4(1): 3-16.

4 Bezier, P., 1972. *Numerical Control: Mathematics and Applications.* John Wiley, U.S.A.

5 Boehm, W, 1980. Inserting new knots into B-spline curves. *Computer-Aided Design*, 12(4): 199-201.

6 Boehm, W., 1981. Generating the Bezier points of B-spline curves and surfaces. *Computer-Aided Design*, 13(6): 365-366.

7 Boehm, W., 1985. Triangular spline algorithms. *Computer Aided Geometric Design*, 2(1): 61-67.

8 Bolton, K. M. 1975. Biarc curves. *Computer-Aided Design*, 7(2): 89-92.

9 Boyse, J. W. and Gilchrist, J. E., 1982. GMSOLID: Interactive modeling for design and analysis of solid. *IEEE CG&A*, 2(2): 27-40.

10 Bradly, M. et al., 1982. *Robot motion planning and control.* The MIT Press, U.S.A.

11 Casale, M. S., 1987. Free-form solid modeling with trimmed surface patches. *IEEE CG&A*, January 1987. pp.33-43.

12 Chang, G., 1982. Matrix formulation of Bezier technique. *Computer-Aided Design*, 14(6).

13 Chiyokura, H. and Kimura, G., 1984. A new surface interpolation method for irregular curve models. *Computer Graphics Forum 3*, 1984. pp.209-218.

14 Chiyokura, H., 1986. Localized surface interpolation method for irregular meshes. In *Advances Computer Graphics*, Kunii, T. L. (ed.), Springer-Verlag, New York.

15 Chiyokura, H., 1988. *Solid modelling with DESIGNBASE.* Addison-Wesley, New York.

16 Choi, B. K., Barash, M. M., and Anderson, D. C., 1984. Automatic recognition of machined surfaces from a 3D solid model. *Computer-Aided Design*, 16(2): 81-86.

17 Choi, B. K., Lee, C. S., Hwang, J. S. and Jun, C. S., 1988a. Compound surface modeling and machining. *Computer-Aided Design*, 20(3): 127-136.

18 Choi, B. K., Shin, H. Y., Yoon, Y. L., and Lee, J. W., 1988b. Triangulation of scattered data in 3D space. *Computer-Aided Design*, 20(5): 239-248.

19 Choi, B. K. and Ju, S. Y., 1989. Constant radius blending in surface modeling. *Computer-Aided Design*, 21(4): 213-220.

20 Choi, B. K. and Jun, C. S., 1989. Ball-end cutter interference avoidance in NC machining of sculptured surfaces. *Computer-Aided Design*, 21(6): 371-378.

21 Choi, B. K. and Lee, C. S., 1990. Sweep surfaces modeling via coordinate transformation and blending. *Computer-Aided Design*, 22(2): 87-96.

22 Choi, B. K., Shin, H. Y., and Yoo, W. S., 1990a. Visually smooth composite surfaces for unevenly spaced 3D data array. to appear in *Computer Aided Geometric Design.*

23 Choi, B. K., Yoo, W. S., and Lee, C. S., 1990b. Matrix representation for NURB curves and surfaces. *Computer Aided Design*, 22(4): 235-240.

24 Choi, Y., 1989. *Vertex-based boundary representation of non-manifold geometric models.* Ph.D. Thesis, Carnegie-Mellon University, U.S.A.

25 Cline, A. K. and Renka, R. L., 1984. Storage efficient method for construction of a Thiessen triangulation. *Rocky Mount. J. Math*, 14(1).

26 Conte, S. D. and deBoor, C., 1980. *Elementary Numerical Analysis.* McGraw-Hill, U.S.A.

27 Coons, S., 1964. *Surfaces for computer aided design.* MAC-TR-41, MIT, U.S.A.

28 Cox, M. G., 1972. The numerical evaluation of B-splines. *Jr. Inst. Maths. Applications*, 10: 134-149.

29 Cubic, 1989. *SWEEP Reference Manual.* Cubic Tech Ltd., Seoul, Korea.

30 DeBoor, C., 1972. On calculating with B-splines. *Jr. Approx. Theory*, 6: 50-62.

31 Dill, J. C., 1981. An application of color graphics to the display of surface curvatures. Proc. SIGRAPH81, *Computer Graphics*, 15(3): 153-161.

32 Dill, J. C. and Rogers, D. F., 1982. Color graphics and ship hull surface curvature. ICCAS Conf. on Computer Appl. in the Automation of Shipyard Operations and Ship Design IV, Annapolis, North-Holland, The Netherlands.

33 Ding, Q. and Davies, B. J., 1987. *Surface Engineering Geometry for Computer Aided Design and Manufacture.* Ellis Horwood, U.K.

34 Duncan, J. P. and Mair, S. G., 1983. *Sculptured Surfaces in Engineering and Medicine.* Cambridge University Press. U.K.

35 Farin, G., 1983. Smooth interpolation to scattered 3D data. In *surfaces in CAGD*, Barnhill and Boehm (eds.), North-Holland Pub. Co., The Netherlands.

36 Farin, G., 1986. Triangular Bernstein-Bezier Patches, *Computer Aided Geometric Design*, 3(2): 83-127.

37 Farin, G., 1988. *Curves and Surfaces for Computer Aided Geometric Design.* Academic Press, U.S.A.

38 Farin, G., Rein, G., Sapidis, N. and Worsey, A. J., 1987. Fairing cubic B-spline curves. *Computer Aided Geometric Design*, 4(2): 91-103.

39 Farouki, R. T., 1986. The approximation of non-degenerate offset surfaces. *Computer Aided Geometric Design*, 3(1): 15-43.

40 Farouki, R. T., 1987. Trimmed-surface algorithms for the evaluation and interrogation of solid boundary representations. *IBM J. Res. Develop.*, 31(3): 314-334.

41 Faux, I. D. and Pratt, M. J., 1981. *Computational Geometry for Design and Manufacture.* Ellis Horwood, U.K.

42 Ferguson, J. C., 1964. Multivariate curve interpolation. *Jr. of ACM*, 11(2): 221-228.

43 Gandhi, A. and Myklebust, A., 1989. A natural language approach to feature based modeling. *ASME Proc. of 15th Design Automation Conf.*, pp.69-78.

44 Gossling, T. H., 1976. The DUCT system of design for practical objects. *Proc. World Congress on the Theory of Machines and Mechanisms*, Millan, Italy.

45 Gregory, J. A., 1983. C^1 rectangular and non-rectangular surface patches. In *Surfaces in CAGD*, Barnhill and Boehm (eds.), North-Holland, The Netherlands.

46 Gregory, J. A., 1986. N-sided surface patches. In *The Mathematics of Surfaces*, J. A. Gregory (ed.), Clarendon Press, Oxford.

47 Gunn, T. G., 1987. *Manufacturing for competitive advantage*, Ballinger Pub. Co., U.S.A.

48 Hoffmann, C. and Hopcroft, J., 1987. The potential method for blending surfaces and corners. In *Geometric Modeling*, Farin, G.(ed.), SIAM, U.S.A., pp.347-365.

49 Holmstrom, L., 1987. Piecewise quadric blending of implicitly defined surfaces. *Computer Aided Geometric Design*, 4(3): 171-189.

50 Hoschek, J., 1984. Detecting regions with undesirable curvature. *Computer Aided Geometric Design*, 1(2): 183-192.

51 Hoschek, J., 1985. Smoothing of curves and surfaces. *Computer Aided Geometric Design*, 2(2): 97-105.

52 Jerard, R. B. and Drysdale, R. L., 1988. Geometric simulation of numerical control machining. *Proc. ASME Int'l Computers in Engineering Conf.*, ASME: 129-136.

53 Kishi, F., 1981. CAD/CAM for the die making industry. *Manufacturing Engineering*, 87(5): 90-92.

54 Kjellander, J., 1983a. Smoothing of cubic parametric splines. *Computer Aided Design*, 15(3): 175-179.

55 Kjellander, J., 1983b. Smoothing of bicubic parametric surfaces. *Computer Aided Design*, 15(5): 288-293.
56 Kjellander, J. and Bjorkenstan, U. C., 1983. Cubic curve fitting using variable segment stiffness for computer aided design. *Computers in Mechanical Engineering*, Nov. 1983, pp.61-66.
57 Klass, R., 1980. Correction of local surface irregularities using reflection lines. *Computer Aided Design*, 12(2): 73-77.
58 Klass, R. and Kaufmann, E., 1988. Smoothing surfaces using reflection lines for family of splines. *Computer Aided Design*, 20(6): 312-316.
59 Klok, F., 1986. Two moving coordinate frames for sweeping along a 3D trajectory. *Computer Aided Geometric Design*, 3(3): 217-229.
60 Kreyszig, E., 1983. *Advanced Engineering Mathematics. 5th Ed.*, John Wiley, U.S.A.
61 Lasser, D., 1986. Intersection of parametric surfaces in the Bernstein-Bezier representation, *Computer Aided Design*, 18(4): 186-192.
62 Lawson, C. L., 1977. Software for C^1 surface interpolation. In *Mathematical Software* III, Rice (ed.), Academic Press, U.S.A.
63 Lawson, C. L., 1984. C^1 surface interpolation for scattered data on sphere. *Rocky Mount. J. Math*, 14(1): 177-201.
64 Lee, C. S., 1990. *Sweep Surface Modeling.* Ph.D. Thesis, KAIST (In Korean).
65 Lee, E., 1987. The rational Bezier representation for conics. In *Geometric Modeling*, G. E. Farin (ed.), SIAM.
66 Lee, R. B. and Fredricks, D. A., 1984. Intersection of parametric surfaces and a plane. *IEEE CG&A*, August 1984, pp. 48-51.
67 Lilly, B. W., Bailey, R. W. and Altan T., 1988. Automated finishing of dies and molds. In *Computer-Aided Design & Manufacture of Dies and Molds*, K. Srinvansan and W. R. Devries (Eds.), ASME, PED 32.
68 Little, F. F., 1983. Convex combination surfaces. In *Surfaces in CAGD*, Barnhill and Boehm (eds.), North-Holland, pp.99-107.
69 Lipschutz, M. M., 1969. *Schaums outline of theory and problems of differential geometry.* McGraw-Hill, New York.
70 Mantyla, M., 1986. Boolean operations of 2-manifolds through vertex neighborhood classification. *ACM Trans. Graphic*, 5(1): 1-29.
71 Mantyla, M., 1988. *An Introduction to Solid Modeling.* Computer Science Press, U.S.A.
72 Markot, R. P. and Magedson, R. L., 1989. Solution of tangential surface an curve intersections. *Computer-Aided Design*, 21(7): 421-429.
73 Maron, M. J., 1982. *Numerical Analysis.* MacMillan Pub. Co., U.S.A.
74 Matos, L. M. C., 1987. A conceptual structure for a robot station programming system. *Robotics* 3.
75 Miller, J. R., 1989. Architectural issues in solid modelers. *IEEE CG&A*, Sept. 1989, pp.72-87.
76 Mortenson, M.E., 1985. *Geometric Modeling.* John Wiley, U.S.A.
77 Mudur, S. P., 1986. Mathematical elements for computer graphics. In *Advances in Computer Graphics I*, G. Enderle (Ed.), Springer-Verlag, New York.
78 NBS, 1986. *Initial Graphics Exchange Specifications.* Version 3.0, April 1986, national Bureau of Standards, U.S.A.
79 Okamoto I., Takahashi A., Sugiura H., Hiramatsu T., Yamada N., and Mori T., 1989. Computer aided design and evaluation system for stamping dies in Toyota, 1989 Society of Automotive Engineers Conference Proceeding: 5.79-5.89.
80 Paul, R. P., 1981. *Robot Manipulators.* The MIT Press, U.S.A.
81 Peng, Q. S., 1984. An algorithm for finding the intersection lines between two B-spline surfaces. *Computer-Aided Design*, 16(4): 191-196.

82 Petersen, C. S., 1984. Adaptive contouring of three dimensional surfaces. *Computer Aided Geometric Design.*, 1(1): 61-74.
83 Poechl, T., 1984. Detecting surface irregularities using isophotes. *Computer Aided Geometric Design*, 1(2): 163-168.
84 Pratt, M. J. and Geisow, A. D., 1986. Surface/surface intersection problems. In *The Mathematics of Surfaces*, Gregory, J. A. (ed.), Oxford Univ. Press, U.K.
85 Renz, W., 1982. Interactive smoothing of digitized point data. *Computer Aided Design*, 14(5): 267-269.
86 Requicha, A. A. G. and Voelker, H. B., 1983. Solid modeling: Current Status and research directions. *IEEE CG&A*. 3(7): 25-37.
87 Requicha, A. A. G. and Voelker, H. B., 1985. Boolean operations in solid modeling: Boundary evaluation and merging algorithms. *Proc. IEEE*, 73(1): 30-44.
88 Richards, W. T., 1985. Automated forging design. *22nd Annual Meeting and Tech. Conf. Proc., AIMT*, pp 357-364.
89 Riesenfeld, R. F., 1973. *Applications of B-spline Approximation to Geometric Problems of CAD*. Ph.D. Thesis, Syracuse University, U.S.A.
90 Rockwood, A. P. and Owen, J. C., 1987. Blending surfaces in solid modeling. In *Geometric Modeling*, Farin, G. (ed.), SIAM, U.S.A., pp 367-383.
91 Rogers, D. F. and Adams, J. A., 1976. *Mathematical Elements for Computer Graphics*. McGraw-Hill, New York.
92 Rossignac, J. R. and Requicha, A. A. G., 1984. Constant radius blending in solid modeling. *Computers in Mechanical Engr.*, July 1984, pp 65-73.
93 Samet, H., 1982. Neighbor finding techniques for images represented by quadtrees. *CG & IP*, 18: 35-57.
94 Sarraga, R. F., 1987. G^1 interpolation of generally unrestricted cubic Bezier curves. *Computer Aided Geometric Design*, 4(1): 23-39.
95 Sederberg, T. W., Anderson, D. C., and Goldman, R. N., 1984. Implicit representation of parametric curves and surfaces. *Computer Vision, Graphics, and Image Processing*, 28: 72-84.
96 Shirman, L. A. and Sequin, C. H., 1987. Local surface interpolation with Bezier patches. *Computer Aided Geometric Design*, 4(4): 279-295.
97 Spath, H., 1974. *Spline Algorithms for Curves and Surfaces*. Utilitas Mathematica Pub., Canada.
98 Stead, S. E., 1984. Estimation of gradients from scattered data. *Rocky Mount. J. Math.*, 14(1): 265-279.
99 Takeuchi Y., Sakamoto M., Abe Y., and Orita R., 1989. Development of a personal CAD/CAM system for mold manufacture, Annals of CIRP, 38(1): 429-432.
100 Tang, J. P. and Oh, S. I., 1988. *AFS: An automated forging design system*. Technical report, Battelle Columbus Div., Ohio, U.S.A.
101 Thomas, S. W., 1984. *Modeling volumes bounded by B-spline surfaces*. Ph.D. Thesis, The University of Utah.
102 Varady, T. and Pratt, M. J., 1984. Design techniques for the definition of solid objects with freeform geometry. *Computer Aided Geometric Design*, 1(3): 207-226.
103 Versprille, K. J., 1975. *CAD Applications of the Rational B-spline Approximation Form*. Ph.D. Thesis, Syracuse University, U.S.A.
104 Waggenspack, W., Anderson, D. C., Abhyankar, S. S. and Bajaj, C., 1987. *Identification and Parameterization of Reducible and Degenerate Curves and Surfaces*. CADLAB Report CAD-87-008, Purdue University, U.S.A.
105 Weiler, K., 1985. Edge-based data structure for solid modeling in curved surface modeling environment. *IEEE CG&A*, 5(1): 21-40.
106 Welbourn, D. B., 1984. Computer-aided engineering in the foundry. *Proc. British Cast Iron Research Association.*
107 Woodward, C. D., 1988. Skinning techniques for interactive B-spline surface interpolation. *Computer-Aided Design*, 20(8): 441-451.
108 Yoo, W. S., 1987. *VC^1 Composite Surface Fitting Considering Chord-length of Regular Point Set.*, M.S. Thesis, KAIST, Seoul, Korea.

INDEX

D

E

F

N

O

P

Q

R

S